To E. J. Applewhite
with appreciation
and best wishes,

Washington. D.C.   3/6/92

István Hargittai

*Quasicrystals,
Networks, and Molecules
of Fivefold Symmetry*

Quasicrystals, Networks, and Molecules of Fivefold Symmetry

István Hargittai,
Editor
Hungarian Academy of Sciences
and Eötvös University

István Hargittai
Hungarian Academy of Sciences and Eötvös University
H-1431 Budapest, Pf. 117, Hungary

Library of Congress Cataloging-in-Publication Data
Quasicrystals, networks, and molecules of fivefold symmetry / edited
    by István Hargittai.
          p.   cm.
    Includes bibliographical references and index.
    ISBN 0–89573–723–X
    1. Metal crystals. 2. Symmetry (Physics)   I.   Hargittai, István.
QD921.Q37   1990
548—dc20   90-12570
                                                                    CIP

British Library Cataloguing in Publication Data
Quasicrystals, networks, and molecules of fivefold
    symmetry.
    1. Crystallography
    I. Title
    548
    ISBN 0-89573-723-X

© 1990 VCH Publishers, Inc.

Printed in the United States of America
Printing History:
10 9 8 7 6 5 4 3 2 1

Published jointly by:

VCH Publishers, Inc.
220 East 23rd Street
Suite 909
New York, New York 10010

VCH Verlagsgesellschaft mbH
P.O. Box 10 11 61
D-6940 Weinheim
Federal Republic of Germany

VCH Publishers (UK) Ltd.
8 Wellington Court
Cambridge CB1 1HZ
United Kingdom

Contents

Preface

Following their discovery by Shechtman et al. in 1984, the study of quasicrystals has opened a new branch of crystallography and solid-state physics. In this book we hope to provide a cross section, or at least a good sampling, of this area of research as it stands at the end of the 1980s. We have taken a broad view of the topic by emphasizing networks and have included discussions of isolated molecules with fivefold symmetry.

It is hoped that this volume will not only stimulate further research in the field, but, by its broad approach, also will initiate interactions among workers in different areas related to quasicrystals. The book will be of primary interest to graduate students and researchers in crystallography, metallurgy, solid-state physics, materials science, geometry, and molecular sciences.

I am grateful to all the contributors and referees for their cooperation.

Budapest, June 1990

István Hargittai

Contributors

Jean-Paul Allouche
C.N.R.S. U.R.A. 0226
Mathématiques et Informatique
351, cours de la Libération
33405 Talance Cedex, France

Michael Baake
Institut für Theoretische Physik
Universität Tübingen
Auf der Morganstelle 14
D-7400 Tübingen,
Federal Republic of Germany

E. Brendsdal
Division of Physical Chemistry
The University of Trondheim
N-7034 Trondheim-NTH, Norway

J. Brunvoll
Division of Physical Chemistry
The University of Trondheim
N-7034 Trondheim-NTH, Norway

L. A. Bursill
School of Physics
University of Melbourne
Parkville, Victoria 3052, Australia

B. N. Cyvin
Division of Physical Chemistry
The University of Trondheim
N-7034 Trondheim-NTH, Norway

S. J. Cyvin
Division of Physical Chemistry
The University of Trondheim
N-7034 Trondheim-NTH, Norway

Françoise Dénoyer
Laboratoire de Physique des Solides
Associé au C.N.R.S.
Bâtiment 510
Université Paris-Sud
91405 Orsay Cédex, France

R. A. Dunlap
Department of Physics
Dalhousie University
Halifax, Nova Scotia, Canada
B3H 3J5

V. Elser
Department of Physics
Cornell University
Ithaca, New York 14853-2501

Gernot Heger
Laboratoire Léon Brillouin
Laboratoire commun C.E.A.-C.N.R.S.
C.E.N.-Saclay
F-91191 Gif Sur Yvette Cedex, France

L. d'Hendecourt
Groupe de Physique des Solides
Université Paris VII
UA 17 du C.N.R.S.
Tour 23-2, place Jussieu
F-75251 Paris Cédex 05, France

D. J. Klein
Texas A&M University at Galveston
Galveston, Texas 77553-1675

Peter Kramer
Institut für Theoretische Physik
Universität Tübingen
Auf der Morganstelle 14
D-7400 Tübingen,
Federal Republic of Germany

Dietmar Kuck
Fakultät für Chemie
Universität Bielefeld
Postfach 8640
D-4800 Bielefeld 1,
Federal Republic of Germany

Marianne Lambert
Laboratoire Léon Brillouin
Laboratoire commun C.E.A.-C.N.R.S.
C.E.N.-Saclay
F-91191 Gif Sur Yvette Cedex, France

D. W. Lawther
Department of Physics
Dalhousie University
Halifax, Nova Scotia, Canada
B3H 3J5

A. Léger
Groupe de Physique des Solides
Université Paris VII
UA 17 du C.N.R.S.
Tour 23-2, place Jussieu
F-75251 Paris Cédex 05, France

Peng Ju Lin
School of Physics
University of Melbourne
Parkville, Victoria 3052, Australia

Alan L. Mackay
Department of Crystallography
Birkbeck College
University of London
Malet Street
London WC1E 7HX, England

Peter McMullen
Department of Mathematics
University College London
Gower Street
London WC1E 6BT, England

Paul G. Mezey
Department of Chemistry
University of Saskatchewan
Saskatoon, Saskatchewan, Canada
S7N 0W0

Hans-Ude Nissen
Laboratory of Solid State Physics
Swiss Federal Institute of Technology
(ETH)
CH-8093 Zürich, Switzerland

Roald Z. Sagdeev
Space Research Institute
Academy of Sciences of the USSR
Profsoyuznaya 84/32
Moscow 117810, USSR

Olivier Salon
E. N. S. de Cachan
Mathématiques
61, avenue du Président Wilson
F-94230 Cachan, France

Martin Schlottmann
Institut für Theoretische Physik
Universität Tübingen
Auf der Morganstelle 14
D-7400 Tübingen,
Federal Republic of Germany

T. G. Schmalz
Texas A&M University at Galveston
Galveston, Texas 77553-1675

W. Schmidt
PAH Research Institute
Sieker Landestrasse 19
D-2070 Ahrensburg,
Federal Republic of Germany

W. A. Seitz
Texas A&M University at Galveston
Galveston, Texas 77553-1675

Marjorie Senechal
Department of Mathematics
Smith College
Northampton, Massachusetts 01063

V. Srinivas
Department of Physics
Dalhousie University
Halifax, Nova Scotia, Canada
B3H 3J5

Peter Stampfli
Institut für Theoretische Physik
Freie Universität Berlin
Arnimalle 14
D-1000 Berlin 33,
Federal Republic of Germany

L. Verstraete
Groupe de Physique des Solides
Université Paris VII
UA 17 du C.N.R.S.
Tour 23-2, place Jussieu
F-75251 Paris Cédex 05, France

E. J. W. Whittaker
Department of Earth Sciences
Parks Road
Oxford OX1 3PR, England

R. M. Whittaker
The Hellenic College of London
Pont Street
London SW1, England

George M. Zaslavsky
Space Research Institute
Academy of Sciences of the USSR
Profsoyuznaya 84/32
Moscow 117810, USSR

Dieter Zeidler
Institut für Theoretische Physik
Universität Tübingen
Auf der Morganstelle 14
D-7400 Tübingen,
Federal Republic of Germany

1 Crystals and Fivefold Symmetry

Alan L. Mackay

Faust: *"Das Pentagramma macht dir Pein?"*
Mephistopheles: *"Beschaut es recht! Es ist nicht gut gezogen;*
                            Der eine Winkel, der nach aussen zu,
                            Ist, wie du siehst, ein wenig offen."

[Faust: "You find my pentagram embarassing?" / Meph. "Beg pardon, Sir, the drawing's not perfected, / For here, this angle on the outer side / Is left, you notice, open at the joint."]

Wagner: *"Was man an der Natur Geheimnisvolles pries,*
                            Das wagen wir verständig zu probieren,
                            Und was sie sonst organisieren liess
                            Das lassen wir kristallisieren."
                    Johann Wolfgang von Goethe (1749–1832)

["The thing in Nature as high mystery prized, / This has our science probed beyond a doubt; / What Nature by slow process organised, / That have we grasped, and crystallised it out."] (trans. *P. Wayne*)

1. Introduction

The history of fivefold symmetry as a mathematical concept probably starts with the Babylonians, from whom the Alexandrian

1

Greeks acquired much of their mathematics. Before that, people could hardly help noticing the fivefold symmetry of many plants, since their lives depended on the correct identification of food. It is evident from the "Tower of Babel" (a "Ziggurat") that the Babylonians had a strong command of solid geometry and that they also invented "organization," and "administration" itself, as well as writing and computation. The mathematics begins with the five regular solids, the tetrahedron, the cube, the octahedron, the dodecahedron, and the icosahedron [1]. However, hardly anyone has dealt with the geometry of the irregular polyhedra. Grünbaum and Shephard remark that "mathematicians have long since regarded it as demeaning to work on problems related to elementary geometry in two or three dimensions, in spite of the fact that it is precisely this sort of mathematics which is of practical value" [2]. From this stricture we must, of course, except Archimedes, and others among the greatest of mathematicians.

In a way these figures later became a mystical cult, promoted by the followers of Pythagoras (ca. 481–411 BC) under Indian influences. What the Pythagoreans really discovered was group theory, and of course this can be applied to a vast range of circumstances. In early classical times the five regular polyhedra became attached to the doctrine of the elements, Earth, Air, Fire, and Water, and with a fifth element, Ether or Essence. This last gave rise to the expression *quintessential*. However, abstract theory is unpopular and people prefer material representations, in the way that saints are used to personify sundry virtues and vices. This Pythagorean cult still exists and seduces people in many cultures from direct physical examination of nature. Even today the regular polyhedra captivate people. Rubik's cube, which had a phenomenal public success, put into people's hands a representation of an extremely large and complex abstract group [3]. Combining sense with nonsense, Buckminster Fuller became a guru for the "dome folk" [4], who actually took to living in symmetrical polyhedra.[1] Indeed, an obsessive concern for symmetry may even have a biochemical basis [5].

Crystals have become a pseudoscientific cult—*Time* [magazine] [6] shows someone "treating bronchitis with an amethyst crystal"—but perhaps they always were since, for example, "amethyst" (a variety of quartz), as the name suggests, was believed to stave off drunkenness.[2] Science exists, barely tolerated by an irrational majority who are still in thrall [7] to religions that depend on supernatural sanctions.[3]

Kepler (1571–1630), who also cast horoscopes in response to the demands of his society, tried hard to see the structure of the Solar System as a nesting

---

[1] Curiously, Fuller seems never to have discovered the skew tessellations of the icosahedron evinced in the $T=7$ polyoma virus.

[2] When we characterize this as hocus-pocus we should not forget the etymology of that expression. ✷

[3] This may be seen from the inscription "D. G. Reg. F. D." [By the grace of God Queen, defender of the faith], still to be seen on every British coin, whereby the legitimacy of the monarchy is ascribed to the supernatural and this myth is, reciprocally, defended by the incumbent.

✷ ... *HOC EST CORPVS FILII*

of the five Platonic solids. He "held that the five polyhedra had been used by God to determine the distances between the six planets." He said "In the mysteries of the Pythagoreans, the five figures were distributed among the planets in this way, not among the elements as Aristotle believed." It is a mark of his greatness that he was eventually able to discard this a priori model, the *Mysterium Cosmographicum* (1596), involving ideas deriving from the crystal spheres of Claudius Ptolemaius, and to enunciate the three laws of planetary motion that bear his name. Unfortunately Kepler lived too early to learn that the five regular polyhedra are in fact (that is, as we now believe!) eigenfunctions of Schrödinger's equation and thus do describe the symmetries of the planetary electrons surrounding an atomic nucleus.

William Davidson (1593–1669), the first British professor of chemistry (albeit an Aberdonian at the Jardin des Plantes in Paris), stood almost at the watershed, suggesting in his book (1641) that the shapes of the Platonic solids had something to do with the shapes of crystals and of leaves, but he could not quite make the connection. For the purposes of this article at least, we regard Platonism as an idealistic view of this real world as being only the projection of some ideal world, and the view that, by manipulation of symbols for the concepts in the ideal world, we can acquire that secure knowledge which is the power of science promised by Francis Bacon— *scientia ipsa potestas est.*

Such views are, of course, consonant with the continuance of various religions, which still promise that there is another world distinct from this material world, in spite of the skepticism voiced through the ages by persecuted minorities. We will not follow here the use of crystals or of the ideas of symmetry, even the "five for the symbols at your door," for religious or magical purposes. Magic is a disease of symbols and there is also a pathology of science.

Many mathematicians regard the world of mathematics as this ideal world, where knowledge a priori from whole numbers, without having to do experiments, is promised. In *Principia Mathematica* Russell and Whitehead thought to establish a sound framework for mathematics, but their achievement was denied by the appearance of Gödel's Theorem. Edington's *Fundamental Theory* sought to derive the universe without observation. Bertrand Russell (*Autobiography*) wrote:

> With equal passion I have sought knowledge. I have wished to understand the hearts of men. I have wished to know why the stars shine. And I have tried to apprehend the Pythagorean power by which number holds sway above the flux. A little of this, but not much, I have achieved.

Pythagoreanism was evident in the *Timaeus* of Plato (428–348 BC), which is largely concerned with the five regular solids. George Sarton goes so far as to say of the *Timaeus* that its "influence was largely an evil one" [8]. Charles de Bovelles (ca. 1479–1566) [9] seems to have been responsible, along with Nicholas of Cusa (ca. 1401–1464), for transmitting "the use of geometrical symbolism to provide mathematical guidance to the divine" from antiquity to Kepler.

Nevertheless, in a sense, various mathematical objects do exist in the noosphere, quite apart from their materializations. The Mandelbrot set must be considered as having lain undiscovered in the noosphere until sufficient computer power had become available to make it visible, just as viruses lay in the biosphere until the advent of the electron microscope permitted them to be seen [10]. Thus, the Platonic view of the existence of a library of ideal objects is not altogether perverse.

The most recent history of fivefold symmetry may be compared to the escape from the idealistic Platonic influence toward a more realistic computational attitude of the Babylonians, whom Derek de Solla Price considered to have a numerical and computational, rather than a geometric or logical or linguistic picture of nature. The Babylonians achieved accuracies not reached again until the Renaissance, because they could accumulate their astronomical observations over 500 years, instead of the 10 years for which individual astronomers could afford to wait. They were, apparently, not worried by the irrationality of $\sqrt{2}$ (or of the Golden Number), but could calculate the numerical value (using their hexadecimal system, which we still employ in measuring angles) to far more places than was physically necessary. On the other hand, the Greeks were shocked to find that the diagonal of a square could not be represented as the ratio of two integers. Our own scientific culture has been similarly shocked by the recent discovery of Shechtman [11] that extended assemblies of atoms could have fivefold symmetry as evidenced experimentally by their diffraction patterns.

2. "Graininess"

Philosophically, determinism has been confused with computability. Just because there are laws it does not follow that the consequences of these laws can be computed with accuracy, or even at all. Quite recently, it has been found that even at the heart of straightforward Newtonian mechanics there is chaos and indeterminacy. Even the behavior of a simple pendulum with Newtonian damping, driven at a fixed frequency, cannot be predicted with indefinitely great accuracy for all regimes. The opening shot of a game of snooker involves many-body interactions and its outcome is designedly indeterminate. It has also been found that "predictability of the orbits of the inner planets, including the Earth, is lost within a few tens of millions of years" [12].

It is uncongenial to think that an infinite number of digits, for example, the square root of two, that is, infinite information, can be stored in a small perfect square formed by some material structure (as the ratio of the diagonal to the edge). Edward Fredkin has developed this idea [13]. Space–time may have an ultimate graininess, just as energy is quantized, even if it is at the level of the Planck length (10^{-35} m). Thus we conclude that the material square is not perfect, but only good enough, just better than its competitors. Since everything in the universe is affected by everything else and hierarchic levels are only approximately separable, symmetry is only approximate and no structures are perfect. The Greeks had a mechanic tradition, as can be seen from the Anti-Kythera machine, which was an astronomical calculator with some 25 metal gear-wheels and a complexity equivalent to that of a modern mechanical clock [14], but this tradition was overlaid by the culture of the scholastic tradition that recreated Greek ideals for Western Europe.

However, the two Greek cultures, the scholastic and the mechanical, were agreeably combined in the great book "On Growth and Form" by D'Arcy Thompson [15], which presaged and indeed stimulated much modern work on the conversion of observational biology into a real science. Since Thompson hardly mentioned atoms, the work was obsolete when it was being written, but the text remains as a continual challenge for modernization. It deals with the shapes of biological structures at a level above that of atoms. In the 1930s this book certainly inspired the Club for Theoretical Biology at Cambridge to found molecular biology (the words were probably W. T. Astbury's coinage) and biochemical morphology.

3. The Pentagon and Dodecahedron

Euclid [of Alexandria] (fourth century BC) working at the Museum of Alexandria produced his masterpiece *The Elements* in 13 books [16]. Those relationships that corresponded to the use of only certain geometric instruments were allowed (the straight edge and the compass). The regular pentagon was constructed in theorems 4.10 and 4.11 of *The Elements*. However, *The Elements* was extended by others and the fifteenth book ended with the construction: "In an icosahedron to inscribe a dodecahedron" and indeed the fourteenth and fifteenth books are mostly about the dodecahedron and the icosahedron. The fourteenth book was in fact written by Hypsicles (fl. first century BC) and the fifteenth by a pupil of Isodoros in the sixth century. Proclus (412–485) characterized *The Elements* as a treatise on the five regular solids for adepts.

Regular polygons with 3,5 (and thus 15) sides can be constructed with ruler and compass (under the artificial rules chosen by Euclid). These sides can be bisected indefinitely to give 4,6,8, etc. Later, Gauss showed that regular polygons with 17, 257, 65537 ... sides can also be constructed within these rules. Thus $\cos(2\pi/17)$ can also be expressed in terms of surds.[4]

---

[4] In 1796 Gauss found

$$16 \cos(2\pi/17) = -1 + \sqrt{17} + \sqrt{2}(17 - \sqrt{17})$$
$$+ \sqrt{68 + 12\sqrt{17} - 4\sqrt{2}(17 - \sqrt{17}) - 8\sqrt{2}(17 + \sqrt{17})}.$$

Seki Takakazu had solved the same equation numerically in 1683 (personal communication from Sugimoto Toshio, September 1989).

4. What Is Special about Fivefold Symmetry?

As we will see, as a consequence of the existence of a lattice, only the axes of rotational symmetry of orders 2, 3, 4, and 6 occur in the space groups to which all proper crystalline substances belong. It is thus possible for molecules with sixfold axes to crystallize so that these six-fold axes coincide with six-fold axes in the crystal space-group, and similarly for axes of order 2, 3 and 4, but this is not possible for molecules will axes of order 5, 7, 8, . . . One major attraction of fivefold symmetry is thus that it is forbidden (in crystals based on lattices). Is this a rule of Nature or of man?

Five is, in a sense, the first "interesting" integer in that given five points in space all cannot be identically related to all the others as the four points in a tetrahedron or the three points in a triangle can be. If five points are in a planar ring then each has neighbors at three different distances. Novelty develops. The Pentagram is more interesting than the Trinity (feuding over the latter and over incommensurability—affecting the date of Easter—split the Western world in AD 325). Five mutually equidistant points define a simplex in four-dimensional space. Five points in three-dimensional space have 10 distances between them, but only 9 coordinates are needed, and thus there is one necessary relationship between the 10 distances. The extent to which this relationship is not satisfied can be interpreted either as the result of experimental errors in the measurement of the distances or as curvature of the whole space into the fourth dimension. Near Johannesburg there is a testing ground (Honeydew) where five points are laid out about 10 km apart and the 10 distances are known to a few millimeters. This is used for testing tellurometers with the assumption that space is three- and not four-dimensional. That is, discrepancies are interpreted as errors and not as curvature of space into the fourth dimension.

5. Crystallinity

The law of constancy of angles (for crystals of the same species) was found by Nicholas Stenson [Steno] (1638–1686) and described in his book *De Solido Intra Solidum* (1669).[5]

Stenson's question as to what the structure inside a crystal is that gives it the regular faces was answered by Haüy's proposal of "molécules inté-grantes." This is now seen to be correct and crystals are indeed made of

---

[5] In October 1988, Niels Stenson, later Bishop of Hamburg, was beatified by the Roman Catholic Church. This was more on account of his apostasy from Protestantism than for his science.

identical molecules or other units stacked in straight rows and columns in three dimensions. There was a hint of this in Democritos who apparently suggested that the surface of a cone would necessarily be stepped. The theory of crystal symmetry was one of the triumphs of nineteenth-century science and has become the foundation of solid-state physics and the science of materials but, just when it was almost perfect, a split in the fabric has appeared with the discovery of quasicrystals with fivefold symmetry [11].

5.1. *The Crystal Lattice*

The theory of space groups of symmetry turns on the existence in a crystal of a lattice. A lattice is an infinite array of points in space that represents the translations under which the crystal structure is invariant. The first step in deriving the 230 space groups is to identify the possible lattices. There are 14 crystal lattices with different symmetries and they are called the 14 Bravais lattices after A. Bravais (1850), but it might be noted that they had been found earlier by M. L. Frankenheim (1842) who reported 15 lattices [17]. It is not clear whether Frankenheim used a different definition or made a mistake and counted the same lattice twice. However, there ought to be a Frankenheim prize for people who got a discovery nearly right!

The symmetry operations include axes of order 2, 3, 4, and 6.

It is simply proved, in the first chapter of every textbook on crystallography and on solid-state physics, that a lattice cannot have identical lattice points and be invariant under the operation of fivefold symmetry axes. Thus, fivefold axes do not enter into the theory of the symmetry of crystals. This omission has now been rectified. Nothing is actually wrong with the old theories but additions must be made.

6. The 230 Crystallographic Space Groups

The 14 Bravais lattices result if we ask what different symmetries an infinite number of identically situated points can have. If we ask how an infinite number of identical asymmetric objects may be arrayed in space so that the surroundings of all objects are the same, then the answer generates the 230 crystallographic space groups. (We must also consider that an object and its mirror image are identical.) The symmetry operations include the same axes of symmetry as for the point groups. The 230 space groups were found independently by three people, Fedorov, Schoenflies, and Barlow, in the 1870s.

No theory is as secure as is the theory of the space groups [19]. This gave some crystallographers the illusion of acquiring absolute knowledge a priori. The iconoclastic aim of overthrowing this model theory seemed quite unrealistic but there has always been the possibility of outflanking the restrictions and developing, under the inspiration of D'Arcy Thompson, a "crystallography" more suited for the progress of biology.

Given the axioms, substantial numerical deductions about Nature could be made. The problem was one of absolute, logical equality or of practical, experimental equality. Essentially the Greek proof that $\sqrt{2}$ could not be expressed as a rational fraction and the Babylonian attitude of calculating it to as many places as necessary differed sharply. The transition in thinking is one from absolute equality to quasiequality.

After all, space group theory applies only to infinite crystals and in fact no crystal is infinite. _*

7. The Penrose Pattern

Various people tried to examine the patterns that could be generated if some of the requirements of strict identity were relaxed. It turned out that a particular pattern, now called the Penrose pattern (after Roger Penrose, Mathematics Institute, Oxford), has been the key to an immense development in crystallographic theory [18]. Before 1984 it had no practical applications. The pattern can be generated in at least five distinct ways:

7.1. Local Rules

Penrose himself was concerned to make a jig-saw tiling, of identical pieces of as few kinds as possible, where the local rules of combination would force nonperiodicity. He succeeded in reducing the number of types of tile to two. He preferred the "kites and darts" forms, but showed that other shapes were possible. Crystallographers have preferred to use two rhombs of angles 72 and 36°, but to correspond to the kites and darts the two ends of the rhombs should be marked to distinguish them.

Penrose thought that if tiles were joined one after the other then it might sometimes be necessary to go back and change some tiles already placed, so that an indefinitely large area might be covered. Steinhardt has recently produced an algorithm by which an infinite area can be generated from local rules, but the rules are not truly local and the whole perimeter of the pattern may have to be scanned.

7.2. Recursive Subdivision

Penrose also showed that his two tiles could be recursively dissected so that a tiling with an infinite number of tiles would be generated. The self-similarity of this class of pattern is important because phase transformations are dealt with in this way in the renormalization theory of Wilson and Fisher.

* INFINITY IS THE ENEMY OF ANY SYSTEM.

7.3. Hierarchic Rules

A very similar pattern was idependently produced by Mackay [20], starting with the hierarchic packing of pentagons and a rule for filling in the resulting gaps between the pentagons. The pattern used two triangular tiles, each half of the rhombs mentioned above.

7.4. Pentagrids

N. de Bruijn showed that the Penrose tiling could be produced and generalized by drawing a grid of five sets of equally spaced lines that defined intersections of two kinds at each of which one of the tiles could be put. This defined the connectivity of the tiling and the tiles were then spaced out appropriately. This construction was very powerful and could be generalized to any symmetry. In the case of fivefold symmetry, if the lines were not equally spaced but with two different spacings (in the ratio of the golden number), then the tiles could be laid directly without further adjustment.

We find too the relationship, for symmetries of order $2n+1$, where there are n different rhombic tiles,

$$\{2 \sin[n\pi/(2n + 1)]\} \{2 \sin[(n - 1)\pi/(2n + 1)]\} \cdots 2 \sin[\pi/(2n + 1)] = \sqrt{n},$$

for example

$$2 \sin(2\pi/5)\ 2 \sin(\pi/5) = \sqrt{5}.$$

The areas of the rhombic tiles (of unit edge) are given by the sines of their vertex angles and their frequencies in an extended pattern are proportional to their areas.

7.5. Shadows of Higher Dimensional Lattices

It was found also, principally by P. Kramer, that the planar Penrose tiling could be produced by projection of a five-dimensional hypercubic lattice in an appropriate direction and that the three-dimensional tiling could be obtained from a six-dimensional hypercubic lattice.

7.6. Three-Dimensional Generalization

The two-dimensional Penrose tiling can be generalized to three dimensions using acute and obtuse rhombohedral tiles suggested by R. Ammann. Each face is a rhombus of angle $\arctan(2) = 63.45°$. The key figure here is the rhombic triacontahedron (with 30 faces) composed of 10 of each kind of rhombohedron. The whole Penrose pattern can be seen to consist of such rhombic triacontahedra, partially overlapping by the sharing of certain tiles. These polyhedra were known to Kepler himself, who must have experimented with packing them together. Such "Keplerian building games" were described by Kowalewski in 1938 [21], although because of the war his book cannot have circulated far.

8. Fivefold Symmetry in Crystals

The above theoretical constructions proceeded quietly until the discovery, by Dan Shechtman, published in 1984 [11], of an Al/Mn alloy that gives diffraction patterns with full icosahedral symmetry. This led to the opening of a new era in crystallography where the trend to see everything as either good crystals or poor crystals has been reversed. We now recognize as regular (rule-given) many other structures that are not crystals. These were, of course, known before, but attitudes have changed. Some thousand papers have followed directly from the stimulus of this discovery [22, 23].

The syllogism goes like this: (1) Every crystal must belong to one or other of the 230 crystallographic space groups. (2) Each of the 230 space groups gives diffraction effects that have the symmetry of one or other of the 32 crystallographic point groups (which embody the symmetry axes of orders only 2, 3, 4, and 6). (3) Some of these materials (the alloys now known as quasicrystals) have diffraction effects that have the symmetry of the icosahedral point group, which is not one of the 32 crystallographic point groups. (Other alloys show other forbidden symmetry axes of orders 8, 10, and 12.) (4) Ergo, the material cannot be a crystal but must be some other regular structure.

Experimental work has succeeded in producing more and more ordered quasicrystals that have external faces showing the symmetry of the icosahedral point group and that give diffraction patterns as sharply defined as those of the best crystals.[6]

It has now been realized that many structures other than orthodox crystals give diffraction patterns with sharp spots. These structures are those which can be represented as three-dimensional shadows of higher-dimensional lattice structures. These structures are called quasi-crystals because they are quasi-periodic. A true crystal, periodic in three dimensions, can be expressed as the linear sum of a series of sinusoidal density waves which have wavelengths which are integral fractions of three non-coplanar base vectors.

$$\rho(x, y, z) = \Sigma_{h,k,l} \; F_{hkl} \cos 2\pi(hx + ky + lz - \alpha_{hkl})$$

h, k, and l are integers. In a quasi-periodic structure or quasi-crystal the density can be expressed as the sum of a finite number of terms where h, k and l are not necessarily integers. Each sine wave corresponds to one spot in the diffraction pattern.

---

[6] The best recipe seems to be $Al_{65}Cu_{20}Fe_{15}$; arc melt the pure elements and anneal for 48 h at 840°.

It was not realized earlier that infinitely extended structures can be devised where the diffraction pattern has fivefold symmetry (or icosahedral symmetry in three-dimensions) but where the structure itself is non-periodic and without fivefold symmetry. The Penrose tiling is an algorithm (one of several) for producing such a pattern.

8.1. Twinned Crystals with Fivefold Symmetry

Twinned crystals with pentagonal symmetry have been observed and reported repeatedly since the earliest times [24–27]. These are mostly based on the approximation of the angle 70.52° between the octahedral 111 faces in the cubic system to 72°, which is one-fifth of a revolution. Five tetrahedra, bounded by 111 faces, may be twinned together to give a pentagonal bipyramid with its fivefold axis along the common [110] axis. This process may be continued so that 20 tetrahedra may be twinned together by the same rule to give an icosahedral cluster [28, 29]. Twinned crystals may often be recognized because of the frequent production of reentrant angles between faces. The fit is good because the 60° edge angle of the tetrahedron corresponds to the center-to-vertex vectors of the icosahedron, which are $\arctan(2) = 63.43°$ apart.

Similarly, regular dodecahedra nearly fill space and four can meet at a point, since the 108° angles between edges fit closely into the 109.47° between the spikes of a tetrahedral calthrop. This is the basis of the proposal by Tilton [30] of the existence of "vitrons" to account for the noncrystalline structure of glass. Essentially, he postulated dodecahedral cages of SiO_4 tetrahedra. Construction following this rule can be continued only for a short distance before strains build up and some dislocation or disclination has to occur. This type of model is still that most favored. The analogous structures, "hydrons," may be seen in clathrate hydrates (for example [31]). These clathrates are now of great economic importance as containers of methane in conditions of permafrost where they have to be melted to release the gases included.

9. Escaping from the Classical Lattice

If we wish to preserve in an extended structure, the fivefold symmetry possible for finite objects, then we have four possible lines of escape from the restrictions. These are *Hierarchy*, *Curved Space*, *Higher Dimensions*, and *Abandonment of Atomicity*.

9.1. Hierarchy

In what respect do crystals differ most decisively from biological structures?

A structure outside the framework of the crystal lattices can be produced in the following way [20]. In order to explain why the sky is black at night (Olbers' Paradox), Charlier suggested [32] that the universe had a hierarchic structure. The mean density thus tended to zero as the gaps increased. Our problem, dealing with the packing of atoms, is to devise a scheme that keeps the density up to a finite value by systematically filling in the gaps. It turned out that the Penrose tiling is such a scheme.

9.2. Positively Curved Space

If we insist on packing pentagons then the space that appears between them can be removed by curving the space and we obtain a dodecahedron with 12 pentagonal panels. Such folding cardboard figures were first developed by Albrecht Dürer. Dürer's diagram for constructing a truncated icosahedron [33] shows a figure that has recently returned to prominence as the shape of the international football (such a shape was known in Greek times) and as the cluster of 60 carbon atoms found by Harold Kroto [34] from graphite vaporized by a laser pulse. Kroto postulates a curving process driven by the occasional occurrence of fivefold rings in the carbon layers.

The idea of curved space has proved very fruitful and important. In virus particles, like the domes of Buckminster Fuller, the average coordination number of each particle is less than six and, as more particles are added, a sheet of positive Gaussian curvature is obtained which eventually closes to give finite particles of icosahedral symmetry. These units then have to pack according to some different rule, usually to give crystals in two or three dimensions.

Negative curvature can be obtained if the coordination number (for planar coordination) is greater than 6, but the topic will not be discussed here [35].

9.3. Higher Dimensionality

A tetrahedron (a simplex in three dimensions) has symmetry operations of order four, although it contains axes of rotational symmetry of orders only three and two. (The operations are those of the permutation of the vertices or fourfold inversion axes.) Similarly, a simplex in four dimensions with five vertices has symmetry operations of order five [36]. After the discovery of the projection method of generating Penrose tilings this symmetry was recalled and the geometry of N-dimensional crystallography has been greatly developed.

9.4. Abandonment of Atomicity

A further way of allowing fivefold symmetry in the plane (or in the solid) is to have fivefold axes everywhere and allow them to be indefinitely close to each other. Here, the population of the space by atoms of finite size must be abandoned, and this approach, due to David Mermin and his colleagues [37, 38], will not be followed here. Axes of any order can be allowed if we abandon atomicity. Fractal structures, continued to infinitely fine structure, have some of the necessary properties, but we wish here to stay with physical structures made of atoms.

9.5. Quasiequivalence

If we look at the crystal structure of solid hexamethyl benzene, where the planar molecule[7] has an axis of sixfold symmetry, we see that it is triclinic and thus the six CH_3 groups of the molecule are not exactly equivalent in the crystal. This is a very common circumstance and often the symmetry of the isolated molecule is not expressed in the symmetry of the crystal when many molecules are aggregated. Usually it is the local symmetry that must give way to the global ordering necessary for reducing the free energy to a minimum. Strict equivalence has to give way to approximate or quasiequivalence.[8]

Since we cannot have an infinite lattice, then atoms in a crystalline array are not absolutely equivalent, but only quasiequivalent.

9.6. Pentagonal and Icosahedral Molecules

If we look up the structure of crystalline cyclopentane we find that it is hexagonal with two molecules per unit cell. This is clearly impossible. Gas electron diffraction data from cyclopentane as a vapor was obtained very early in the history of crystallography [39]. It is now believed that the molecule of cyclopentane is slightly puckered.

On a larger scale, the β-subunits in the heavy riboflavine synthetase of *Bacillus subtilis* form pentamers [40] as do molecules of human serum amyloid P component [41]. In all these cases the expression of the fivefold symmetry is frustrated and the orthodox crystal symmetries win. Many important viruses (including that involved in AIDS) form icosahedral particles which then crystallize [10, 45, 44].

10. Quasicrystals

There has been, in fact, a steady progress toward the development of a biological crystallography. Two events in the early 1950s presaged this. At Birkbeck College, Aaron Klug and Rosalind Franklin began the study of the icosahedral arrangement of units in the spherical

---

[7] This was the first pictorial demonstration, by Kathleen Lonsdale, of the shape of the benzene ring.

[8] The word "quasi" entered the English language via Norman French and, from the customary usages of the legal profession is, in England, pronounced "kwei-sai" (OED).

viruses and how these icosahedra then formed crystals. This led to a study of the packing of icosahedra, in the course of which Bernal drew our attention to the possibilities of hierarchic packing mentioned by Charlier [42]. It was immediately clear that this was a way to fulfill the crystallographic requirement of identical surroundings but outside the framework of the 14 Bravais lattices [28]. Caspar and Klug [43] formulated the principle of quasiequivalence. On the surface of a sphere only 60 asymmetric objects can be situated so that they are all exactly equivalent. If more than 60 are arranged on a hexagonal lattice and this lattice is wrapped round an icosahedron, by cutting out sectors so that 12 lattice points have only fivefold instead of sixfold symmetry, then the units can only be quasiequivalent. Linus Pauling demonstrated the occurrence of icosahedral coordination in metals and also [46] (with Robert Corey) postulated the α-helix in chains of amino acids. Both these objects had symmetries incompatible with the site symmetries available in crystals. In particular, the α-helix was a structure with two incommensurable periodicities (pitch and the spacing of equivalent atoms along the chain), which meant that for a long chain (like keratin) crystallization was frustrated. The "double helix" of DNA, discovered by Watson and Crick, also has such incommensurability.

The term "quasicrystal," as coined by Levine and Steinhardt [47], refers back to the mathematical term "quasiperiodic," which means a pattern that can be built up as the sum a finite number of periodic waves. If these waves have integrally related periodicities then we get a real crystal, if not integrally related, then a quasicrystal. The term "quasilattice" was used by Mackay [48] in a sense associated with the usage of Caspar and Klug. In considering a crystal (or other structure) as the linear sum of a number of sine waves (Fourier synthesis), the relative phases of the waves (how peaks in one wave are placed with respect to peaks in another) are of the greatest importance.

10.1. Icosahedral Symmetry in Diffraction Patterns

When the X-ray (or electron, or neutron) diffraction pattern of a crystal is recorded then a three-dimensional lattice, called the weighted reciprocal lattice, is obtained. Each point in this lattice corresponds to a sine wave of density in the real structure. However, knowledge of the relative phases of the waves is lost. It follows that if, for example, an X-ray diffraction pattern shows fivefold symmetry, the original structure may be one of any of the infinite number of structures to be obtained by varying the phases of the Fourier waves. Structures that give the same diffraction pattern are called homometric.

Crystallographers had not had occasion earlier to consider the diffraction properties of those three-dimensionally periodic structures that were projections from some higher dimensionally periodic structure. They were surprised, therefore, to find a simple theorem showing that such structures should give diffraction patterns with sharp spots [49].

10.2. Nuclear Clusters

Linus Pauling has been continuously the main partisan for the recognition of the occurrence of the icosahedron in crystal and other structures. He also disturbed orthodox crystallography by inventing the α-helix, an irrational helix of protein resides (about 3.4 residues per turn), which was incompatible with strict crystallinity [46]. He was, moreover, the pioneer in showing the presence of icosahedral coordination polyhedra in metallic structures such as β-tungsten and particularly in $Mg_{32}(Al, Zn)_{49}$, which has proved to be closely connected with the quasicrystal phases. Paradoxically, but perhaps in character, Pauling led the main opposition to the recognition of quasicrystals as a new phenomenon and has insisted that they were twins. Pauling also proposed [50] that atomic nuclei were icosahedral clusters of nucleons, but this theory seems to have sunk without trace.

Levitov [51] has shown that the additional symmetry axes, 5, 8, 10, and 12 can occur in quasicrystals. The numbers are those for which the cosines of the angles $2\pi/N$ is of the form $a + b\sqrt{c}$, where a, b, and c are integers. These symmetry axes are in fact those found in quasicrystals, but Levitov's analysis uses physical rather than mathematical considerations. However, Lagrange has a theorem to the effect that such algebraic numbers, the roots of quadratic equations, can be expressed as continued fractions, which are periodic. This rule can be exploited to give self-similar rectangles of which the Golden Rectangle [52] is only the simplest (and the most slowly converging).

11. Neopythagoreanism

As a technical and intellectual challenge, characterized as the "Mount Everest of alicyclic chemistry," Paquette and his co-workers [53] have expended immense chemical virtuosity in creating the molecule dodecahedrane. In 1981 they had constructed dimethyl dodecahedrane and determined its molecular shape from X-ray crystal structure analysis. The dimethyl dodecahedrane is approximately f.c.c. and the dodecahedrane is expected to be f.c.c. It has an extraordinarily high melting point (430 ± 10°C). At the time of the last report dodecahedrane had not been crystallized; perhaps it might form a liquid with local icosahedral ordering although the gap between melting point and decomposition point was rather small. We wait with the greatest interest to hear about its properties.

12. Conclusion—Frustration

The main significance of fivefold symmetry for science is that it furnishes us with an explicit example of frustration, which has proved a most fertile concept in the physics of condensed matter and has been developed particularly by the French school of physicists. Neither we nor nature can have everything simultaneously—not all things are possible, even with or without God. We have only the freedom of necessity. "Nature must obey necessity" as Shakespeare (*Julius Caesar* IV:iii), Democritos, Monod, Bernal, and many others have also recognized. Science probes the limits of necessity and, in the case of fivefold symmetry, has found a corridor that leads us through to a new territory.

REFERENCES

1. H. S. M. Coxeter, *Regular Polytopes*. Macmillan, New York, 1948 (2nd. ed., 1963).

2. B. Grübaum and G. C. Shephard, in *A Handbook of Applicable Mathematics*, W. Ledermann and S. Vajda (eds.), Vol. VB, p. 728. Wiley, New York, 1985.

3. H. Zassenhaus, "Rubik's cube: a toy, a Galois tool, group theory for everybody," *Physica*, **114A,** 629–637 (1982).

4. S. Baer, *Zome Primer*. Zomeworks Corporation, Albuquerque, NM, 1970.

5. J. L. Rapoport, "The biology of obsessions and compulsions," *Sci. Am.*, **260,** 64 (March 1989).

6. *Time* [Magazine], 19 Jan. 1987, p. 66.

7. M. Gardner, *New York Review of Books*, 30 June 1988, 43–45.

8. G. Sarton, *Introduction to the History of Science*, I.113. Carnegie Institution, 1927.

9. P. M. Sanders, "Charles de Bovelles's treatise on the regular polyhedra (Paris, 1511)," *Ann. Sci.*, **41,** 513–566 (1984).

10. J. T. Finch and A. Klug, "Structure of Poliomyelitis virus," *Nature (London)*, **183,** 1709–1714 (20 June 1959).

11. D. Shechtman, I. Blech, D. Gratias, and J. W. Cahn, "Metallic phase with long-range order and no translational symmetry," *Phys. Rev. Lett.*, **53,** 1951–1953 (1984).

12. J. Laskar, "A numerical experiment on the chaotic behavious of the solar system," *Nature (London)*, **338,** 237–238 (1989).

13. R. Wright, *Three Scientists and their Gods*. p. 26. Times Books, New York, 1988.

14. D. J. de Solla Price, *Gears from the Greeks: The Anti-Kythera Mechanism—A Calendar Computer from ca. 80 BC*, New York, 1975. But see also A. G. Bromley, *Centaurus*, **29,** 5–27 (1986).

15. D'Arcy W. Thompson, *On Growth and Form*, Cambridge University Press, Cambridge, 1917.

16. *Euclide's Elements*, Isaac Barrow (ed.). London, 1732.

17. J. J. Burckhardt, *Die Symmetrie der Kristalle*. Birkhäuser, Basel, 1988.

18. B. Grünbaum and G. C. Shephard, *Tilings and Patterns*. Freeman, New York, 1987.

19. *The International Tables for Crystallography*, Vol. A, p. 777.

20. A. L. Mackay, "Generalised crystallography," *Izv. Jugoslav. Centra Krist.* **10**, 15–36 (1975).

21. G. Kowalewski, *Der Keplersche Körper und andere Bauspiele.* Verlag K. F. Köhlers Antiquarium, Leipzig, 1938.

22. P. J. Steinhardt and S. Ostlund, *The Physics of Quasicrystals.* World Scientific, Singapore, 1987.

23. A. L. Mackay, "A bibliography of quasicrystals," *Int. J. Rapid Solid.*, **2**(2), Suppl. S1–S41 (1987).

24. W. E. van der Veen, Z. *Kristallogr.*, **52**, 512 (1912/13).

25. R. L. Schwoebel, *J. Appl. Phys.*, **37**(6), 2515–2516 (1966).

26. K. Kobayashi and L. M. Hogan, "Fivefold twinned silicon crystals grown in an Al-16% Si melt," *Phil Mag. A.*, **40**(3), 399–407 (1979).

27. L. A. Bursill and R. L. Withers, "On the multiple orientation relationships between haematite and magnetite," *J. Appl. Crystallogr.*, **12**, 287–294 (1979). (See *J. Appl. Cryst.*, **13**, 346–353 (1980).

28. A. L. Mackay, "A dense non-crystallographic packing of equal spheres," *Acta Crystallogr.* **15**, 916–918 (1962).

29. S. Ino and S. Ogawa, "Multiply twinned particles at earlier stages of gold film formation on alkali halide crystals," *J. Phys. Soc. Jpn.*, **22**, 1365–1374 (1967).

30. L. W. Tilton, "Role of vitrons in alkali silicate binary glasses," *J. Res. NBS*, **60**(4), 351–364 (1958).

31. G. P. Johari and H. A. M. Chew, "O-H stretching in ice clathrate," Nature (*London*), **303**, 604 (1983).

32. C. V. L. Charlier, "How an infinite world may be built up," *Arkiv Mat. Astron. Fy.*, 22, 1–35 (1922).

33. E. Schröder, *Dürer Kunst und Geometrie.* Akademie-Verlag, Berlin, 1980.

34. H. W. Kroto, J. R. Heath, S. C. O'Brien, R. F. Curl, and R. E. Smalley, "C_{60}: Buckminsterfullerene," *Nature* (*London*), **318**, 162–163 (1985). H. W. Kroto and K. McKay, "The formation of quasi-icosahedral spiral shell carbon particles," preprint (1987).

35. A. L. Mackay, "Non-Euclidean crystallography," pp. 347–371. *Colloq. Mathematica Societatis Janos Bolyai*, **48**. Intuitive Geometry, Siofok, 1985.

36. A. L. Mackay and G. S. Pawley, "Bravais lattices in four-dimensional space," *Acta Crystallogr*, **16**, 11–19 (1963).

37. D. S. Rokhsar, D. C. Wright, and N. D. Mermin, "The two dimensional quasicrystallographic space groups with rotational symmetries less than 23-fold," *Acta Crystallogr.* A (1987).

38. N. D. Mermin, D. S. Rokhsar, and D. C. Wright, "Beware of 46-fold symmetry: The classification of two-dimensional quasicrystallographic lattices," *Phys. Rev. Lett.*, **58**(20), 2099–2101 (1987).

39. I. Hargittai and M. Hargittai (eds.), *Stereochemical Applications of Gas-Phase Electron Diffraction.* VCH, New York, 1988.

40. R. Ladenstein, H. D. Bartunik, M. Schneider, and R. Huber, "Structure analysis of an icosahedral multi-enzyme complex: heavy riboflavin synthetase from *Bacillus subtilis. Angew. Chem.*, **27**(1), 79–88 (1988).

41.   S. P. Wood, G. Oliva, B. P. O'Hara, H. E. White, T. L. Blundell, S. J. Perkins, I. Sardharwalla, and M. B. Pepys, "A pentameric form of human serum amyloid P component," *J. Mol. Biol.*, **202**, 169–173 (1988).

42.   J. D. Bernal, "Generalised crystallography," *The Origin of Life*, p. xv and Appendix 3. London, 1967. *Sov. Phys.*—Crystallogr., **13**, 811–831 (1969).

43.   D. L. D. Caspar and A. Klug, "Physical principles in the construction of regular viruses," *Cold Spring Harbor Symp. Quant. Biol.*, **27**,1 (1962).

44.   L. Liljas, "The structure of spherical virsues," *Prog. Biophys. Mol. Biol.*, **48**, 1–36 (1986).

45.   A. Acharya, E. Fry, D. Stuart, G. Fox, D. Rowlands, and F. Brown, "The three-dimensional structure of foot-and-mouth disease virus at 2.9A resolution," *Nature (London)*, **337**, 709–716 (1989).

46.   L. Pauling and R. B. Corey, *Proc. Natl. Acad. Sci. U.S.A.* **37**, 729 (1951).

47.   D. Levine and P. J. Steinhardt, "Quasi-crystals: A new class of ordered structures," *Phys. Rev. Lett.* **53**, 2477–2480 (1984).

48.   A. L. Mackay, "De Nive Quinquangula—On the pentagonal snowflake," *Sov. Phys. Cryst. Kristallog.*, **26**(5), 909–918 (1981).

49.   M. V. Jarić, *Introduction to Quasicrystals*, Vol. 1 of *Aperiodicity and Order*. Academic Press, Boston, 1988.

50.   L. Pauling, "The close-packed-spheron theory and nuclear fission," *Science*, **150**, 297–305 (1965)

51.   L. S. Levitov, "Why only quadratic irrationalities are observed in quasi-crystals," *Comm. Math. Phy.* Earlier as "Local rules for quasi-crystals," *Zh. E. T. F.*, **93**, 1832–1847 (1987).

52.   H. S. M. Coxeter, *Introduction to Geometry*, p. 164. Wiley, NY, 1961.

53.   L. A. Paquette et al., "Total synthesis of dodecahedrane," *J. Am. Chem. Soc.*, **105**, 5456 (1983). L. A. Paquette, "Plato's solid in a retort: The dodecahedrane story," in *Strategies and Tactics in Organic Synthesis*, pp. 175–200. Academic Press, New York, 1984. L. A. Paquette, D. W. Balogh, R. Usha, D. Kountz, and G. G. Cristoph, "Crystal and molecular structure of a pentagonal dodeca-hedrane," *Science*, **211**, 575–576 (1981). See also *Angew. Chem.*, **28/3** (1989).

2 Generalizing Crystallography: Puzzles and Problems in Dimension 1

Marjorie Senechal

1. Prologue

In Eudoxia, which spreads both upward and down, with winding alleys, steps, dead ends, hovels, a carpet is preserved in which you can observe the city's true form. At first sight nothing seems to resemble Eudoxia less than the design of that carpet, laid out in symmetrical motives whose patterns are repeated along straight and circular lines, interwoven with brilliantly colored spires, in a repetition that can be followed throughout the whole woof. But if you pause and examine it carefully, you become convinced that each place in the carpet corresponds to a place in the city and all the things contained in the city are included in the design, arranged according to their true relationship, which escapes your eye distracted by the bustle, the throngs, the shoving. All of Eudoxia's confusion, the mules' braying, the lampblack stains, the fish smell is what is evident in the incomplete perspective you grasp; but the carpet proves that there is a point from which the city shows its true proportions, the geometrical scheme implicit in its every, tiniest detail.

It is easy to get lost in Eudoxia: but when you concentrate and stare at the carpet, you recognize the street you were seeking in a crimson or indigo or magenta thread which, in a wide loop, brings you to the purple enclosure that is your real destination. . . .

An oracle was questioned about the mysterious bond between two objects so dissimilar as the carpet and the city. One of the two objects—the oracle replied—has the form the gods gave the starry sky and the orbits in which the worlds revolve; the other is an approximate reflection, like every human creation.

For some time the augurs had been sure that the carpet's harmonious pattern was of divine origin. The oracle was interpreted in this sense, arousing no controversy. But you could, similarly, come to the opposite conclusion: that the true map of the universe is the city of Eudoxia, just as it is, a stain that spreads out shapelessly, with crooked streets, houses that crumble one upon the other amid clouds of dust, fires, screams in the darkness.

—From Invisible Cities, by Italo Calvino [1]

2. What Is a Generalized Crystal?

For centuries, people have been searching for geometric patterns that exhibit the universal scheme of crystal structure in its tiniest detail. This search led eventually to the theory of infinite repeating patterns, which turned out to be a reliable guide to understanding the geometric properties of "normal" crystals. Does there exist a simple, aesthetically satisfying scheme that can help us understand the complex, nonperiodic structures we are studying today?

Since the discovery of X-ray diffraction by crystals in 1912—the Rosetta Stone of the solid state—crystallography has gradually broadened its scope to include patterns that are much more general than ideal periodic structures. Today, as Alan Mackay has pointed out [2], "Crystallography is only incidentally concerned with crystals . . . crystallography is rapidly becoming the science of structure at a particular level of organization, being concerned with structures bigger than those represented by simple atoms but smaller than those of, for example, the bacteriophage. It deals with form and function at those levels, particularly with the way in which large-scale form is the expression of local force."

The problems that are the subject matter of contemporary mathematical crystallography are not only very broad, they are also very deep. We know very little about the way in which local force expresses large-scale form: as the geometer S. S. Chern recently noted [3], the relation between local and global order is one of the mysteries of geometry. From the point of view of statistical mechanics, this problem has not been resolved even for periodic crystals!

A periodic crystal is a structure that can be abstractly modeled by a lattice (the lattice points represent atoms or molecules). We can consider the periodicity of the lattice to be its large scale form, however that periodicity was actually achieved. A "generalized crystal," then, ought to be a structure that can be modeled by something we can call a generalized lattice, whatever that may mean. Of course, every real crystal is a generalized crystal: no crystal has a structure that is a perfect lattice. Crystallographers deduce the ideal lattice structure of a real crystal by studying the diffraction patterns it produces, using their knowledge of how distortions affect diffraction. Figure 1, a series of plane patterns and their optical diffraction patterns, suggests some of the relations between diffraction patterns and the structures that produce them.

We are looking for something more general than these distorted lattices, yet something we can still call crystalline. Where should we look? Since 1981, when Alan Mackay produced a diffraction pattern by passing a laser beam through a mask of holes located at the vertices in a Penrose tiling [4], it has been assumed that the key feature of a generalized crystal is non-crystallographic symmetry. Notice, in Figure 1, that even though the distortions of the lattices are in some cases considerable, the rotational symmetries of the diffraction patterns are not other than twofold, threefold, fourfold, or sixfold. These are the symmetries permitted by the so-called "crystallographic restriction," a theorem that states that if a two- or three-dimensional pattern is periodic then any rotational symmetry in the pattern will have one of those orders. The diffraction patterns of the Penrose tiling and quasicrystals, which are sharp and clear, exhibit fivefold symmetry, which is not on the above list. Therefore the Penrose tilings and quasicrystals are not periodic, but, like periodic crystals, they produce nice diffraction patterns.

It is probable that if the diffraction patterns produced by these objects had fivefold symmetry but were hazy, no one would have paid much attention to them. Their noncrystallographic symmetry per se is not their "quasicrystalline" feature; they were dubbed quasicrystals because of the bright spots in their diffraction patterns. Thus a working definition of a generalized crystal might be a nonperiodic point pattern that has this diffraction property.

The question then is, what geometric features of the pattern are responsible for the bright spots? The answer to this question is still unknown, even in the simplest, one-dimensional case. In this article, I will attempt to show something of the scope of the problem, and why it is a difficult one.

The reader who wants to learn more about the mathematics of generalized crystals is in for an intellectual adventure. The literature is widely scattered in mathematics and physics journals, and almost none of it has been written with the general reader in mind. Three fundamental papers on one-dimensional generalized crystals are [5], [6], and [7]. Other important articles dealing with various aspects of the problem include [8], [9], [10], and [11]. For an overview of the quasicrystal problem, see [12]; for background in number theory, tiling theory, and substitution sequences, consult [13], [14], and [15].

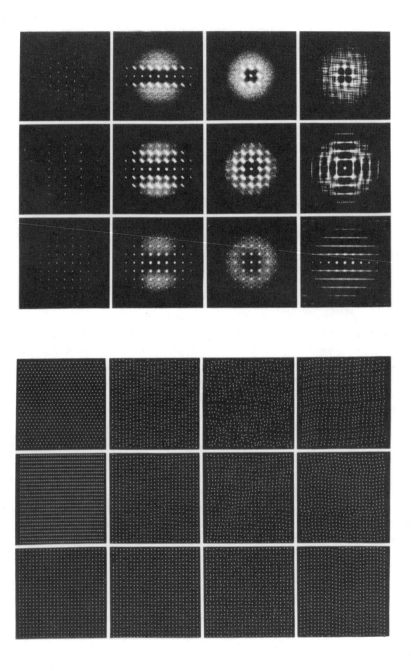

Figure 1. Ideal and distorted lattices (below) and their diffraction patterns (above). From G. Harburn, C. A. Taylor, and T. R. Welberry, *An Atlas of Optical Transforms*, Cornell University Press, Ithaca, 1975.

3. A Misleading Example: The Fibonacci Sequence

The most famous example of a one-dimensional generalized crystal is the Fibonacci sequence. This sequence, which plays a role in the theory of Penrose tilings, can be generated by a substitution rule,

$$a \rightarrow ab, \quad b \rightarrow a, \tag{1}$$

where a and b are adult and baby rabbits, or more abstractly, lengths of intervals (in this case $a = 1$, $b = 1/\tau$). As we apply this rule over and over again, ad infinitum, in the line-segment interpretation we construct in the limit an infinite sequence of points on the line with two spacings, 1 and $1/\tau$. The sequence is not periodic, but it can be coordinatized; it turns out to be the set of points

$$u_n = \frac{n}{\tau} + \frac{1}{\tau + 1} \left[\frac{n}{\tau} \right], \tag{2}$$

where $[x]$ is the greatest integer less than or equal to x, and $\tau = (1 + \sqrt{5})/2$ is the positive root of the quadratic equation $x^2 - x - 1 = 0$. Notice that

$$u_{n+1} - u_n = \frac{1}{\tau} + \frac{1}{\tau + 1} \left(\left[\frac{n + 1}{\tau} \right] - \left[\frac{n}{\tau} \right] \right) = \frac{1}{\tau} \text{ or } 1, \tag{3}$$

according as $[(n + 1)/\tau] - [n/\tau]$ is 0 or 1. A portion of the Fibonacci sequence is shown in Figure 2.

We can interpret the substitution rule as a linear transformation with matrix

$$\begin{pmatrix} 1 & 1 \\ 1 & 0 \end{pmatrix}.$$

The vector $(1, 1/\tau)$ is an eigenvector of this matrix, so the transformation fixes the line $y = x/\tau$. Now think of $\{u_n\}$ as a sequence of points on this line. The substitution rule tells us that adjacent segments of lengths $1/\tau$ and 1 can be grouped together to form larger segments. The eigenvalue belonging to $(1, 1/\tau)$ is τ, so the new lengths are τ and 1. Grouping again, the lengths are $\tau + 1$ and τ, and so forth; at each stage the ratio of the lengths is $\tau{:}1$. This means that the Fibonacci sequence is *self-similar*. The self-similarity property is very special: had we used any other length ratio for a and b, we could still use this rule to create a substitution sequence, but the larger tile lengths would not be proportional to the smaller ones.

The Fibonacci sequence satisfies our definition of a generalized crystal because there are bright spots in its diffraction pattern (Fig. 3).

Figure 2. A portion of the Fibonacci sequence.

$$0 \quad \tfrac{1}{\tau} \quad 1+\tfrac{1}{\tau}$$

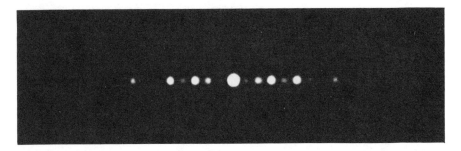

Figure 3. A diffraction pattern produced by the Fibonacci sequence. From S. Y. Litvin, A. B. Romberger, and D. B. Litvin, "Generation and experimental measurement of a one-dimensional quasicrystal diffraction pattern," *Am. J. Phys.* **56**, 72–75 (1988).

The Fibonacci sequence is very intriguing (indeed many people have become obsessed with it), but as an example of a generalized crystal it is misleading in many ways. As we will see, although it is nonperiodic it has an "average lattice" and so is not essentially different from the patterns in Figure 1. Moreover, its many interesting features are surprisingly independent: there are other generalized crystals that share some but not all of them. This makes any expectation of finding a scheme that encompasses all generalized crystals much more problematic.

Let us look at the Fibonacci sequence more closely. Notice, first of all, that the diffraction pattern in Figure 3 does not have noncrystallographic symmetry. Of course, being one-dimensional, the only rotational symmetry that it could possibly have is twofold. But this obvious remark shows clearly that *noncrystallographic symmetry is incidental to our problem.*

We will now show the the Fibonacci sequence (2) can be generated in at least four different ways. In fact, each method gives rise to a large class of interesting sequences. Some of the sequences are generalized crystals according to our definition; we will prove this in Section 4. But, as we will see in Section 5, it is by no means clear whether and how these classes are related to one another.

Method 1. We have already shown how the Fibonacci sequence can be generated by the substitution rule $a \to ab$, $b \to a$. This is a special case of a general procedure in which we construct substitution sequences using n segment lengths and any substitution rule. For example, suppose we use two lengths d and r. We can start the process with a segment of length d, say. We first substitute α segments of length d and β segments of length r for this segment. Next, we substitute α ds and β rs for each segment of length d, and γ ds and δ rs for each segment of length r. We continue this substitution process ad infinitum.

We can codify the process in a linear transformation T, where

$$T(d) = \alpha d + \beta r, \qquad T(r) = \gamma d + \delta r. \tag{4}$$

The matrix for T is

$$\begin{pmatrix} \alpha & \gamma \\ \beta & \delta \end{pmatrix}.$$

(As we have seen, for the Fibonacci sequence $\{u_n\}$, $\alpha = 1$, $\beta = 1$, $\gamma = 1$, and $\delta = 0$.) The sequence, which is the "limit" of infinitely many interations of T, will be nonperiodic for appropriatechoices of d and r. In the special case that

$$\begin{pmatrix} d \\ r \end{pmatrix}$$

is a right eigenvector belonging to the leading eigenvalue of T, the sequence will be self-similar.

Method 2. The very same sequence can be obtained by projecting a set of lattice points in the plane onto a line (Fig. 4). The lattice points to be projected lie in the strip of the plane between the lines $y = (1/\tau)x$ and $y = (1/\tau)x + \tau$. This is a special case of a more general construction that can be used to produce the two- and three-dimensional Penrose tiles and many other interesting patterns. We start with the integer lattice in a space of any dimension (these are the points with integer coordinates with respect to a Cartesian frame), and choose any irrational line (a line whose only

Figure 4. The Fibonacci sequence can be obtained by projection.

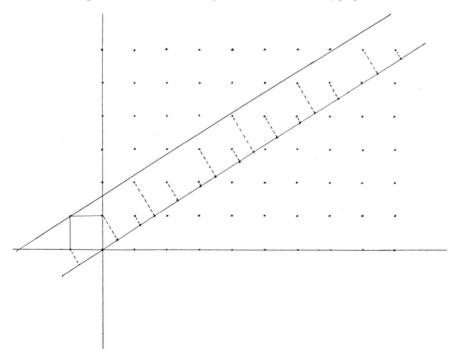

Figure 5. The Fibonacci sequence has an average lattice (slanted lines).

lattice point is the origin). We construct an infinite cylinder of finite radius with this line as axis, and project the points of the lattice that lie in the cylinder onto the axis or a line parallel to it. The projected points will form a sequence that is discrete and nonperiodic.

Method 3. The Fibonacci sequence can be obtained by perturbing the one-dimensional lattice whose coordinates are $(3/\tau - 1)n$ by a varying but bounded amount. To show this, we use the fact that for any real number $x, x = [x] + \{x\}$, where as above $[x]$ is the greatest integer $\leq x$, and $\{x\}$ is the "fractional" part of x. Then after multiplying (2) by τ, rearranging, and using the well-known relations $\tau^2 = \tau + 1$ and $1/\tau = \tau - 1$, we find that

$$\tau u_n = (3 - \tau)n - (1/\tau)\{n/\tau\}. \tag{5}$$

This means that the sequence (2) has an "average lattice" of spacing $3/\tau - 1$, with a disturbance or modulation of at most $1/\tau^2$ (Fig. 5).

Obviously this too is just one representative of a large class of sequences with average lattices,

$$x_n = n\alpha - \beta\{n\gamma\}. \tag{6}$$

The variation will be small if β/α is small; in any case it will be bounded.

Method 4. The Fibonacci sequence can be generated by a "circle map." We wind a linear lattice $n\alpha$ around a circle of circumference 1, and note which lattice points fall within an arc or "window" of some specified length Δ. We construct a new sequence by defining

$$y_n = A \qquad \text{or} \qquad y_n = B, \tag{7}$$

according to whether $n\alpha$ does or does not fall within the "window." The symbols A and B can be interpreted in many ways: they might be the numbers 0 and 1, or segment lengths l_1 and l_2, or vectors $\mathbf{u}$ and $\mathbf{v}$. For the Fibonacci sequence, the lattice has spacing $\alpha = 1/\tau$, the window arclength Δ is also equal to $1/\tau$, and A and B are line segments of lengths 1 and $1/\tau$ (Fig. 6).

These four methods appear to be very different, and to some extent they are very different. We see that each of them is also very general.

4. Computation

In this section, we will describe some of the techniques that can be used to predict, a priori, whether a pattern will produce a diffraction pattern with bright spots. In particular, we will show that some subclasses of the classes of sequences discussed in the previous section are generalized crystals.

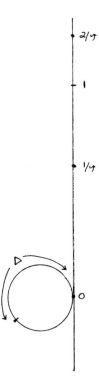

Figure 6. The Fibonacci sequence can be generated by a circle map.

In optics courses it is shown that we can predict the diffraction pattern produced by a set S of points in space by computing the Fourier transform of the "density function" which describes the location of the points of the set. If S is one-dimensional, its points can be labeled $r_i, i \in Z$, and the density function of S has the form

$$\rho(x) = \sum_n \delta(x - r_n), \qquad (8)$$

where $\delta(x - r_n)$ is a Dirac delta function located at the point $x = r_n$. Of course, neither $\rho(x)$ nor $\delta(x)$ is a function in the usual sense. The Dirac delta is the limit of a sequence of bell-shaped Gaussian curves of decreasing widths and increasing heights:

$$\delta(x) = \lim_{a \to 0} \frac{1}{\sqrt{2\pi a^2}} \exp\left(\frac{-x^2}{2a^2}\right). \qquad (9)$$

The Gaussian describes the density of a unit mass on the real line; as we decrease its width, we increase the amount of mass near the origin. In the limit, all of the mass is located at the origin. Thus

$$\delta(x) \, = \, 0 \quad \text{if} \quad x \neq 0 \quad \text{and} \quad \int_{-\infty}^{\infty} \delta(x) \, dx \, = \, 1. \tag{10}$$

Accordingly, we can think of $\delta(x)$ either as a limit of a sequence of functions or as a discrete measure. A rigorous theory that permits us to handle sums of delta functions can be constructed under either interpretation. The diffraction pattern produced by S is given by the Fourier transform of $\rho(x)$,

$$\hat{\rho}(t) \, = \, \sum_{n} \exp(2\pi i t r_n). \tag{11}$$

For the diffraction pattern to show bright spots, there must be values t_i of t for which $\hat{\rho}(t_i)$ is a delta function. In other words, $\hat{\rho}(t)$ must contain a sum of deltas (it may also contain a continuous component). The patterns in the top row of Figure 1 suggest that when the points of S form a linear lattice, so do the bright spots in its diffraction pattern. In the one-dimensional case, this means that if

$$\rho(x) \, = \, \sum \delta(x \, - \, \theta n) \tag{12}$$

for some real number θ, then $\hat{\rho}(t)$ should have the same general form.

That $\hat{\rho}(t)$ does have this form is in fact a famous theorem, the elegant *Poisson summation formula*, which states that

$$\sum \exp(2\pi i t n \theta) \, = \, \sum \delta(t \, - \, n/\theta). \tag{13}$$

Since the left-hand side of the formula is the Fourier transform of $\rho(x) = \Sigma \, \delta(x - n\theta)$, this says that the transform describes a lattice with spacing $1/\theta$.

The sets S we are considering are more general than lattices, but they are like lattices in that their density functions are sums of delta functions (such sets are sometimes called *Dirac combs*). So we want to characterize those Dirac combs whose Fourier transforms are again Dirac combs, or at least contain them. A complete characterization still eludes us, but for appropriately chosen parameters, the four classes of sequences described in Section 3 provide examples of such sets S.

Projection (and circle-map) sequences are generated by systematically selecting some of the points of a lattice and then transforming the selected points. For all projected sequences, and for circle map sequences with appropriate parameters, the density function $\rho(x)$ is a product of the density function of the lattice and of the function that does the selecting (this is a characteristic function, which assumes the value 1 at the points it wants to retain and 0 everywhere else). Because $\rho(x)$ is a product, we can use standard theorems about products of transforms (albeit in rather nonstandard ways) to compute the Fourier transforms of the sequences themselves. Projected sequences are always generalized crystals; for the circle map sequences, the existence of deltas in the Fourier transform depends (in rather complicated ways) on how α and Δ are related and how A and B are interpreted.

Sequences with average lattices of the type

$$x_n = n\alpha - \beta\{n\gamma\} \tag{14}$$

are always generalized crystals. In this case we can evaluate

$$\hat{\rho}(t) = \sum \exp[2\pi i(n\alpha - \beta\{n\gamma\}t)] \tag{15}$$

directly. Note that

$$\exp[2\pi i(n\alpha - \beta\{n\gamma\}t)] = \exp(2\pi i n\alpha)\exp(-2\pi i\beta\{n\gamma\}t) \tag{16}$$

and, since $\{x\}$ is a periodic function, we can expand the second factor in a Fourier series:

$$\exp(-2\pi it\beta\{n\gamma\}) = \sum_{m=-\infty}^{\infty} c_m(t) \exp(2\pi i mn\gamma), \tag{17}$$

where $c_m(t)$ is the mth Fourier coefficient. After substituting (17) into (16), (16) into (15), and rearranging, we have

$$\hat{\rho}(t) = \sum_m c_m(t) \sum_n \exp[2\pi i n\gamma(t + m)] \tag{18}$$

which, if we apply the Poisson summation formula to the term on the right, is equal to $\sum_n \sum_m c_m(t)\delta[n - \gamma(t + m)]$, or

$$\hat{\rho}(t) = \sum_n \sum_m c_m(t)\delta[t - (n/\gamma - m)]. \tag{19}$$

This means that $\hat{\rho}(t)$ is a sum of delta functions located at the points $t = n/\gamma - m$, for all integers m and n. Notice that this set of points is dense on the real line, and that these deltas have "weights" that are the Fourier coefficients $c_m(t)$.

The computation of the Fourier transforms of sequences produced by substitution rules is quite different, and not all substitution sequences are generalized crystals. The properties of the substitution matrix T determine whether and where deltas are appear in $\hat{\rho}(t)$.

As we iterate T, we obtain increasingly long finite sequences of points that are the endpoints of the line segments we have substituted for the line segments in the previous stage. Assume that at stage n the right-most endpoint has coordinate r_n. Equation (11) suggests that we should be able to evaluate the Fourier transform at $t = t_i$ by studying the growth of the finite sum

$$S_N(t_i) = \sum_{n=1}^{N} \exp(2\pi i t_i r_n) \tag{20}$$

as N increases without bound. Since $|\exp(2\pi i t_i r_n)| = 1$, the sum in (20) is at most N. We can conclude that $\hat{\rho}(t_i)$ will be a delta function if the sum grows like N, that is, if

$$\lim_{N\to\infty} \frac{1}{N} \left| \sum_{n=1}^{N} \exp(2\pi i t_i r_n) \right| = c, \tag{21}$$

where c is a positive constant.

We can evaluate the limit in Eq. (21) because, at each stage of the substitution process, r_n is a linear combination of the nth powers of the eigenvalues of T. Thus, what happens in the limit depends on the sizes of the eigenvalues. If T is a 2×2 matrix, it has two eigenvalues, λ_1 and λ_2. If $|\lambda_1| > 1$ and $|\lambda_2| < 1$ then r_n behaves, asymptotically, like λ_1^n, since the powers of λ_2 become arbitrarily small. In this case,

$$\hat{\rho}(t_i) = \lim_{N \to \infty} \sum_{n=1}^{N} \exp(2\pi i t_i \lambda_1^n) \tag{22}$$

and $\hat{\rho}$ will behave like a delta function at t_i if, as n increases, $t_i \lambda_1^n$ approaches integral values.

The numbers λ that have this property for some real number $t_i \neq 0$ are known as Pisot–Vijayaraghavan, or P–V, numbers. We are in luck: the P–V numbers are also precisely the numbers κ such that $|\kappa| > 1$ and κ is an eigenvalue of an integer matrix whose other eigenvalues have absolute value less than one. Thus if $|\lambda_1| > 1$ and $|\lambda_2] < 1$, there are values of t for which $\hat{\rho}(t)$ is a delta function and the substitution sequence is a generalized crystal.

It turns out that for all four classes of generalized crystals, the points t at which the delta functions in the transform are located are not isolated, but are everywhere dense on the line: the situation that we found in the average lattice case is typical. We see bright spots in the diffraction pattern because some of the deltas have weights which make them appear sharper than the others.

5. Searching for the Carpet

We have seen that each of the four sequence-generating methods that we have discussed includes at least a large subclass of generalized crystals. Is this due to some underlying type of geometric order common to all of them? This is one of the most important questions of generalized crystallography.

All of these methods produced sequences that are discrete and in which there is an upper bound on the distance between successive points. In each case there are also other interesting local patterns. But it is not yet clear whether these local patterns are essentially the same in all four cases, nor is it known which properties of the local geometry are responsible for the Dirac deltas.

Be that as it may, there is some overlap among these classes of sequences. It would be very nice if we could conclude that they really form one large class, or even if we could be sure that every one-dimenensional generalized crystal belongs to one of these four classes. Unfortunately (or fortunately, if you enjoy studying sequences), the situation is much more complicated. For example:

- Some projected sequences have average lattices. For example, this will be the case if the width of the "cylinder" is chosen to be the diagonal of the unit square of the plane lattice (Fig. 7). These particular projected sequences can also be generated by circle maps. But if we choose the radius of the cylinder differently, there may be no average lattice and no circle map. Also, in general, a projected sequence will not be invariant under any substitution rule.

- We can construct sequences with average lattices by using other kinds of modulations, not only {x}. It is not known which of these more general sequences can be obtained by projections, circle maps, or substitutions.

- Circle maps can be used to generate sequences that do not have average lattices! Whether they do or do not depends again on the values that have

Figure 7. When the radius of the cylinder is equal to the length of a diagonal of a unit square, a projected sequence will have an average lattice.

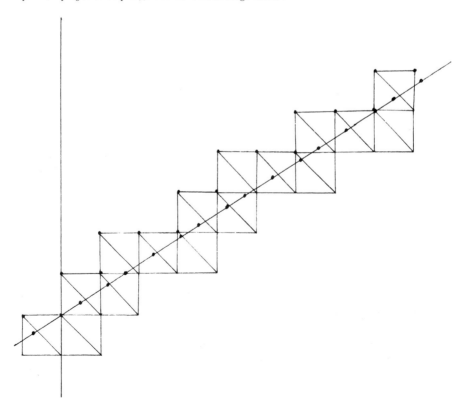

been assigned to the parameters. When an average lattice does exist, it may require a more complex construction than that in Figure 7.

- Sequences generated by substitution rules are *contained* in projected sets if the leading eigenvalue is a P–V number, but it is not known whether they are projected sets per se. It is also not known which of them have average lattices, nor which can be generated by circle maps.

- If it should turn out that there are some subtle but crucial geometric properties shared by the four classes of sequences that are responsible for their beautiful diffraction patterns, it will probably also be the case that these four classes by no means exhaust the list of one-dimensional generalized crystals.

Although a casual reading of the quasicrystal literature might suggest that the one-dimensional case was settled long ago, careful readers will be properly suspicious of all definitive pronouncements. There appears to be no easy solution to the problem of characterization; perhaps there is no general characterization. As Poul Anderson once remarked [16], *"I have yet to see a problem, no matter how complicated, which, when looked at in the right way, did not become still more complicated."*

REFERENCES

1. I. Calvino, *Invisible Cities*, W. Weaver (trans.), Harcourt Brace Jovanovich, San Diego and New York, 1974, "Cities and the Sky 1," 96–97.

2. A. L. Mackay, "Generalized crystallography," *Izvj. Jugosl. centr. krist., Zagreb*, **10**, 15–36 (1975).

3. S. S. Chern, Lecture, University of Massachusetts, Amherst, spring 1989.

4. A. L. Mackay, "Crystallography and the Penrose pattern," Physica **114A**, 609–613 (1982).

5. S. Aubry, C. Godrèche, and J. M. Luck, "Scaling properties of a structure intermediate between quasiperiodic and random," *J. Stat. Phys.*, **5**, 1033–1075 (1988).

6. E. Bombieri and J. E. Taylor, "Quasicrystals, tilings, and algebraic number theory: Some preliminary connections," *Cont. Math.* **64**, *Am. Math. Soc.*, 241–264 (1987).

7. A. Katz and M. Duneau, "Quasiperiodic patterns and icosahedral symmetry," *J. Phys.* **47**, 181–196 (1986).

8. N. G. De Bruijn, "Quasicrystals and their Fourier transform." *Kon. Nederl. Akad. Wetensch. Proc. Ser A* **89** (Indagationes Mathematicae 48), 123–152 (1986).

9. N. G. De Bruijn, "Updown generation of Beatty sequences," preprint.

10. C. Godrèche and C. Oguey, "Construction of average lattices for quasiperiodic structures by the section method," *J. Phys. France* **51**, 21–37 (1990).

11. R. Porter, "The applications of the properties of Fourier transforms to quasi-crystals," M. Sc. Thesis, Rutgers University, 1988.

12. M. Senechal and J. E. Taylor, "Quasicrystals: The view from Les Houches," *The Math. Intell.* vol. 12, no. 2, 54–64, (1990).

13. J. W. S. Cassels, *An Introduction to Diophantine Approximation.* Cambridge University Press, Cambridge, 1965.

14. B. Grünbaum and G. S. Shephard, *Tilings and Patterns*. Freeman, New York, 1987.

15. M. Quffélec, *Substitution Dynamical Systems—Spectral Analysis*. Lecture Notes in Mathematics 1294, Springer-Verlag, New York, 1987.

16. P. Anderson, *New Sci.*, **25** (1969). (Quoted in A. L. Mackay, *Scientific Quotations: Harvest of a Quiet Eye*. New York Institute of Physics, New York, 1977.)

3 Dynamic Origin of Fivefold Symmetry

**Roald Z. Sagdeev and
George M. Zaslavsky**

1. Introduction

"Chaos" is the random motion of a system without any random forcing. How is that possible? From the scientific community many can directly give a more of less understandable formal answer to the question as well as examples from different branches of scientific and nonscientific fields. Twenty years ago that was impossible; instead of "chaos" scientists introduced more cautious terms such as "irregularity," "nonperiodicity," and "quasirandomness." The word closest to "chaos" was "stochasticity."

"Chaos" comes from the instability of orbits. A small perturbation in the initial condition appears to make exponential growth of the average distance between the perturbed and unperturbed orbit. This property is known as local instability of a system in its phase space.

The phase space of an arbitrary system can be considered like a geometric object filled with chaotic or regular orbits. With some level of reliableness the area of chaotic motion forms a finitely measured region in the full phase space. This region can be studied both as a geometric and topologic object.

Consider a Hamiltonian close to integrable. The Hamiltonian of such a system can be written in the form

$$H = H_0 + \epsilon V \tag{1}$$

where H_0 is the unperturbed Hamiltonian for an integrable system, ϵV is the potential of perturbation, and ϵ is a small parameter of the perturbation.

The orbits related to H_0 are regular in full phase space. For very general conditions the perturbation leads to destruction of regular motion in some zones of phase space and to the onset of chaotic motion in these zones. It is proved for some cases that the area of destruction of irregular orbits tends to zero as $\epsilon \to 0$ [1]. So, the area of chaos forms in the phase space a small measured object that induces a pattern in phase volume.

For some simple cases the structure of the small area of chaos can be eliminated in the limit $\epsilon \to 0$ [2]. These are the stochastic layers and stochastic webs. The stochastic layers replace the small region close to the unperturbed separatrices that are destroyed due to a perturbation. The stochastic web is formed by a connected set of stochastic layers.

Existence of the webs was predicted by Arnold [3] from the topologic consideration of the phase space of the system (1). Infinite transport of particles is possible inside channels of the web. That is the so-called Arnold diffusion, which usually appears if a number (N) of degrees of freedom of a system is

$$N > 2. \tag{2}$$

The property of interest is the ergodicity of transport along the web. If the initial condition of an orbit is taken inside the web then this is the only orbit that will fulfill the area of the web. Such a random walking process of only one particle with an appropriate initial condition inside the web's area may be considered as a process of visualizing the stochastic web.

The symmetry properties of a dynamic system should be implied by the symmetry properties of the web of the system. Such systems with crystal and quasicrystal symmetry for their webs were introduced [4, 5]. Different properties of webs with symmetry and quasisymmetry are described by Zaslavsky et al. [2], and geometric properties of webs with quasicrystal symmetry are mentioned by Arnold [6]. In these kind of systems the webs appear even for $N = 1\frac{1}{2}$ [compare with (2)]. Consider the manifold G_t of all points from full phase space of a system at the moment t. Let $\hat{T}$ be the shift operator that acts on G_t and makes the shift of a system orbit at time T along its trajectory

$$G_{t+T} = \hat{T}G_t. \tag{3}$$

Equation (3) determines the group of motion acting on G. There are different invariants of this group. For example, the manifold of all stagnation points of G forms a trivial invariant manifold, which can be considered a realization of some symmetry.

Dynamic induction of space-realizing symmetries was discussed in detail by Weyl [7]. New understanding of this idea arises from the existence of areas of chaos in the phase space of a dynamic system. Regions of chaotic motion are invariant under the arbitrary dynamic shift T along the orbits. So it may be considered the realization of some symmetry. But it is important to note that the real chaotic submanifold is not homogeneous in phase space; it has many holes (islands). Therefore an area of chaos creates an invariant pattern in phase space. If chaos is weak then chaotic zones can form some tiling of phase space, and the symmetry of the tiling is governed by the symmetry of the dynamic properties of the system. These ideas were realized [4, 5] in the construction of the generator of the tiling of a plane with arbitrary q-fold symmetry (see also [8]).

One of the most important points in understanding the dynamic origin of symmetry is that the arbitrary orbit fulfills the connected invariant area of chaos (the connected web) because of the ergodic property of the chaotic orbit. This shows that one chaotic orbit leads to a tiling of the phase space with symmetry of the stochastic web. This tiling may be of fivefold symmetry, which is isomorphic to the well-known Penrose tiling [9, 10] or to the not-so well known Islamic ornaments from the Alhambra (Grenada) [11].

2. Web Mapping as Dynamic Generator of Crystal and Quasicrystal Symmetry

Consider the dynamic system of two variables (u, v) with Hamiltonian

$$H = \frac{1}{2}(u^2 + v^2) - K \cos v \sum_{n=-\infty}^{\infty} \delta(t - n\,\alpha) \tag{3}$$

where α and K are parameters. Equation (3) describes a linear oscillator with frequency one, forcing by a periodical set of δ-functions kicks of intensity K. The period of the kicks is equal to α. The equations of motion are

$$\dot{u} = \frac{\partial H}{\partial v} = v + K \sin v \sum_{n=-\infty}^{\infty} \delta(t - \alpha n) \tag{4}$$

$$\dot{v} = -\frac{\partial H}{\partial u} = -u.$$

It is possible to replace differential Eqs. (4) by the difference equations (mapping) due to existence of δ-functions in (4). Let $(\bar{u}, \bar{v})$ and (u, v) be the two pairs of variables just before two successive kicks. Then the exact connection between $(\bar{u}, \bar{v})$ and (u, v) is

$$\hat{M}_\alpha \begin{cases} \bar{u} = (u + \sin v) \cos \alpha + v \sin \alpha \\ \bar{v} = -(u + \sin v) \sin \alpha + u \cos \alpha. \end{cases} \tag{5}$$

The transformation $\hat{M}_\alpha$ may be considered as a product of two inversions:

$$\hat{M}_\alpha = R(\alpha)P(K)$$

where

$$P(K): \begin{pmatrix} u \\ v \end{pmatrix} = \begin{pmatrix} u + K \sin v \\ v \end{pmatrix}$$

$$R(\alpha): \begin{pmatrix} u \\ v \end{pmatrix} = \begin{pmatrix} u \cos \alpha & v \sin \alpha \\ -u \sin \alpha & v \cos \alpha \end{pmatrix}.$$

$P(K)$ is transformation with nonlinear perturbation proportional to K and $R(\alpha)$ is linear rotation on angle α.

The resonance rotation appears under the condition

$$\alpha = 2\pi/q \tag{6}$$

where q is integer. The more general case of $\alpha = 2\pi$ (rational) will not be considered here. Corresponding mapping

$$\hat{M}_q = \hat{M}_{\alpha=2\pi/q} \tag{7}$$

will be called web mapping. The reason is that the phase portrait of the system (u, v) generated by $\hat{M}_q$ has an area of chaos in the form of infinite connected webs. For example, if $q = 4$ then the mapping $\hat{M}_4$ is

$$\hat{M}_4: \bar{u} = v, \qquad \bar{v} = -u - K \sin v,$$

and the web has fourfold crystal symmetry (Fig. 1). All points of Figure 1a belong to the only orbit of (5) or (6) for $q = 4$. Their plots are done for a finite time. The larger the time of plotting, the larger will be the size of the web.

The web is not the only area of chaos. There is an infinite number of other disconnected small areas of chaos that are separated from the web

Figure 1. Fourfold symmetric stochastic web (a) and detail (b).

a   b

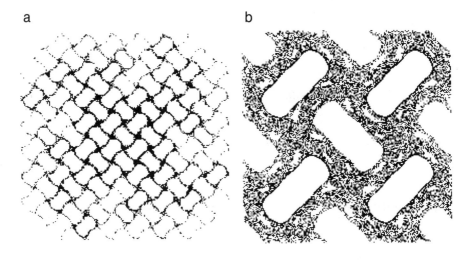

by invariant regular orbits. Many gaps with very complicated phase portraits inside exist in the web. So the web characterizes only some special infinite connected part of the phase plane fulfilled with chaotic orbits. If K is small then the webs are very thin. The thickness of the web is exponentially small [2, 5]

$$\delta \sim \exp(-\text{const}/K). \tag{8}$$

For small enough K many details of the fine structure of the web such as small gaps and small meanderings of the boundaries of the web can be ignored. This infinite connected web realizes the fourfold symmetry tiling of the plane.

The main property of (7) is the existence of the connected webs for arbitrary q-fold symmetry. Examples of webs with fivefold quasicrystal symmetry are given on Figure 2. Not all details of the webs are displayed because of a finite time for observations. But a set of parallel direct lines, rotated five times, is clearly seen. The grid of lines may be connected with a so-called Ammann lattice [12] (see details in [2]). That is the principal property of fivefold symmetry.

Use of the mapping, $\hat{M}_\alpha$, offers a great advantage over many other methods to determine quasicrystal symmetry. It is an explicit form for generating such symmetries by using the only dynamic orbit inside the web, which has the needed symmetry controlled by the only parameter q. An example for $q = 7$ is given in Figure 3.

Figure 2. Webs with fivefold symmetry.

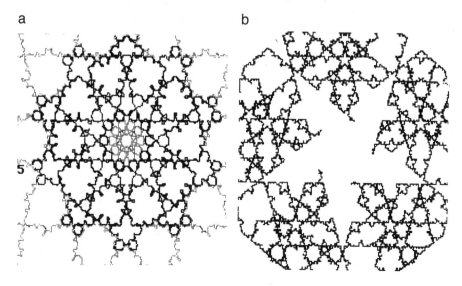

a

b

5

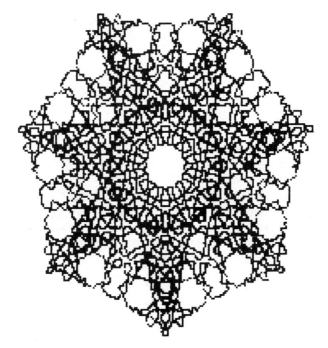

Figure 3. Web with sevenfold symmetry.

3. Skeletons of the Webs

A roughening procedure to delete many small details of webs can be introduced as an averaging of the dynamic system (3) over high frequencies. This leads to the stationary Hamiltonian system

$$\overline{H}_q = \sum_{j=1}^{q} \cos(\mathbf{Re}_j) \qquad (9)$$

where

$$\mathbf{R} = (u, v)$$

is the vector of position on the phase plane (u, v), and

$$\mathbf{e}_j = (\cos \frac{2\pi}{q} j, \sin \frac{2\pi}{q} j), \qquad (j = 1, \ldots, q)$$

are the unit vectors that form the regular star. The connection between (3) and (9) is not so simple, but in some sense Eq. (9) can be considered the first approximation to (3), made explicit for a constant multiplier [2, 5].

The Hamiltonian $\overline{H}_q$ describes a stationary dynamic system with one degree of freedom and without any chaos. Lines of constant level $\overline{H}_q =$

const = E can be considered and their topology on the phase plane (u, v) depends on the value of E.

For example, for $q = 4$

$$\overline{H}_4 = \cos u + \cos v. \tag{10}$$

Solution of the equation

$$\cos u + \cos v = \text{const} \neq 0$$

gives an infinite number of identical closed curves that tiles the full plane with the symmetry of the square lattice. But the lines of the singular value $\overline{H}_4 = 0$ form the real square lattice on the plane (u, v). All elements of this lattice are separatrices and all their intersections are saddle points. So the tiling generated by the Hamiltonian (10) is of fourfold symmetry. The width of the square lattice that appears for $\overline{H}_4 = 0$ is zero. But if we consider small layer δH in the vicinity of $\overline{H}_4 = 0$ and consider the whole family of lines of constant levels with $\overline{H}_4 \in (-\delta H, \delta H)$ then the square web of the width of order δH replaces the square lattice of zero width.

The last consideration is very important in understanding how skeletons with quasicrystal symmetry appear. Only for $q = 3, 4$, or 6, the saddle points that produce the connected web are situated on the same plane as $\overline{H}_q = $ const. But for any other values, $q \neq 3, 4$, or 6, there is no value of $\overline{H}_q$ for which the system of separatrices forms an infinite connected network. But such a connected network appears as a web of finite width if we consider the family of all orbits belonging to the values of

$$\overline{H}_q \in (\overline{H}_{q0} - \delta H, \overline{H}_{q0} + \delta H). \tag{11}$$

Here the value $\overline{H}_{q0}$ is some special value of $\overline{H}_q$ for which the distribution of a number of saddle points versus $\overline{H}_q$ reaches its maximum [2].

For $q = 5$ the $\overline{H}_{q0} \approx 1$. The correspondent skeleton with fivefold symmetry is shown in Figure 4. The analogous skeleton can be obtained for $q = 7$, 8, and 12 (Fig. 5). These figures show that Hamiltonian $\overline{H}_q$ can be considered as a dynamic generator of skeletons both for crystal and for quasicrystal symmetry.

A comment of special significance involves the Fourier spectrum of patterns. Let S_Γ be the manifold of all points that belong to a pattern from area Γ. Let

$$\delta(\mathbf{R}) = \begin{cases} 1, & \mathbf{R} \in S_\Gamma \\ 0, & \mathbf{R} \notin S_\Gamma, \end{cases}$$

where $\mathbf{R}$ is vector of arbitrary point from Γ. In other words $\delta(\mathbf{R})$ is a characteristic function of the given pattern. Then the Fourier spectrum of the pattern is defined as

$$S(\mathbf{k}) = \lim_{V(\Gamma) \to \infty} \frac{1}{(2\pi)^2} \frac{1}{V(\Gamma)} \int_\Gamma e^{i\mathbf{k}\mathbf{R}} \delta(\mathbf{R}) \, d\mathbf{R}, \tag{12}$$

Figure 4. Skeleton with fivefold symmetry.

where $\mathbf{V}(\Gamma)$ is the volume of the phase space area Γ. If we apply (12) to some pattern then we will receive a new two-dimensional Fourier spectrum pattern. Zaslavsky et al. [2] show that the results of the Fourier spectrum pattern in the case $q = 5$ for the web (Fig. 2a) and for the skeleton (Fig. 4) almost absolutely coincide with the X-ray spectrum of real quasicrystal from Shechtman et al. [13].

4. Decorations and Similarity

All the webs described have the property of self-similarity. That is a special group property mentioned by Weyl [7] and discussed in more detail in Shubnikov [14]. Stochastic webs are fractals and their similarity is a typical property of fractals [15]. So the detailed shape of the webs is very complicated. The more precisely we wish to determine the shape of its boundaries, the more complicated is the pattern observed. This property is inherent in any stochastic layer, and can hardly be discussed in terms of

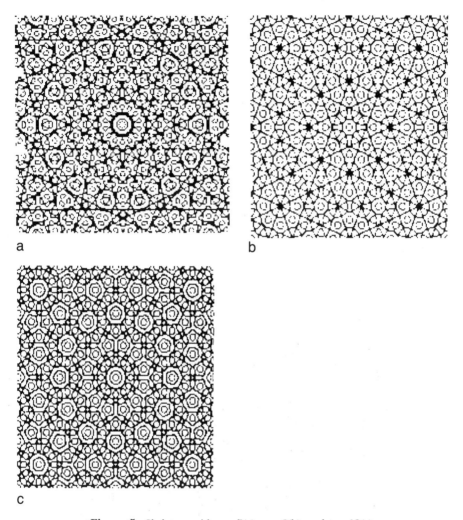

Figure 5. Skeletons with $q = 7$(a); $q = 8$(b); and $q = 12$(c).

any exact symmetry. The interaction of translational and rotational symmetries should disrupt both, even if the coupling constant K is small. But defects of a symmetry may be weak and therefore symmetry will exist in some approximate and perhaps poorly defined sense.

We understand at an intuitive level that a certain degree of coarsening of a web will make it more regular, i.e., more "symmetric." Sometimes it is useful to move away from attempts to define some "pure" symmetry and to use in its place a "quasisymmetry." This is the role played by the mapping $\hat{M}_q$, which may be thought of as a generator of tilings with a "quasisymmetry" of the "quasicrystal" type. In this sense the smoothed patterns specified by

Hamiltonian $\overline{H}_q$ are more "regular." When we go from a web to smoothed skeletons, there is some smoothing out of kines, and certain elements disappear. Thus the skeletons may be considered as a decoration of the webs.

Skeletons may be decorated also. The simple example (Fig. 6) shows the way to get the Penrose tiling from the skeleton on Fig. 4 [s].

Skeletons can be considered stencils to create different ornaments. Figure 7a gives an example from Islamic art in Tbilisi (Georgia, USSR). It consists of decagons as the main element of the ornament. But this kind of tiling does not possess fivefold symmetry. That is clearly seen from Figure 7b, in which the lines of the ornament are plotted over the skeleton from Figure 5 treated as the stencil. The correct way of including such decagons into fivefold symmetry tiling is shown in Figure 7c.

The last example (Fig. 8) shows the sevenfold symmetry tiling that includes only three kind of rhombuses [2].

Figure 6. Penrose tiling from skeleton with fivefold symmetry.

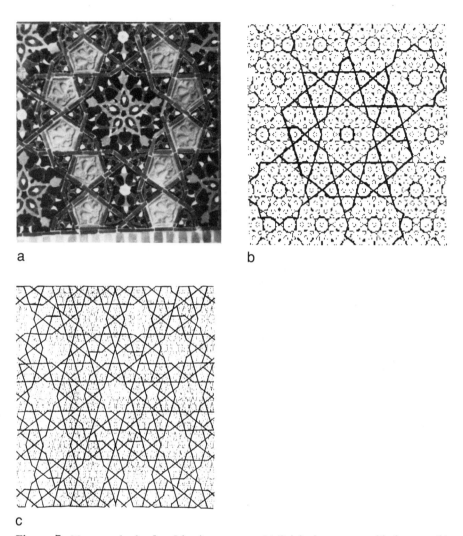

Figure 7. Metamorphosis of an Islamic ornament. (a) Original ornament with decagon. (b) Decoding the ornament by adjusting it to a skeleton. (c) New ornament with the same decagon but with fivefold symmetry.

5. Conclusions

Dynamic generating of different quasicrystal symmetries shows the special role of chaotic orbits and invariant regions of chaos in the phase space of a system. The topology of dynamic systems is controlled by singular elements of the system. This property can be considered as an indication of the way to create various kinds of symmetries. Sometimes a geometric version of this process of constructing the given symmetry is much more complicated than the dynamic way. That is because of the possibility of weak chaos. A thin area of chaos such as stochastic layers and

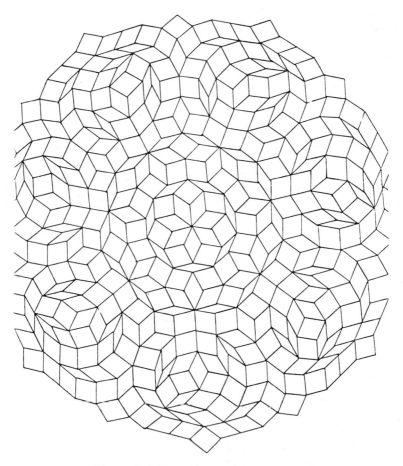

Figure 8. Tiling with sevenfold symmetry.

stochastic webs fits small regions between elements of tiling. So the packing of the phase space by the elements of regular motion will be not absolutely dense. Chaos creates the possibility of making a quasisymmetric tiling.

REFERENCES

1. V. I. Arnold, *Mathematical Methods of Classical Mechanics*. Springer, New York, 1978.
2. G. M. Zaslavsky, R. Z. Sagdeev, D. A. Usikov, and A. A. Chernikov, "Minimal chaos, stochastic webs and structures of quasicrystal symmetry." *Sov. Phys.—Uspekhi*, **31**, 887 (1988).
3. V. I. Arnold, "On the instability of dynamical systems with many degrees of freedom," *Dokl. Akad. Nauk USSR*, **142**, 758 (1962).

4. G. M. Zaslavsky, M. Yu. Zakharov, R. Z. Sagdeev, D. A. Usikov, and A. A. Chernikov, "Stochastic web and diffusion of particles in a magnetic field," *Sov. Phys. JETP*, **64**, 294 (1986).

5. G. M. Zaslavsky, M. Yu. Zakharov, R. Z. Sagdeev, D. A. Usikov, and A. A. Chernikov, "Generation of ordered structures with a symmetry axis from a Hamiltonian dynamics," *JETP Lett.*, **44**, 451 (1986).

6. V. I. Arnold, "Remarks on quasicrystallic symmetries," *Physica D*, **33**, 21 (1989).

7. H. Weyl, *Symmetry*. Princeton Univ. Press, Princeton, 1952.

8. A. A. Chernikov, R. Z. Sagdeev, D. A. Usikov, and G. M. Zaslavsky, "Symmetry and chaos," *Comp. Math. Appl.* **17**, 17–32 (1989), and in *Symmetry 2: Unifying Human Understanding*, I. Hargittai (ed). Pergamon Press, Oxford, 1989.

9. R. Penrose, "The role of aesthetics in pure and applied mathematical research," *Bull. Inst. Math. Appl.*, **10**, 266 (1974).

10. M. Gardner, "Mathematical games," *Sci. Am.*, **236**, 110 (1977).

11. D. Wade, *Pattern in Islamic Art*. Studio Vista, London, 1976.

12. B. Grünbaum and G. S. Shephard, *Tiling and Patterns*. Freeman, New York, 1987.

13. D. Shechtman, I. Blech, D. Gratias, and J. W. Cahn, "Metallic phase with long-range orientational order and no translational symmetry," *Phys. Rev. Lett.*, **53**, 1951 (1984).

14. A. V. Shubnikov, "Symmetry of similarity," *Sov. Phys. Crystallogr.*, **5**(4), 469 (1961), and in *Crystal Symmetries*, I. Hargittai and B. K. Vainshtein (eds.), pp. 365–371. Pergamon Press, Oxford, 1988.

15. B. Mandelbrot, *The Fractal Geometry of Nature*, Freeman, New York, 1977.

4 Computer-Simulated Images of Icosahedral, Pentagonal, and Decagonal Clusters of Atoms

Peng Ju Lin and L. A. Bursill

1. Introduction

Many authors have published high-resolution electron microscopic (HREM) images of icosahedral and decagonal alloys (see, for example, some collected papers on aperiodic and quasicrystalline alloys [1, 2]). However it has been disappointing that analysis of such images of aperiodic alloy structures has failed to establish definitive structural information. The images often reveal interesting arrays of black and white spots (see, for example, Fig. 1a, taken from a sample of Al_6Mn [3]). Pentagonal and decagonal rings of white (or black) intensity occur that map readily onto two-dimensional Penrose tiling patterns [3, 4]. Agreement between image and tiling patterns is not fortuitous, since the images of course are themselves decorated tilings possessing local pentagonal and decagonal symmetries, whereas the symmetry of the optical transform of the images or the power spectra must be decagonal [5]. To deduce the three-dimensional quasicrystalline structure is a typical nontrivial inversion problem for diffraction theory [5]. In the following we investigate, by computer simulation, HREM images for a large set of icosahedral and decagonal clusters, containing 50–1500 atoms. The aim of this work was to probe the model sensitiveness of HREM images with respect to such aperiodic structures, given optimized electron optical imaging parameters.

This work has led to a new experimental study of both crystalline and quasicrystalline Al(Si)Mn alloys [6]. Thus carefully chosen electron optical

49

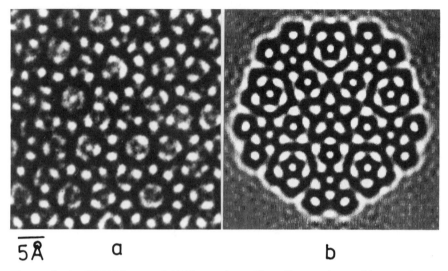

$\overline{5\,Å}$ a   b

Figure 1. (a) HREM image of Al_6Mn quasicrystalline alloy specimen; this area shows a localized pseudo-fivefold symmetry axis. (b) Computer-simulated image for cluster model Q (see Fig. 6b below). Large white spots match a, but fine detail does not.

conditions established by computer simulation were then used to obtain high-quality experimental images, which could be analyzed in a meaningful way.

Small icosahedral clusters of atoms have intrinsic scientific interest in terms of the energetics and stability of ultrafine particles.

The catalytic activity of such small clusters of atoms makes them interesting too, from the point of view of applied chemistry and chemical engineering.

In Section 2 we introduce briefly the subject of ultrafine metal cluster structure research, placing it in context with respect to the subject of quasi-crystallinity. Subsequent sections present a survey of structural models investigated (Section 3), a description of the computer simulation technique for image analysis (Section 4), the results of the computer simulations testing model sensitiveness of the HREM images (Section 5), and finally our conslusions (Section 6).

2. Ultrafine Atomic Clusters and Quasicrystals

The study of the symmetry and structure of clusters of 12, 13, 55—up to several hundreds of atoms lies at the core of understanding of relative stability of icosahedral alloys, both crystalline and quasicrystalline, as well as of glassy metals. Frank [7] considered an icosahedral cluster of 12 atoms about a central sphere as energetically preferable to cubic or hexagonal close-packed coordination spheres (e.g., cuboctahedral), at least for simple Lennard–Jones pair potentials. He used this argument to explain the remarkable degree of supercooling possible in simple liquid metals due to the abundance of icosahedral clusters. It was a modern study of this problem that led Levine and Steinhardt into the subject of quasicrystals [8]. Frank's work later extended to the glass–crystalline transition as a supercooled liquid approaches a Frank–Kasper phase [9], introducing the concept of a disordered disclination network [10]. Extended icosahedral correlations were also proposed in small "amorphon" cluster models of the structure of metallic glasses [11]. Pauling [12] provided a beautifully clear statement of the transition from clusters having icosahedral symmetry in the liquid state into complex crystalline alloy phases; see especially his illustrations [12, Figs. 11–14].

The structures of metallic particles used as catalysts lie somewhere between those of translationally periodic crystalline arrangements and the aperiodic aggregates proposed for glasses. Hoare and Pal [11] have described a range of coordination polyhedra, often invoking the concept of multiple internal twinning, forerunner of the controversy between proponents of the reality of Penrose tiling patterns [13] versus Pauling's claim for multiply twinned crystalline phases [14]. Such multiply twinned particles may grow from a nucleus of four atoms in a tetrahedral close-packed arrangement, onto which additional atoms occupy the centers of each face, producing very stable clusters of 7 or 13 atoms, which become the precursors of the pentagonal bipyramid or icosahedron, respectively [11]. These shapes may be maintained by the addition of shells of atoms [15], which, for small clusters, are thought to be relatively more stable than a cluster of the same number of atoms having a face-centered cubic or hexagonal close-packed structure. It would seem likely that extended arrays of pentagonal bipyramidally coordinated clusters should lead naturally to decagonal quasicrystalline alloys, in the same way that icosahedral or rhombic triacontahedral units should lead to icosahedrally symmetric quasicrystals [16]. In addition to decorations of two- or three-dimensional Penrose tilings [17], it is possible to conceive of hierarchical packings of icosahedral units [18] or essentially randomly positioned icosahedral units [19] cemented together by Al or Mn atoms; in the latter case a large proportion of the atoms have bond orientational order (which may be perfect in a Penrose tiling), but there are also a significant proportion of "fill-in" atoms, which are not necessarily part of an icosahedral cluster.

3. Description of Some Icosahedral and Pentagonal Test Objects

A Cartesian set of axes was used for the atomic coordinates and cell edges. These are shown schematically in Figure 2, where the relationship to a six-axial system is indicated. Note that the 12 vertices of a

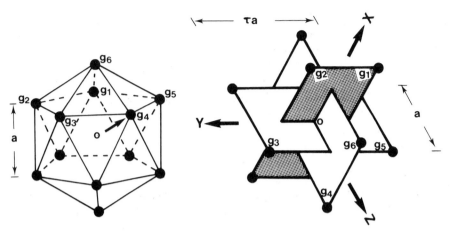

Figure 2. Relationship between six-axial basis set g_i ($i = 1, \ldots, 6$) and Cartesian axes X, Y, Z.

regular icosahedron may be decomposed into sets of vertices of three orthogonal golden rectangles, shown in Figure 2. Thus $g_1 = (\tau, \bar{1}, 0)$, $g_2 = (\tau, 1, 0)$, $g_3 = (0, \tau, 1)$, $g_4 = (\bar{1}, 0, \tau)$, $g_5 = (0, \tau, 1)$, $g_6 = (1, 0, \tau)$. Figure 3 shows pentagonal rings containing five elementary units. These may stack along a fivefold projection axis with the second and succeeding levels either eclipsed (Fig. 3a), forming columns of pentagonal prisms (Fig. 3b), staggered (Fig. 3c), or forming pentagonal antiprisms (Fig. 3d). Capping antiprisms top and bottom (Fig. 3e) gives rise to pentagonal clusters consisting of face-shaped pairs of truncated icosahedra (Fig. 3f). Such structural units may be used to decorate two-dimensional Penrose tilings, as shown in Figures 4a–c. Successive layers may be added, to produce pentagonal prisms (Fig. 4d), pentagonal antiprisms (Fig. 4e), or coherent mixtures of prisms and antiprisms (Fig. 4f). All the models in Figures 3 and 4 possess a single fivefold or $\overline{10}$ axis.

In Figure 5 a series of structural models is developed, starting with a pentagonal ring of icosahedra in Figure 5a. Successive icosahedra may be added in corner-shared (Fig. 5b) or edge-shared fashion (Fig. 5c). The latter forms a much denser cluster (Fig. 5d), possessing $\overline{10}$ point symmetry. Figures 5e–g show alternative ways of filling the central position within pentagons, designed to allow comparison of changes in the relative occupancies of pentagonal centers.

These same modeling principles may be extended for decorated Penrose tilings, as shown in Figures 6a–c. Again any number of stacking variants may be invoked along the unique fivefold or $\overline{10}$ projection axis, illustrated by Figures 6d–h. The periodicity along the projection axes may be chosen to model various decagonal phases [20]. This set of models was designed to vary the relative amounts of edge- and corner-sharing of pentagonal

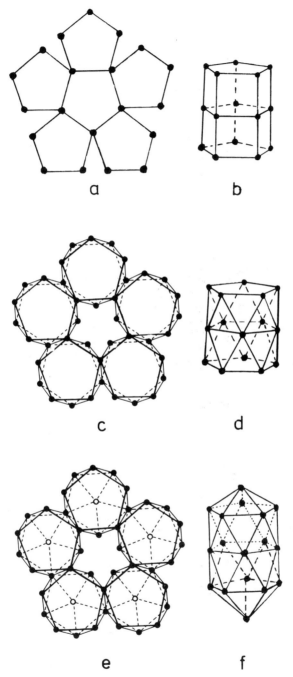

Figure 3. (a) Pentagonal cluster (eclipsed layers) giving rise to (b) pentagonal prismatic columns along fivefold axis. Capping top and bottom (c) gives rise to (f), face-shared truncated icosahedra. (c) Pentagonal cluster (staggered layers) giving rise to (d) pentagonal antiprismatic columns along fivefold axis.

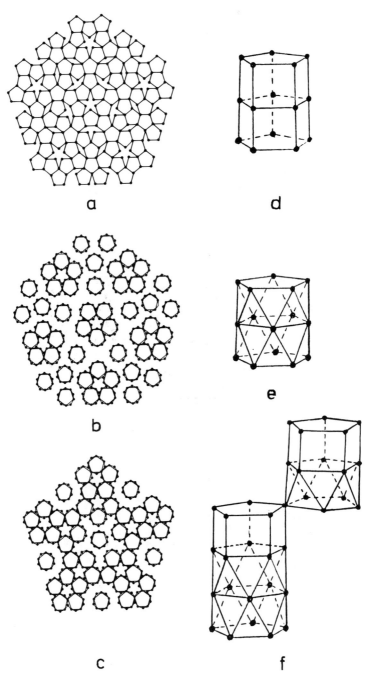

Figure 4. (a) Two-dimensional Penrose tiling (eclipsed stacking) giving rise to (d) pentagonal prismatic columns. (b) Two-dimensional cluster (staggered stacking) giving rise to (e) pentagonal antiprisms. (c) Third and fourth layers of a cluster containing a mixture of pentagonal prisms (a) and antiprisms (e) with sequence given in (f).

54

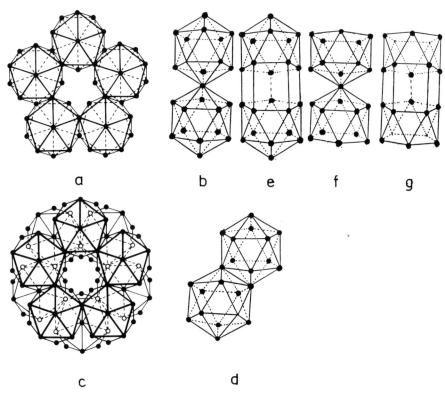

a   b   e   f   g

c   d

Figure 5. Pentagonal cluster of five icosahedra surrounding a central stellated icosahedron. (b)–(g) Five different ways of extending (a) along a unique fivefold axis giving rise to different occupancies at sites shown in projection (a). (c) Cluster of 10 icosahedra surrounding a central tesselated icosahedron; the nature of the edge-sharing of icosahedra shown here is clarified in (d).

prisms, antiprisms, or icosahedra. Thus we were able to investigate the model sensitivity of the computed images to varying occupancies of ring or centering positions along the projection axis.

Hiraga et al. [18] proposed a three-dimensional structure for an icosahedral phase (so-called hierarchial model) in which 12 unit icosahedra (edge a_0; see Fig. 7a) aggregate by edge-sharing to form a larger icosahedron of edge length $a_1 = (2 + 1/\tau)a_0$ (Fig. 7c) where τ is the golden mean. Similarly, further generations of icosahedral clusters may be envisaged (Fig. 7c) with edge length $a_n = (2 + 1/\tau)\, a_0$. The fundamental icosahedral units may be centered or noncentered (Fig. 7a, b). It is worth noting that it is necessary to introduce a further tesselation of the central icosahedron for each hierarchical generation (Fig. 7c), to avoid leaving voids in this model. There are also some unrealistically short interatomic distances (a_0/τ), that would have to be avoided in a real structure. However, it was interesting to compare

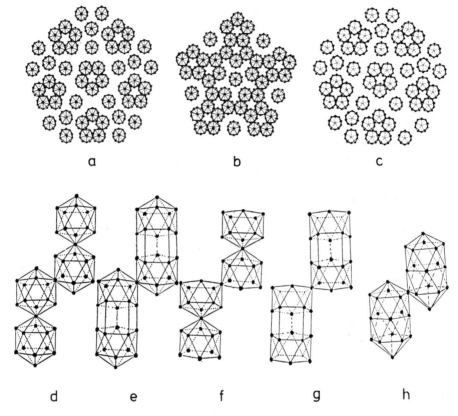

Figure 6. (a)–(c) Extended two-dimensional Penrose tilings of icosahedra (or truncated icosahedra), which may be stacked according to the ways shown in (d–h) to produce a variety of atomic densities viewed along a fivefold (or 1̄0) projection axis.

computer simulations for these giant molecular clusters with those for three-dimensional Penrose tilings.

Unit rhombic triacontahedra were constructed, following Mackay [16], out of 10 oblate and 10 prolate rhombohedra ($\alpha = 63.43°$ and $116.57°$, respectively, Fig. 8a, b). A perspective view is given as Figure 8c. A second-generation cluster was then constructed, shown projected along a fivefold axis in Figure 8d. Note that only the external facets form an object with true icosahedral symmetry. Permutations of the basic rhombohedral units would readily lower the true symmetry of the cluster to pseudoicosahedral for a real structure. We have chosen a symmetrical model (Fig. 8d) for purposes of computer simulations.

All the models used in simulations are summarized in Table 1, where a shorthand symbol for each distinct type may be found. Note that in the

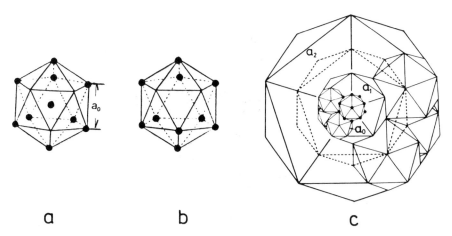

Figure 7. Filled (a) or empty (b) icosahedral units used to construct first and second generations (c) of larger icosahedral units according to Hiraga's hierarchical model [18].

simulations the unit interatomic distance was chosen as $a_0 = 2.8$Å, as appropriate for Al–Al, for example. For the triacontahedral model the unit rhombohedra had edge length $a_R = 4.6$ Å.

4. Computer Simulation Technique

The calculations were based on the physical optics approach to electron diffraction and imaging due to Cowley and Moodie [21] using the multislice technique introduced by Goodman and Moodie [22]. All the nonlinear N-beam dynamic scattering, as well as lens aberrations and Fresnel propagation effects in the objective lens of an HREM instrument were included. Periodic continuation methods, as developed for atomic resolution images [23–25] were obtained using MUM software (Melbourne University Multislice). This allows HREM images of finite aperiodic objects to be simulated reliably. Thus diffuse scattering, due to finite size effects, as well as internal defects in packing, may be readily included in the calculations. The atomic clusters (described above) were placed at the center of a large pseudocrystalline unit cell (dimensions are given in Table 1), when there was minimum overlap of waves scattered by adjacent clusters. The image of one such unit cell then exhibits contrast virtually identical to that of an isolated cluster. The techniques were first applied to small icosahedral clusters of gold atoms by Barry, Bursill, and Sanders [26]. The present report is restricted to five- or $\overline{10}$-fold projection axes of icosahedral, pentagonal, and decagonal clusters of atoms. The electron optical parameters refer to a JEOL-2000EX instrument with spherical aberration coefficient

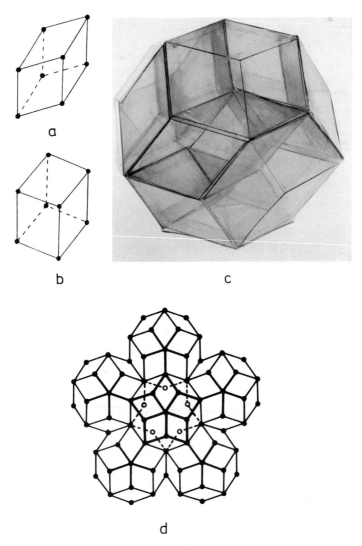

Figure 8. Rhombic triacontahedron (a) composed of 10 oblate (b) and 10 prolate (c) rhombohedra. Second-generation cluster of rhombic triacontahedra (d) forming a small segment of a three-dimensional Penrose tiling.

$C_s = 0.94$ mm and chromatic aberration defocus spread 50 Å. Through-focal series of images were obtained for the useful range of experimental values typical for this instrument (i.e., $-1000 \leqslant \Delta f$ Å $\leqslant 400$). Note that the so-called Scherzer optimum focus condition occurs for $\Delta f = -600$ Å. The effective resolution of the image simulations was limited to 2.3 Å, the value expected for interpretable structure images for the JEOL-2000EX instrument [27].

Table 1. Summary of Cluster Structures

Figure Number	Code	Number of Atoms	Cell Dimensions $(a \times b \times c \text{ Å}^3)$
3a, 3b	A	40	11.86 × 11.86 × 4.8
3c, 3c	B	40	11.86 × 11.86 × 4.8
5a, 5b	C	95	11.86 × 11.86 × 10.65
5e	D	90	11.86 × 11.86 × 10.65
5f	E	85	11.86 × 11.86 × 10.65
5g	F	80	11.86 × 11.86 × 10.65
5c, 5d	G	90	11.86 × 11.86 × 9.17
3e, 5f	H	70	11.86 × 11.86 × 9.17
8c	I	42	34.76 × 34.76 × 16.9
8d	J	202	34.76 × 34.76 × 16.9
7a, 7c	K	116	11.86 × 11.86 × 13.8
7b, 7c	L	104	11.86 × 11.86 × 13.8
7c	M	1292	31.05 × 31.05 × 36.50
7c	N	1148	31.05 × 31.05 × 36.50
4a	O	920	55 × 55 × 10.64
6h	P	1052	55 × 55 × 10.64
4b	Q	1080	55 × 55 × 13.30
6d	R	1502	55 × 55 × 15.96
6e	S	1452	55 × 55 × 15.96
6f	T	1356	55 × 55 × 15.96
6g	U	1260	55 × 55 × 15.96

5. Results

5.1. Small Clusters

Figure 9 gives a matrix of computed images for 11 different ent clusters as a function of objective lens defocus ($-1000 \leqslant \Delta f \text{ Å} \leqslant 400$). Note there are interesting changes of contrast with both objective lens defocus and structural model. As well as changes in internal detail, the apparent shape changes significantly with defocus. Fresnel fringe detail extends outward in the vicinity of a cluster, which effect becomes dominant for Δf values far from the Scherzer optimum image condition ($\Delta f = -600$ Å). Fresnel effects also superpose on fine details within the cluster, since the latter contain surface facets (cluster–vacuum interfaces) that are characteristic of the cluster and may produce significant phase contrast in the images. These optical details do not directly represent atomic detail of course, although for some clusters they may provide a distinctive "fingerprint" image. Note in Figure 9 (models C–F) that pentagonal rings of five white spots are relatively stable for $-600 \leqslant \Delta f \text{ Å} \leqslant -300$, although the pentagons change slightly in edge dimension. Black spot contrast within the cluster does not necessarily represent the projected potential or atomic charge density in the cluster. Similarly, white spots do not necessarily represent

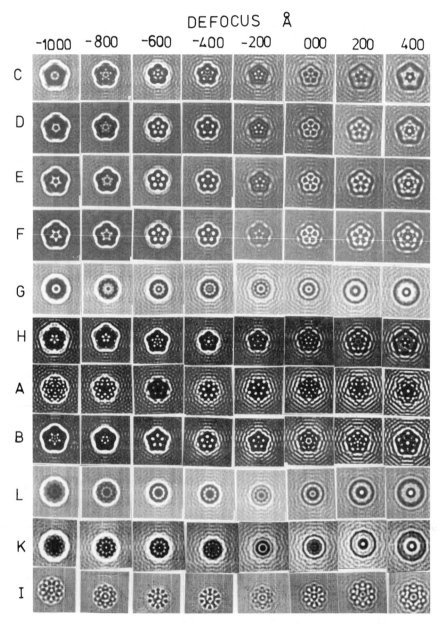

Figure 9. Matrix of computer-simulated images for 11 different small cluster models for objective lens defocus $-1000 \leqslant \Delta f \, Å \leqslant 400$. Compare different models for Scherzer optimum image condition ($\Delta f = -600$ Å), then compare this with different defoci.

60

relatively empty tunnels through the cluster; rather, they represent sites at the center of pentagonal rings. The intensity and size of such white spots may vary with the projected occupancy of such sites. In general, at Scherzer defocus ($\Delta f = -600$ Å), the higher the projected atomic density, the larger and darker the black spots, or the smaller and weaker the light spots. Thus these cluster images are structure model sensitive, but it may not be a trivial matter to solve the inversion problem (i.e., to convert image intensity into atomic density or projected potential). However, corner-shared pentagonal rings (A) should be distinguishable from clusters containing face-sharing pentagonal prisms (B) or edge-sharing icosahedral structural units (G). In particular, the latter shows just-resolved rings of 10 black or white spots, depending on defocus.

The first-generation icosahedral clusters, whether centered (K) or non-centered (L) show relatively well-resolved 10-rings; the ring diameter varies with defocus. For the Scherzer condition the ring diameter of 7.8 Å is consistent with the icosahedral ring diameter. The triacontahedral cluster (J: see Fig. 12, below) behaves in a distinct fashion, compared with the first-generation hierarchical cluster (K). The former gives a 13.2 Å diameter 10-ring of black spots (at Scherzer), which undergoes a contrast reversal to 10- and 5-rings of white spots as defocus varies away from Scherzer. As well, we see a changing diameter for the 10-ring of white spots. Clearly, if any structural information is to be available directly from the image, it is essential to identify the Scherzer optimum defocus condition.

5.2. Two-Dimensional Penrose Tiling Models

Figure 10 gives a matrix of images for seven two-dimensional Penrose tilings as a function of objective lens defocus ($-1000 \leqslant \Delta f$ Å $\leqslant 400$). As for Figue 9 the images are model sensitive to a significant extent—at the Scherzer optimum defocus condition, for example ($\Delta f = -600$ Å). When the through-focal series are compared, even greater discrimination between models is readily apparent, surpassing that exhibited by the smaller clusters. The model Q shows a remarkable similarity to an experimental image (Fig. 1b). Individual pentagonal prismatic or antiprismatic columns may be identified as a white spot and larger dark spots correspond to edge-shared icosahedral units, for $\Delta f = -400$Å. All the images may be identified as decorated two-dimensional Penrose tilings, at least in the interior away from edge effects.

It is interesting to ask if the degree of model sensitiveness of the images may be increased by availability of higher instrumental resolution. Figure 11 shows a through-focus-through resolution matrix of images for model Q. Briefly we conclude that fine detail certainly increases with higher resolution, for all defocus settings. However most of this increased detail is of little use, as it is not simply related to the projected atomic positions in the

DEFOCUS Å

-1000 - 800 -600 -400 -200 000 200 400

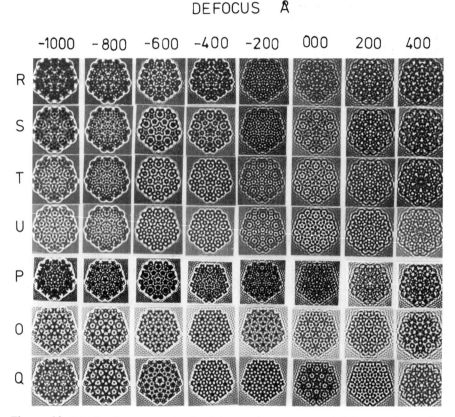

Figure 10. Matrix of computer-simulated images for seven extended Penrose clusters for objective lens defocus $-1000 \leq \Delta f$ Å ≤ 400. Compare model sensitiveness of images for Scherzer defocus ($\Delta f = -600$ Å) then also compare for deviations in Δf.

corresponding cluster (e.g., Fig. 4b). However, some aesthetically pleasing results were obtained.

5.3. Second-Generation Hierarchical Model

Complete through-focal series of images for the centered and noncentered hierarchial models are illustrated in Figure 12 a and b ($-1000 \leq \Delta d$ Å ≤ 1400). Comparison of the two models for any given defocus (e.g., $\Delta f = -600$ Å) will reveal quite distinct images, even for such a relatively small change of structure. However, for 2.3 Å resolution there is virtually no direct structural information, in terms of one-for-one relationship with projected charge density. Furthermore, increases in instrumental resolution (at 200 kV) are likely to make interpretation more, rather than less, difficult. Comparison of hierarchical with rhombic triacontahedral clusters is however quite revealing.

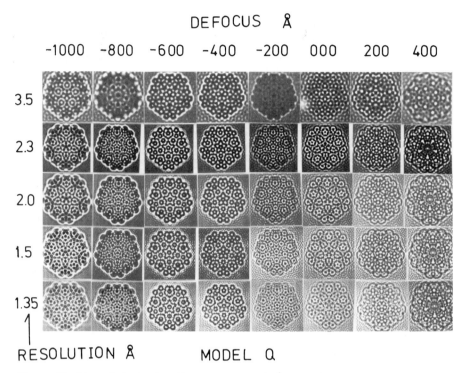

Figure 11. Through-focal series of images ($-1000 \leq \Delta f$ Å ≤ 400) for model Q for instrumental resolution in the 3.5–1.35 Å range. Note increasing detail for the latter, with, however, no increase in interpretable structural information.

5.4. Rhombic–Triacontahedral Clusters

A through-focal series ($-1000 \leq \Delta f$ Å ≤ 400) is given for model J in Figure 12c. All the images are quite distinct from those of the hierarchial model (Fig. 12 a, b) but resemble those of the two-dimensional Penrose tilings (Fig. 10). The image at Scherzer ($\Delta f = -600$ Å) most closely resembles the structure, but the effective thickness of the cluster was only 26 Å, which is really almost two-dimensional. Six interlocking decagonal rings of black spots may be found for $\Delta f = -600$ Å or -500 Å, whereas six interlocking rings of white spots are found for $\Delta f = -1000$, -900, -400, -100, and 100 Å. This set of images is strictly pseudopentagonal due to our choice of occupancies in the calculation. Departures from fivefold symmetry become more apparent for specific Δf values.

6. Conclusion

The computer simulations above have allowed us to explore the limitations of HREM image analysis for structure determination of quasicrystalline clusters or even local structural elements in extended

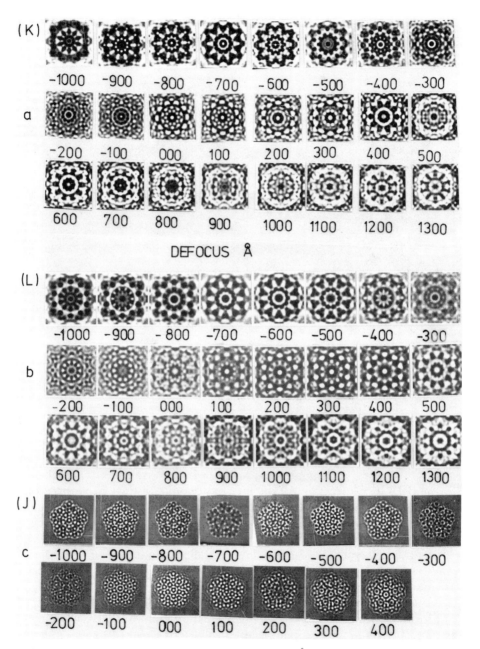

Figure 12. Through-focal series of images ($-1000 \leq \Delta f \,\text{Å} \leq 400$) for (a) centered and (b) noncentered second-generation hierarchical clusters. Note model sensitiveness of the images, without, however, any obvious inversion to map out the structure of the cluster (cf. fig. 7). (c) Through-focal series ($-1000 \leq \Delta f \,\text{Å} \leq 400$) of computer-simulated images of the rhombic triacontahedral cluster (cf. Fig. 8). Images are quite distinct from (a) and (b).

64

aperiodic objects. Clearly, there is a very significant degree of model sensitiveness available, in the sense that the image may provide a fingerprint for certain structural elements, provided electron optical parameters such as specimen thickness and objective lens defocus have been determined. It is equally clear, however, that direct inversion from image to structure is not a realistic possibility. Whereas many possible structures may be eliminated, it may not be possible to arrive at unique structural solutions.

Based on the foregoing experience, it occurred to us that a reasonable procedure would be to record experimental images of known complex icosahedral alloys, in the crystalline phase, then use the computer simulations to identify "fingerprint" imaging conditions whereby certain structural elements could be reliably identified in images of aperiodic quasi-crystalline or amorphous specimens. This principle has been used by Song in subsequent work [6].

From the present results it is worth noting that the experimental image (Fig. 1a) could be reproduced to a large extent (Fig. 1b) by cluster model Q (Table 1) apart from some model-dependent details. That image suffered from the experimental uncertainty that it could not be proved a priori whether the specimen (Al_6Mn) was icosahedral or decagonal phase, nor were the thickness or defocus conditions established. In the subsequent work on Al(Si)Mn crystalline and icosahedral alloys, strenuous efforts have been made to correct these limitations [6].

Acknowledgments

This work was supported by the Australian Research Council. Peng Ju Lin is grateful for the award of a National Research Fellowship.

REFERENCES

1. D. Gratias and L. Michel (eds.), International Workshop on Aperiodic Crystals, Les Houches, March 1986, in *J. Phys. Colloq.* **C3**, Suppl. 7 (1986).

2. P. J. Steinhardt and S. Ostlund (eds.), *The Physics of Quasicrystals*. World Scientific, Singapore, 1987.

3. L. A. Bursill and Peng Ju Lin, "Penrose tiling observed in a quasicrystal," *Nature (London)*, **316**, 50–51 (1985).

4. P. Guyot and M. Audier, "TEM study of the relationships between decagonal and icosahedral phases in Al-Mn," *J. Microsc.* **S10**, 575–582 (1985); "Al_4Mn quasicrystal atomic structure, diffraction data in Penrose tiling," *Phil. Mag.* **B53**, L43–L51 (1986).

5. J. M. Cowley, *Diffraction Physics*, 2nd ed, Chap. 2. North-Holland, Amsterdam, 1981.

6. Guang Li Song, Ph.D. thesis, University of Melbourne, 1990.

7. F. C. Frank, "Supercooling of liquids," *Proc. R. Soc. London, Ser. A*, **215**, 43–46 (1952).

8. D. Levine and P. J. Steinhardt, "Quasicrystals: A new class of ordered structures," *Phys. Rev. Lett.* **53**, 2477–2480 (1984).

9. F. C. Frank and J. S. Kasper, "Complex alloy structures regarded as sphere packings, Part I: Definitions and basic principles," *Acta Crystallogr.* **11**, 184–190 (1958); "Part II: Analysis and classification of representative structures," **12**, 483–489 (1959).

10. D. R. Nelson, "Liquids and glasses in spaces of incommensurate curvature," *Phys. Rev. Lett.* **50**, 982–985 (1983); "Order, frustration and defects in liquids and glasses," *Phys. Rev. B* **28**, 5515–5535 (1983).

11. M. R. Hoare and P. Pal, "Physical cluster mechanics: Statics and energy surfaces for monatomic systems," *Adv. Phys.* **20**, 161–196 (1971).

12. L. Pauling, *The Nature of the Chemical Bond*, 3rd ed., pp. 425–428. Cornell University Press, Ithaca, NY, 1960.

13. J. W. Cahn, D. Gratias, D. Shechtman, A. L. Mackay, P. A. Bancel, P. A. Heiney, P. W. Stephens, A. I. Goldman, and A. A. Berezin, "Pauling's model not universally accepted," *Nature (London)*, **319**, 102–104 (1986); P. A. Heiney, P. A. Bancel, and P. M. Horn, "Comment on 'So-called icosahedral and decagonal quasicrystals are twins of an 820-atom cubic crystal,'" *Phys. Rev. Lett.*, **59**, 2119–2120 (1987).

14. L. Pauling, "So-called icosahedral and decagonal quasicrystals are twins of an 820-atom cubic crystal," *Phys. Rev. Lett.* **58**, 365–368 (1987).

15. A. L. Mackay, "A dense non-crystallographic packing of equal spheres," *Acta Crystallogr.* **15**, 916–918 (1962).

16. A. L. Mackay, "Crystallography and the Penrose pattern," *Physica*, **114A**, 609–613 (1982).

17. V. Elser and C. J. Henley, "Crystal and quasicrystal structures in Al-Mn-Si Alloys," *Phys. Rev. Lett.* **55**, 2883–2886 (1985).

18. K. Hiraga, M. Hirabayashi, A. Inoue, and T. Masumoto, "Icosahedral quasi-crystals of a melt-quenched Al-Mn alloy observed by high-resolution electron microscopy," *Sci. Rep. Res. Inst. Tohokua Univ.* **A32**, 309–314 (1985).

19. D. Schechtman and I. A. Blech, "The microsctructure of rapidly-solidified Al$_6$Mn," *Metall. Trans.* **16A**, 1005–1012 (1985).

20. K. K. Fung, C. Y. Yang, Y. Q. Zhou, J. G. Zhou, W. S. Zhan, and B. G. Shen, "Icosahedrally related decagonal quasicrystal in rapidly cooled Al-14 at.%-Fe alloy," *Phys. Rev. Lett.* **56**, 2060–2063 (1986).

21. J. M. Cowley and A. F. Moodie, "The scattering of electrons by atoms and crystals: Parts I–III," *Acta Crystallogr.* **10**, 609–619 (1957); **12**, 353–359 (1959); **12**, 360–367 (1959). See also J. M. Cowley, "The electron optical imaging of crystal lattices," *Acta Crystallogr.* **12**, 367–375 (1959).

22. P. Goodman and A. F. Moodie, "Numerical evaluation of *N*-beam wave functions in electron scattering by the multislice method," *Acta Crystallogr.* **A30**, 280–293 (1974).

23. D. S. Maclagen, L. A. Bursill, and A. E. C. Spargo, "High-resolution image

calculations of defects using the method of periodic continuation," *Phil. Mag.* **35**, 757–780 (1977).

24. A. R. Wilson, L. A. Bursill, and A. E. C. Spargo, "Fresnel diffraction effects in high-resolution images (< 3 Å): Effect of spherical aberration on the Fresnel fringe," *Optik*, **52**, 313–336 (1978–1979).

25. Peng Ju Lin and L. A. Bursill, "Electron optical study of polar surface facetting of ruby and sapphires," *Optik*, **81**, 167–174 (1989).

26. J. C. Barry, L. A. Bursill, and J. V. Sanders, "Electron microscope images of icosahedral and cuboctahedral (f.c.c. packing) clusters of atoms," *Austr. J. Phys.* **38**, 437–448 (1985).

27. L. A. Bursill, J. C. Barry, and J. L. Hutchinson, "Electron optical imaging of diamond," *Optik*, **65**, 271–293 (1983).

5 X-Ray Diffraction Study of Slowly Solidified Icosahedral Alloys

Françoise Dénoyer, Gernot Heger and Marianne Lambert

1. Introduction

Since the discovery by Shechtman et al. [1] of a metallic phase with long-range icosahedral orientational symmetry, considerable experimental and theoretical work has been devoted to the study and the understanding of quasicrystalline alloys. In 1986 large single quasicrystalline grain samples of Al–Cu–Fe, with typical sample size of a few hundred micrometers, were prepared by slow solidification [2]. To this time, structural investigations had been restricted to the study of rapidly quenched alloys. In this last case, the size of quasicrystalline particles (typically a few micrometers) limited structural studies to electron microscopy measurements on small quasicrystalline grains or X-ray or neutron diffraction measurements on powder samples. (For a review of rapidly quenched quasicrystalline alloys studied by the X-ray powder diffraction method see, for example, the article by Dunlap et al. (this volume). In the present paper we report recent results obtained by single-crystal X-ray diffraction methods (as opposed to powder methods), on Al–Cu–Li and Al–Cu–Fe alloys: the two systems in which large, single, quasicrystalline grain samples are available. In the case of the Al–Li–Cu alloy, the results are interpreted on the basis of an icosahedral quasicrystal model, whereas in the case of the Al–Cu–Fe alloy the results are interpreted in terms of crystalline aggregates with a rhombohedral crystalline cell with $a = 32.16$ Å and $\alpha = 36°$. In these two ternary alloys, an overall icosahedral symmetry is found to be angularly preserved.

2. Aluminum–Lithium–Copper Alloy

Large single grains of the icosahedral phase with a composition near Al_6Li_3Cu have been prepared by slow solidification at the Péchiney Research Center in Voreppe, France [2, 3]. The pseudospherical single grains are formed by natural "golden rhombus" facets resulting in a rhombic triacontahedron morphology. Figure 1 shows a sample of about 300 μm diameter glued onto a glass fiber with a twofold axis A aligned along the fiber direction. We observe a macroscopic fivefold symmetry at the vertex formed by five rhombic facets perpendicular to the twofold axis A.

Monochromatic Laue transmission photographs of very long exposure time were taken with the twofold axis, A, perpendicular to the incident beam direction [4, 5]. In a systematic manner the sample orientation was varied in turning around A with steps of $2°$. The results obtained in this study are typical of the diffraction pattern expected for a single icosahedral quasicrystalline grain. Figure 2 shows the patterns recorded for special orientations of the sample corresponding to the cases of two-, three-, and fivefold axes parallel to the incident beam. In this way, we were able to prove that the *overall icosahedral symmetry is angularly verified*. The striking general features of these patterns are as follows.

1. A high density of diffraction peaks. Because of the sensitivity of the method, we are able to distinguish peaks with intensities covering a

Figure 1. Photograph of an Al_6Li_3Cu sample glued onto a glass fiber. The triacontahedral morphology of the quasicrystal is visible, with a fivefold axis perpendicular to the plane of the figure.

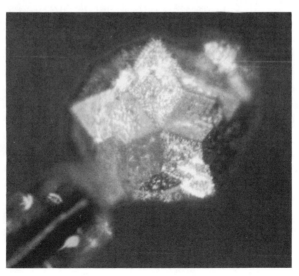

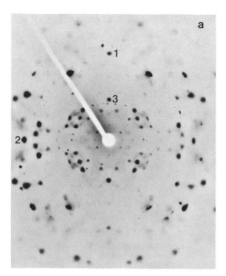

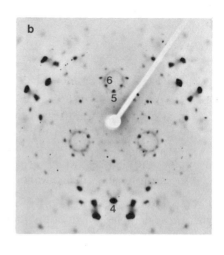

Figure 2. Al_6Li_3Cu: monochromatic transmission X-ray Laue diffraction patterns obtained with (a) a twofold axis, (b) a three-fold axis, and (c) a fivefold axis of the sample parallel to the incident beam; Cu K_α radiation (λ = 1.54 Å) has been selected using the (002) reflection of a pyrolitic graphite mon-ochromator. For some typical reflections, N and M values, multiplicities, and indexes are given below, following the indexing method proposed by Cahn et al. [6]

No.	N	M	Multiplicity	Indices
1	24	36	60	0/2 2/4 0/0
2	24	36	20	2/4 0/2 0/0
3	10	9	60	1/0 0/3 0/0
4	20	32	30	0/0 2/4 0/0
5	8	4	12	2/0 0/2 0/0
7	6	9	20	1/2 0/1 0/0
8	8	8	60	0/2 0/2 0/0
9	18	25	60	1/4 0/1 0/0
10	26	41	60	3/4 0/1 0/0

The $\lambda/2$ contamination is negligible, except for a few diffraction peaks of very small intensity; the point labeled 6 corresponds to the center of a diffuse circle; it can be indexed as 1/1 1/2 0/0 ($N = 7$, $M = 11$, multiplicity = 12). Its very small intensity is not due to a violation of the extinction rules given in Ref. 6 for icosahedral primitive quasilattice (i.e., $h + k' = 2n$, $k + l' = 2n$, $l + h' = 2n$) but to a very small $\lambda/2$ contamination. A small contamination from residual crystalline phases is revealed in the form of punctuated powder rings. These spots give directly the order of magnitude of the experimental resolution.

71

range of a factor of more than 1000. Some peaks are strongly anisotropic and are even composed of several distinct components.

2. In addition to the diffraction peaks, the patterns show clear evidence for complex diffuse scattering that obeys icosahedral symmetry. Diffuse features of several kinds have been observed [4], the most spectacular ones being circles, which are clearly visible when the threefold axis of the sample is aligned parallel to the incident beam (see Fig. 2b).

To compare directly the diffraction peak distribution in reciprocal space with the calculated diffraction pattern for an icosahedral quasicrystal as given by Cahn et al. [6], overexposed monochromatic precession photomicrographs were taken [5]. Figure 3 presents the results obtained for the zero level in reciprocal space with the two-, three-, and fivefold axes parallel to the precession axis. Once again, all peaks are enlarged or composed of distinct components, and different shapes are observed for different symmetry directions. Each pattern is investigated by utilizing schematic drawings with a differentiation between peak intensities. The following observations can be made.

1. All reflections can be indexed on the basis of the reciprocal quasilattice, using the notation introduced by Cahn et al. [6] with a quasilattice constant $d_0 = 19.20$ Å. Following the results of these authors, we conclude that the selection rules for a six-dimensional (6D) primitive hypercubic lattice are fulfilled. The 6D lattice parameter is given by $A = d_0/[2(2 + \tau)]^{1/2} = 7.138$ Å.

2. Important deviations between our experimental intensities and those calculated from the inverse distance of the points in the 6D primitive reciprocal lattice from the cut plane [6] are found. Some reflections that are predicted to be very strong by the model [e.g., (4/6 2/4 0/0) or (3/4 3/5 2/3)] are not visible on the patterns of Figure 3a and c; others [e.g., (4/6 4/6 0/0) or (2/4 2/4 0/0)] are visible but are very weak (cf. Fig. 3a). On the other hand, many more weak diffraction peaks appear in the photographs as a result of the very long exposure. The intensity modulations are certainly related to the atomic ordering, although this is not yet well known. Quantitative measurements of peak intensities for structure determination have been performed on a four-

---

Figure 3. Al_6Li_3Cu: monochromatic X-ray precession photographs obtained with an incident beam wavelength $\lambda = 0.711$ Å (Mo K_α radiation) for the zero-level reciprocal planes with (a) a twofold axis, (b) a threefold axis, and (c) a fivefold axis parallel to the precession axis: Each pattern is explored in the form of a schematic drawing with differentiation between peak intensities: strong, medium and weak following the size of the circles. By using solid circles ●, dotted circles ○, and combined open and solid circles ☉, we can compare the experimental results and the calculated intensities (see Ref. 6): ● ☉ both correspond to observed reflections, ☉ stands for peaks that are not present in the calculated patterns [6], and ○ denotes peak positions present in the calculated patterns for which no intensity was observed.

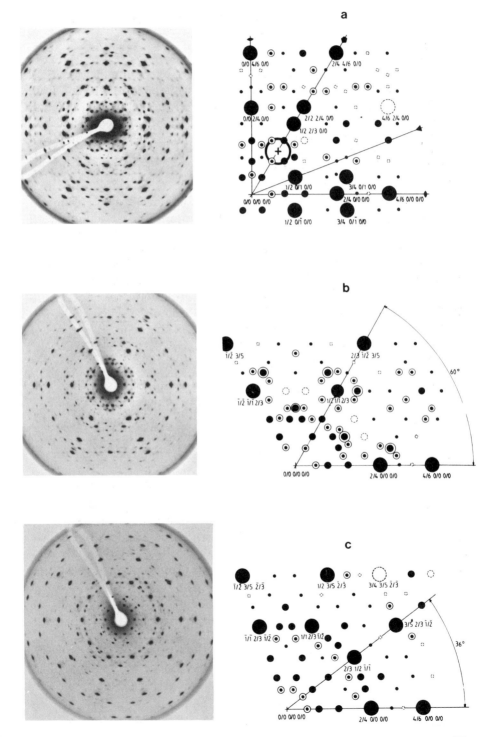

73

circle diffractometer by Pannetier et al. [7] on this same sample and by Elswijk et al. [8]. The latter authors have refined the structure on the basis of the three-dimensional Penrose packing.

In perfect agreement with the monochromatic Laue results, the precession photograph of Figure 3a shows diffuse scattering in the form of rings. From a systematic study, we were able to prove that these rings result from spherical shells of diffuse scattering in reciprocal space. The most intense shells are located along the fivefold axis. These 12 spherical shells are centered at points of the reciprocal lattice that are forbidden by selection rules [(1/1 1/2 0/0) in Fig. 3a] and are located at a distance corresponding to 3.86 Å in real space. From the radius of the circle, a characteristic distance of 16.3 Å can be deduced. There is some indication for similar diffuse scattering located at larger distances from the origin in reciprocal space, although the intensity is much lower [centers of circles at, e.g., (1/1 3/4 0/0) or (2/3 3/5 0/0)]. No shell centered at the origin of the quasilattice was observed. We have suggested [5] that the only tentative explanation for these observations is the introduction of displacement disorder. A qualitative model has been made on the basis of the "strip model" [9] by introducing a fluctuating strip. This is in good agreement with our experimental results.

To conclude this section we emphasize five of the main advantages of single crystal diffraction methods compared to powder methods especially in the case of quasicrystalline alloys.

1. With single crystal methods it becomes possible to demonstrate that the overall icosahedral symmetry is accurately angularly verified.

2. Because of quasiperiodicity there exist a lot of very weak diffraction peaks, which can be observed and localized in the reciprocal quasilattice. In powder diffraction methods, these very weak diffraction peaks appear in an important modulated background. This "background" has a physical meaning and usually it is not taken into consideration when a structure determination is made.

3. It is not possible from powder methods to distinguish between some diffraction peaks of wave vector $\mathbf{Q}_1$, $\mathbf{Q}_2$ with multiplicities $\mu_1 = \mu_2$ that have the same absolute value $|\mathbf{Q}_1| = |\mathbf{Q}_2|$. This is the case for many diffraction peaks and can lead to some confusion in the indexing of powder data.

4. Since deviations to the perfect structure are inherent in many icosahedral phases (e.g., phason-like disorder), it is important to measure the complete shape anisotropy (or splitting) of the diffraction peaks in the reciprocal quasilattice. This is possible only using single crystal diffraction methods, and four-circle diffractometer measurements give more quantitative information than photographic methods [10, 11].

5. Another manifestation of disorder can be found in the form of very weak anisotropic diffuse scattering, and the photographic single crystal

technique is a suitable method for the detection and localization of this scattering in the reciprocal space.

3. Aluminum–Copper–Iron Alloy

The Al–Cu–Fe icosahedral phase was first reported by Tsai and Masumoto [12, 13] as a stable icosahedral phase. The indexing of a powder X-ray diffraction pattern by Ishimasa et al. [14] has revealed that the scaling (or self-similarity ratio) along the five-, three-, and twofold axes was the golden mean, τ, instead of τ^3. More recently, Ebalard and Spaepen [15] have identified the quasilattice with a face-centered 6D-hypercubic lattice, and this structure has been described by Devaud-Rzepski et al. [16] as an ordered F-superstructure of the usual primitive quasilattice with at least two motifs in the 6D representation. This new quasicrystalline phase has been observed in micrometer sized grains of rapidly quenched alloys as well as in as-cast alloys. The structural studies therefore have been limited to electron diffraction measurements on single-domain grains and X-ray diffraction measurements on powder samples.

Recently, large single grain samples of Al–Cu–Fe have been grown by slow solidification. The sample studied here was prepared at the Péchiney Research Center. An alloy of composition $Al_{65}Cu_{20}Fe_{15}$ was cast at 1200°C under an argon atmosphere followed by cooling to room temperature under vacuum at a rate of less than 1°C/min. When the ingot was broken, it exhibited many prismatic dendrites and just a few brilliant dodecahedral particles. An electron probe microanalysis [17] has shown the chemical composition of these latter particles, presumed to be of icosahedral structure, to be homogeneous and to correspond to $Al_{63.5}Cu_{24}Fe_{12.5}$. Figure 4 shows a typical particle with a diameter of a few hundred micrometers, which is well adapted for precise investigation of the reciprocal space by single crystal X-ray diffraction method (i.e., a study of the type reported for Al_6Li_3Cu in Section 2).

To compare directly the diffraction peak distribution in reciprocal space with the calculated diffraction pattern for an icosahedral quasicrystal with a face-centered 6D-hypercubic lattice, as given by Devaud-Rzepski et al. [16], we have made monochromatic X-ray precession photographs with very long exposure time [18]. The results obtained with an incident beam wavelength of $\lambda = 1.542$ Å (Cu K_α radiation), for the zero level in reciprocal space are shown in Figure 5a–c for the five-, three-, and twofold axes parallel to the precession axis. Each pattern can be directly compared with the results obtained for an icosahedral Al_6Li_3Cu quasicrystal (see Fig. 5a'–c'). The intensity distribution of the reflections is quite different, but the icosahedral orientational symmetry is found to be correct without any deviation in the two cases. The measured d values (d_{meas}), for the Al–Cu–Fe sample, obtained from peak positions along the two-, three-, and fivefold axes are

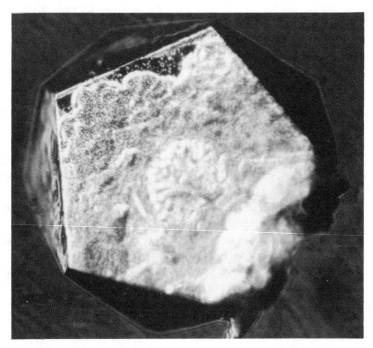

Figure 4. Photomicrograph of a typical particle of composition $Al_{63.5}Cu_{24}Fe_{12.5}$ found inside the ingot with pentagonal dodecahedral morphology.

given in Tables 1a–c. A list of some of the main calculated reflections has been published in Ref. 16; their indices $h/h'\ k/k'\ l/l'$ ($l/l' = 0/0$) and their corresponding $N = h^2 + k^2 + l^2 + h'^2 + k'^2 + l'^2$, $M = h'^2 + k'^2 + l'^2 + 2hh' + 2kk' + 2ll'$, $Q_{\parallel} = \sqrt{N + M\tau}$ and $Q_{\perp} = \sqrt{\tau(N\tau - M)}$ are also given in the tables. Using a direct 6D F-lattice parameter of $A_F = 2 \times 6.312$ Å, the corresponding d values (d_{calc}) have been calculated. These are listed in Tables 1a–c, from which the following points appear.

1. For many diffraction peaks, good agreement is found between the d_{meas} and the d_{calc} values.

2. A number of diffraction peaks, which have been observed by electron microscopy (see, e.g., Ref. 16), are not observed in spite of the extreme sensitivity of the photographic method. This is the case, for instance, for the reflections $\bar{1}/1\ 1/0$, $1/0\ 0/1$, $2/0\ 0/2$, $0/2\ 2/2$, $2/1\ 1/3$, $0/3\ 3/3$, $\bar{2}/2\ 0/0$, $2/0\ 0/0$, $0/4\ 0/0$, $4/2\ 0/0$, . . .

3. Additional diffraction peaks of weak or medium intensity are observed but cannot be reasonably indexed using a face-centered 6D-hypercubic lattice with parameter $A_F = 2 \times 6.312$ Å. Therefore, although this slowly solidified Al–Cu–Fe alloy shows *an overall orientational icosahedral*

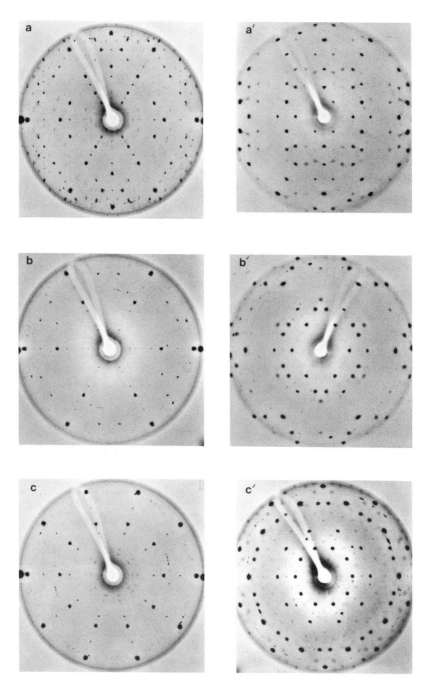

Figure 5. Comparison of monochromatic X-ray precession photographs obtained with an incident beam wavelength of $\lambda = 1.542$ Å (Cu K_α radiation) and the zero-level reciprocal plane with (a, a′) a twofold axis, (b, b′) a threefold axis, and (c, c′) a fivefold axis parallel to the precession axis: (a–c) $Al_{63.5}Cu_{24}Fe_{12.5}$ sample and Al_6Li_3Cu sample.

Table 1a. Indexing of Reflections along Twofold Axis

d_{meas} (Å)	$h/h'\ k/k'$	N	M	$Q_\parallel$	d_{calc} (Å)	$Q_\perp$	Comment	$(HKL), d$ (Å) $a = 32.16$ Å, $\alpha = 36°$
16.09 ± 0.3	$\bar{2}$/2 0/0	8	−4	1.236	13.76	5.236	Moderate−weak	(110), 16.07
	2/0 0/0	4	0	2	8.490	3.236		
8.08 ± 0.05							Weak	(220), 8.036
6.48 ± 0.04							λ/2 (2/2 0/0)	
6.015 ± 0.04	0/2 0/0	4	4	3.236	5.247	2	λ/3 (2/4 0/0)	
4.01 ± 0.04							λ/2 (2/4 0/0)	
3.232 ± 0.02	2/2 0/0	8	12	5.236	3.243	1.236	Strong	(550), 3.216
	0/4 0/0	16	16	6.472	2.624	4		
2.48 ± 0.006	4/2 0/0	20	20	7.236	2.347	4.472	λ/2 (4/6 0/0)	
2.005 ± 0.006	2/4 0/0	20	32	8.472	2.004	0.764	Very strong	(880), 2.010

Table 1b. Indexing of Reflections along Threefold Axis

d_{meas} (Å)	$h/h'\ k/k'$	N	M	$Q_\parallel$	d_{calc} (Å)	$Q_\perp$	Comment	$(HKL), d$ (Å) $a = 32.16$ Å, $\alpha = 36°$
9.95 ± 0.15	1/0 $\bar{2}$/1	6	−3	1.070	18.177	4.535	Weak	(333), 10.016
	0.1 1/1	3	0	1.732	9.803	1.732		
7.53 ± 0.07							λ/2 (1/2 0/1)	
6.09 ± 0.06	1/1 1/0	3	3	2.803	6.059	5.605	Moderate−weak	(555), 6.010
	0/2 2/2	12	0	3.464	4.902			

d_{meas} (Å)	$h/h'\,k/k'$	N	M	$Q_\parallel$	d_{calc} (Å)	$Q_\perp$	Comment	(HKL), d
3.746 ± 0.020	1/2 0/1	6	9	4.535	3.745	1.070	Strong	(888), 3.756
	2/2 2/0	12	12	5.605	3.029	3.464		
	1/3 1/2	15	15	6.267	2.710	3.872		
2.313 ± 0.006	2/3 1/1	15	24	7.337	2.314	0.662	Strong–moderate	(131313), 2.31

Table 1c. Indexing of Reflections along Fivefold Axis

d_{meas} (Å)	$h/h'\,k/k'$	N	M	$Q_\parallel$	d_{calc} (Å)	$Q_\perp$	Comment	(HKL), d (Å) $a = 32.16$ Å, $\alpha = 36°$
16.85 ± 0.3	2/1 $\bar{1}$/1	7	−4	0.726	23.370	4.980	Moderate	(100), 16.90
8.49 ± 0.08	$\bar{1}$/1 1/0	3	−1	1.176	14.444	3.078	Moderate	(200), 8.45
6.83 ± 0.05	1/0 0/1	2	1	1.902	8.927	1.902	λ/2 (1/1 1/2)	
6.32 ± 0.05							λ/3 (1/2 2/3)	
5.56 ± 0.08	0/1 1/1	3	4	3.078	5.517	1.176	Moderate	(300), 5.63
	2/0 0/2	8	4	3.804	4.463	3.804		
4.22 ± 0.03							λ/2 (1/2 2/3)	
3.395 ± 0.03	$\bar{1}$/2 2/1	10	5	4.253	3.992	4.253	Strong	(500), 3.38
	1/1 1/2	7	11	4.980	3.410	0.727		
	0/2 2/2	12	16	6.155	2.759	2.351		
	2/1 1/3	15	20	6.882	2.467	2.629		
	1/3 3/2	23	19	7.331	2.316	5.429		
2.105 ± 0.007	1/2 2/3	18	29	8.057	2.107	0.449	Very strong	(800), 2.113
1.879 ± 0.003	3/1 1/4	27	31	8.784	1.933	4.53	Moderate	(900), 1.879

symmetry, the indexing of its corresponding diffraction pattern is incompatible with a face-centered 6D-hypercubic lattice.

Complementary electron microscopy experiments have been performed on fragments of dodecahedral particles coming from the same ingot [19]. These electron diffraction patterns exhibit large spots and diffuse scattering. Moreover, the fivefold axes were not always strictly equivalent for different fragments from the same dodecahedral particle. The perfect equivalence of the six fivefold axes of the sample studied in the X-ray precession experiments has been carefully demonstrated using a four-circle X-ray diffractometer. For geometrical reasons, only four of the six fivefold axes were measured in the Burger method. In addition $\omega/2\theta$ scans of reflections along the different symmetry axes have revealed a remarkable sharpness of the diffraction peaks. This indicates a high degree of coherence in the spatial interferences, comparable to that encountered in crystals of small mosaic spread. At variance with the Al_6Li_3Cu case [10, 11], no $Q_\perp$ dependence was detected in the width of reflection peaks. This suggest one more reason to suspect the existence of a crystalline state.

Electron diffraction patterns have in fact been interpreted in terms of a periodic microcrystalline structure [17, 20], and the appropriate Bravais lattice has been found to be trigonal with cell parameters $a = 32.08$ Å and $\alpha = 36°$. Corresponding to the results of electron microscopy, the X-ray reflections in Tables 1a–c as observed along the two-, three-, and fivefold axes can be interpreted as arising from the 110, 220, 550, 880, 333, 555, 888, 13 13 13, and 100, 200, 300, 500, 800, 900 reflections, respectively, of the trigonal Bravais lattice of refined cell parameters $a = 32.16$ Å and $\alpha = 36°$. Thus there is a very good agreement with the interpretation based on the electron diffraction study.

From these present results, an immediate question arises: How is space tiled with only *one* rhombohedral cell in such a way as to preserve a mean perfect icosahedral orientationl order? A priori, the task of explaining the microstructual arrangement of such a state would appear to be rather difficult. However, as guide for further investigations, an answer is partially given by two-dimensional models presented in Ref. 21. A consideration of the size of coherent domains helps us to understand the origin of the differences between electron diffraction experiments on small fragments of the sample and X-ray diffraction patterns of macroscopic samples of the same phase [17, 19, 20]. Moreover, it is important to realize that in such a case structural defects should occur at grain boundaries.

Finally, it should be pointed out that the structural model presented here is basically different from that obtained by other authors [12–16], who claimed to have observed a perfect icosahedral structure. This apparent discrepancy, due to different preparation methods (rapidly quenched or slowly cooled alloys), is probably related to the occurrence at high temperature of a phase transition between these "pseudoicosahedral" and perfect icosahedral states [17, 20].

Acknowledgments

The authors would like to thank J. M. Lang and P. Sainfort for supplying Al$_6$Li$_3$Cu samples and J. M. Lang and P. Duroux for the preparation of the Al–Cu–Fe ingot.

REFERENCES

1. D. Shechtman, I. Blech, D. Gratias, and J. W. Cahn, "Metallic phase with long-range orientational order and no translational symmetry," *Phys. Rev. Lett.* **53**, 1951–1953 (1984).

2. B. Dubost, J. M. Lang, M. Tanaka, P. Sainfort, and M. Audier, "Large AlLiCu single quasicrystals with triacontahedral solidification morphology," *Nature (London)*, **324**, 48–50 (1986).

3. J. M. Lang, M. Audier, B. Dubost, and P. Sainfort, "Growth morphologies of the Al–Li–Cu icosahedral phase," *J. Crys. Growth*, **83**, 456–465 (1987).

4. F. Dénoyer, G. Heger, M. Lambert and J. M. Lang, "Diffraction des rayons X par un mono-quasicristal AlLiCu," *C.R. Acad. Sci. Paris*, **304**, Ser. 11, 625–627 (1987).

5. F. Dénoyer, G. Heger, M. Lambert, J. M. Lang, and P. Sainfort, "X-ray diffraction study on uniformly oriented quasicrystals," *J. Phy.*, **48**, 1357–1361 (1987).

6. J. W. Cahn, D. Shechtman, and D. Gratias, "Indexing of icosahedral quasiperiodic crystals," *J. Mat. Res.*, **1**, 13–26 (1986).

7. J. Pannetier, G. Heger, C. Janot, M. Audier, J. M. Lang, and P. Sainfort, in preparation.

8. H. B. Elswijk, J. T. M. de Hosson, S. Van Smallen, and J. L. de Boer, "Determination of the crystal structure of icosahedral Al–Cu–Li," *Phys. Rev.* **38**, 1681–1685 (1988).

9. L. S. Levitov, "Diffuse scattering in quasicrystals," *Europhys. Lett.* **6**, 419–424 (1988).

10. P. A. Bancel, P. A. Heiney, P. M. Horn, and F. W. Gayle, "High-resolution structural study of faceted icosahedral Al$_6$Li$_3$Cu," preprint.

11. G. Heger, F. Dénoyer, J. P. Lauriat, M. Lambert, S. Lefebvre, and M. Bessières, in preparation.

12. A. P. Tsai, A. Inoue, and T. Matsumoto, "Preparation of a new Al–Cu–Fe quasicrystal with large grain sizes by rapid solidification," *J. Mat. Sci. Lett.* **6**, 1403–1405 (1987).

13. A. P. Tsai, A. Inoue, and T. Matsumoto, "A stable quasicrystal in Al–Cu–Fe system," *Jpn. J. Appl. Phys.* **26**, L1505–L1507 (1987).

14. T. Ishimasa, Y. Fukano, and M. Tsuchimori, "Quasicrystal structure in Al–Cu–Fe annealed alloy," *Phil. Mag. Lett.* **58**, 157–165 (1988).

15. S. Ebalard and F. Spaepen, "The body-centered-cubic-type icosahedral reciprocal lattice of the Al–Cu–Fe quasi-periodic crystal," *J. Matr. Res.* **4**, 39–43 (1989).

16. J. Devaud-Rzepski, A. Quivy, Y. Calvayrac, M. Cornier-Quicandon, and D. Gratias, "Antiphase domains in icosahedral Al–Cu–Fe alloy," *Phil. Mag. B* (1989), **60**, 855–869 (1989).

17. M. Audier and P. Guyot, "Microcrystalline AlFeCu phase of pseudo icosahedral

symmetry," AAR Conference, ICTP, Trieste, July 1989, to be published.

18. F. Dénoyer, G. Heger, M. Lambert, M. Audier, and P. Guyot, "X-ray and TEM studies of AlFeCu dodecahedral particles: Characterization of their microcrystalline state of pseudo-icosahedral symmetry," *J. Phys.*, **51**, 651–660 (1990).

19. F. Dénoyer, G. Heger, M. Audier, and M. Lambert, "Etude par diffraction des "quasicristaux" Al–Cu–Fe," French National Colloquium on Quasicrystals, Nancy, March 1989.

20. M. Audier and P. Guyot, "Rhombohedral to icosahedral solid state transformation in the $Al_{65}Cu_{20}Fe_{15}$ alloy," Third International Meeting on Quasicrystals; Incommensurate Structure in Condensed Mater, May 27–June 2, 1989, Vista Hermosa, Mexico, to be published.

21. M. Lambert and F. Dénoyer, "Pavages plans et quasicristaux," *C.R. Acad. Sci. Paris, Série II*, **309**, 1463–1467 (1989).

6

Analysis of X-Ray Powder Diffraction Patterns of Rapidly Quenched Al-Based Alloys on the Basis of Quasicrystalline and Cubic Models

R. A. Dunlap, D. W. Lawther, and V. Srinivas

1. Introduction

Since the report of icosahedral symmetry in rapidly solidified Al-transition metal alloys by Shechtman et al. [1] in 1984, there has been considerable debate concerning the actual microstructure of these materials. They have been viewed as true quasicrystals in the sense suggested by Levine and Steinhardt [2]. These are commonly modeled on the basis of a three-dimensional Penrose tiling [3, 4]. Based on the work of Pauling, they have also been viewed as multiply twinned crystallites with a large unit cell cubic structure [5–9].

On the basis of electron diffraction patterns evidence has been presented that seems to favor the quasicrystal picture [e.g., 10]. The suggested validity of the multiple-twin model has been based on the analysis of X-ray diffraction patterns [5–9]. Although a good description of the X-ray patterns is also obtained on the basis of a projection of six-dimensional icosahedral vectors to three dimensions [11, 12], we are not aware of a critical comparison of these models based on an analysis of X-ray diffraction patterns. In the present work, we begin with a brief discussion of the two major models for analyzing X-ray diffraction patterns of these alloys. We then critically analyze previously published diffraction patterns of rapidly solidified Al-based alloys, as well as those of a number of new single-phase alloys that exhibit the same structure.

2. Structural Models of Quasicrystals

2.1. *Cubic Model*

Pauling [5–9] has proposed a model for the structure of alloys exhibiting fivefold symmetry that is based on a cubic unit cell with a large number of atoms. He has suggested two distinct structures. One consists of unit cells with 820 atoms [6], and is characteristic of the Al-transition metal-based alloys, related metalloid-containing systems, and Pd–Si–U [7]. This structure is proposed to consist of 104-atom icosahedral complexes, each made up of 20 Friauf polyhedra. Eight of these polyhedra are arranged in a cubic β-tungsten structure. Twelve atoms are shared between these polyhedra, yielding the 820 atom cell, which belongs to the space group $O_h^3 Pm3n$ and has a lattice parameter of approximately 23 Å.

The second related structure proposed by Pauling [8] consists of 1012-atom cells. This has been proposed for Al–Li–Cu, Ga–Mn–Zn [8], Ti–Ni–Si [5], and related alloys. In this case, the β-W structure is decorated with eight icosahedral clusters. Two of these are the 104-atom cluster, which appears in the 820 atom structure. The other six clusters consist of 136-atomic icosahedral complexes. These may be derived from the 104-atom cluster by placing an additional atom adjacent to each of the 20 hexagonal faces and each of the 12 pentagonal faces. These eight complexes share a total of 24 atoms, yielding the 1012-atom unit cell, which has a lattice parameter of ~24.5–26.0 Å.

Since the model of Pauling is based on the cubic β-W structure, an analysis of the X-ray diffraction data proceeds along conventional lines. In this case, the scattering vector for a particular reflection, $q = 2\pi/d$, is related to the three Miller indices, n_i, as

$$q = \frac{2\pi}{a_c} \left| \sum_{i=1}^{3} n_i \varepsilon_i \right|, \tag{1}$$

where a_c is the cubic lattice parameter and the ε_i are the three-dimensional basis vectors.

Although this model can satisfactorily describe the observed X-ray diffraction patterns, it is necessary to invoke a model involving microtwinning for the grain growth to explain the presence of fivefold symmetry in the electron diffractograms.

2.2. *Quasicrystalline Model*

The icosahedral symmetry of the quasilattice structure is defined by the construction of six orthogonal basis vectors in six dimensions.

The cut and projection to three dimensions, as described by Cahn et al. [4], define the six three-dimensional scattering vectors, $\boldsymbol{\varepsilon}_\parallel$, which determine the peak locations. An orthogonal projection yields the six three-dimensional vectors, $\boldsymbol{\varepsilon}_\perp$ which define the peak intensities. Following Dunlap et al. [3], we have taken the six $\boldsymbol{\varepsilon}_i = \boldsymbol{\varepsilon}_\parallel$ vectors to be cyclic permutations of $(1, \tau, 0)$ with an appropriate normalization, μ, as

$$
\begin{array}{ll}
\boldsymbol{\varepsilon}_1 = \mu(1, \tau, 0) & \boldsymbol{\varepsilon}_4 = \mu(0, 1, -\tau) \\
\boldsymbol{\varepsilon}_2 = \mu(1, -\tau, 0) & \boldsymbol{\varepsilon}_5 = \mu(\tau, 0, 1) \\
\boldsymbol{\varepsilon}_3 = \mu(0, 1, \tau) & \boldsymbol{\varepsilon}_6 = \mu(-\tau, 0, 1),
\end{array}
\tag{2}
$$

where $\tau = (1 + \sqrt{5})/2 = 1.61803\ldots$ and $\mu = (1 + \tau^2)^{-1/2} = 0.52573.\ldots$

The generalized diffraction condition is given as [3]

$$
q = \frac{\pi\sqrt{5}}{a_q} (1 + \tau^2)^{1/2} \left| \sum_{i=1}^{6} n_i \boldsymbol{\varepsilon}_i \right|,
\tag{3}
$$

where a_q is the quasilattice constant as defined by Elser [11]. This model also satisfactorily describes the X-ray diffraction patterns of those materials and, as well, implicitly explains the fivefold symmetry of the electron diffraction patterns.

2.3. Density of Diffraction Peaks

In general, for a least squares fit to experimental data, the goodness-of-fit, i.e., χ^2, depends not only on the suitability of the model used to fit the data but also on the number of parameters used in the fit. The use of a sufficient number of free parameters allows for the fitting of the data without necessarily validating the fitting function. In a comparison of more than one fitting model applied to a given set of data, the number of free parameters for each model must be taken into consideration.

In a comparison between the suitability of the quasicrystalline and crystalline models for the description of X-ray diffraction patterns, the value of the lattice parameter (or the quasilattice parameter) as well as a goodness-of-fit measure are obtained by a linear fit for the measured q-values. For each reflection, q is taken to be of the form

$$
q = \frac{k}{a} \left| \sum_i n_i \boldsymbol{\varepsilon}_i \right|,
\tag{4}
$$

where the n_i are the Miller indices, $\boldsymbol{\varepsilon}_i$ the basis vectors, a is the (quasi)lattice parameter, and k is a model dependent constant, as described in the previous section. In both cases, the fit of q to $|\sum n_i \boldsymbol{\varepsilon}_i|$ is linear in $1/a$, and this is the single fitted parameter. Although it is customary to iterate the n_i to find the best value of the Miller indices for each reflection, once these have been established the data for either model are fit to a straight line with a single free parameter.

In contrast to most data sets, X-ray diffraction results are described by an independent coordinate parameter, $|\sum n_i \boldsymbol{\varepsilon}_i|$, which is discrete. The ability of a model to fit an arbitrary set of data is therefore dependent on the spacing of the discrete lines allowed by the complete set of n_i. To compare the quasicrystalline and cubic models, we calculate the number of diffraction lines per unit angle for both of these.

2.3.1. Cubic Model   The Bragg condition for diffraction for a cubic structure is given by

$$n\lambda = 2a_c \sin\theta, \tag{5}$$

where n is an integer which satisfies the condition for the Miller indices, n_i,

$$n = \left[\sum_{j=1}^{3} \left(\sum_{i=1}^{3} n_i \varepsilon_{ij} \right)^2 \right]^{1/2}. \tag{6}$$

In this case, the three ε_i are orthogonal and we find that the density of diffraction lines per unit angle,

$$N_c = dn/d\theta, \tag{7}$$

is given by a dn of 1 in $\Sigma\, n_i^2$, so

$$N_c = \frac{d}{d\theta} \left(\frac{4a_c^2}{\lambda^2} \sin^2\theta \right) = \frac{8a_c^2}{\lambda^2} \sin\theta \cos\theta. \tag{8}$$

Some integer values of $\Sigma\, n_i^2$ are not allowed and as a result certain adjacent peaks are twice as far apart as (8) would indicate. Nevertheless, (8) is a good estimate of the maximum peak density. We see that N_c depends on a_c^2, and the use of an arbitrarily large value for a_c allows for the successful fitting of any diffraction pattern. N_c as a function of θ is shown in Figure 1.

Figure 1. Density of lattice reflections per unit degree of the cubic model, N_c, as a function of diffraction angle, θ. Results are for Cu K_α radiation ($\lambda = 1.54$ Å), $a_c = 23.4$ Å.

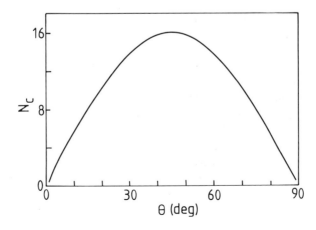

2.3.2. *Quasicrystal Model* The scattering vector, q, is related to the scattering angle as

$$q = \frac{4\pi}{\lambda} \sin \theta. \tag{9}$$

From expression (3) we can relate this to the basis vectors as

$$\left| \sum_{i=1}^{6} n_i \boldsymbol{\varepsilon}_i \right| = \frac{4a_q}{\lambda \sqrt{5}(1 + \tau^2)^{1/2}} \sin \theta. \tag{10}$$

The magnitude of the scattering vector is determined to be

$$\left| \sum_{i=1}^{6} n_i \boldsymbol{\varepsilon}_i \right| = \left[\sum_{j=1}^{3} \left(\sum_{i=1}^{6} n_i \varepsilon_{ij} \right)^2 \right]^{1/2}, \tag{11}$$

where the j indices represent the three three-dimensional components of the three-dimensional projections of the basis vectors.

The derivative of (11) with respect to θ is then written as

$$\frac{d}{d\theta} \left[\sum_{j=1}^{3} \left(\sum_{i=1}^{6} n_i \varepsilon_{ij} \right)^2 \right] = \frac{32 a_q^2}{5\lambda^2(1 + \tau^2)} \sin \theta \cos \theta. \tag{12}$$

The difficulty in evaluating this expression comes from the fact that N_q is not approximately equal to the left-hand side of this equation. This arises because the basis vectors, $\boldsymbol{\varepsilon}_i$, are not orthogonal in three dimensions. It then remains to estimate the derivative on the left-hand side of equation (12).

The sum over j in Eq. (12) is over the three spatial coordinates. Since the components of the $\boldsymbol{\varepsilon}_i$ are permutations of 0, μ, and $\mu\tau$, and the n_i are integers, then each term in the sum over j can be written as

$$\left(\sum_{i=1}^{6} n_i \varepsilon_{ij} \right)^2 = \mu^2 (a_j + b_j \tau)^2, \tag{13}$$

where the a_j and b_j are integers. The sum over j in expression (11) gives

$$\sum_{j=1}^{3} \left(\sum_{i=1}^{6} n_i \varepsilon_{ij} \right)^2 = \sum_{j=1}^{3} \mu^2 (a_j + b_j \tau)^2. \tag{14}$$

The allowed values of the left-hand side of Eq. (14) are determined by the allowed values of a_j and b_j on the right-hand side. This in turn gives the relative spacing of the X-ray diffraction peaks, as in Eq. (12).

While most authors have indexed peaks in the X-ray diffraction patterns using quasicrystalline Miller indices $|n_i| \leq 2$, in some cases [3] indices up to $|n_i| \leq 4$ have been used. In no case is there experimental evidence to indicate that reflections with $|n_i| > 4$ have measurable intensity. For the present investigation, we use $|n_i| \leq 4$, although it should be noted that the results obtained here are relatively insensitive to the maximum values of $|n_i|$ used.

An inspection of Eq. (13) in the context of the definition of the basis vectors in Eq. (2) shows that if $|n_i| \leqslant n_0$ then the a_j and b_j in Eq. (14) can take on the $2n_0 + 1$ integral values between $-2n_0$ and $+2n_0$. This means that there are $[(2n_0 + 1)^2 + 1]/2$ values of $\mu^2(a_i + b_i\tau)^2$, which range from a minimum of 0 to a maximum of $4n_0\mu^2(1 + \tau)^2$.

Table 1 gives allowed values of $\mu^2(a_i + b_i\tau)^2$ for $|n_i| \leqslant 4$, along with the differences, Δ, between adjacent values. As the X-ray diffraction line spacing is related to Δ, we clearly see that there is a highly nonuniform distribution of diffraction peaks. This is in contrast with the situation for the cubic model where the only anomalies occurred for integral values of Σn_i^2, which were disallowed.

As a method of establishing the probable peak separation, we have, in Figure 2, plotted a histogram of Δ values from Table 1. This shows a peak between $\Delta = 0.1$ and $\Delta = 0.2$. Changing n_0 or changing the step size for the histogram does not significantly alter the location of the peak in Figure 2. Therefore, to the accuracy with which we feel that an interpretation of a χ^2 analysis is meaningful, it is sufficient to take the most probable value of Δ of 0.1.

Equation (12) now gives the value of the density of reflections as

$$N_q = \frac{64a_q^2}{\lambda^2(1 + \tau^2)} \sin \theta \cos \theta. \tag{15}$$

A comparison of this with the expression (8) gives

$$\frac{N_q}{N_c} = \frac{8a_q^2}{(1 + \tau^2)a_c^2}. \tag{16}$$

Using typical values of $a_q = 4.61$ Å and $a_c = 23.4$ Å gives

$$N_q/N_c = 0.086. \tag{17}$$

Hence, a comparison of χ^2 for fits to the cubic and quasicrystalline models for the X-ray diffraction patterns of these alloys must take into account the results of Eq. (16).

3. Results and Analysis

A typical X-ray diffraction pattern for a quasicrystalline material is illustrated in Figure 3. The diffraction peaks shown here have been indexed according to the scheme for quasicrystalline materials proposed by Bancel et al. [13]. Details of these peaks, along with the corresponding indices assigned on the basis of Pauling's cubic model, are given in Table 2. A number of previous X-ray diffraction studies of quasicrystals have been

Table 1. Allowed Values of $\mu^2(a_i + b_i\tau)^2$ from Eq. (14) and Corresponding a_j and b_j and Differences [a]

a_j	b_j	$\mu^2(a_j + b_j\tau)^2$	Δ
0	0	0.000	
3	-2	0.015	0.015
2	-1	0.040	0.025
4	-2	0.161	0.121
4	-3	0.202	0.041
1	0	0.276	0.074
2	-2	0.422	0.146
3	-1	0.528	0.106
0	1	0.724	0.196
3	-3	0.950	0.226
1	-1	1.056	0.106
2	0	1.106	0.050
1	-2	1.382	0.276
4	-1	1.568	0.186
4	-4	1.689	0.121
1	1	1.894	0.205
2	-3	2.251	0.357
3	0	2.488	0.237
0	2	2.894	0.406
3	-4	3.332	0.438
2	1	3.618	0.286
1	-3	4.106	0.488
4	0	4.422	0.316
1	-2	4.960	0.538
2	-4	5.528	0.568
3	1	5.894	0.366
0	3	6.512	0.618
2	2	7.578	1.066
1	-4	8.276	0.698
4	1	8.724	0.448
1	3	9.472	0.748
3	2	10.748	1.276
0	4	11.578	0.830
2	3	12.984	1.406
4	2	14.472	1.488
1	4	15.432	0.960
3	3	17.050	1.618
2	4	19.839	2.788
4	3	21.668	1.829
3	4	24.798	3.130
4	4	30.311	5.513

[a] a_j and b_j are not unique but may both differ by a sign.

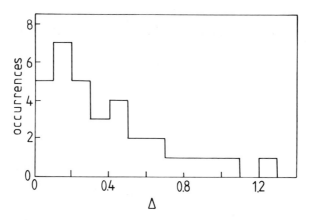

Figure 2. Histogram of Δ values from Table 1 ($|n_i| \leq 4$).

conducted on mixed phase systems. With the exception of the data for $Al_{85}Mn_{14}$ from Bancel et al. [13], the present analysis is restricted to well-ordered, essentially single-phase systems.

For a variety of systems measured q-values of the diffraction peaks were least squares fitted to values calculated by either Eq. (1), for the cubic model, or Eq. (3) for the quasicrystalline model. An iterative method was used to determine the best values of the Miller indices, n_i, and the (quasi)lattice parameter a_c (or a_q) by minimizing the goodness-of-fit parameter:

$$\chi^2 = \frac{1}{N-1} \sum_{k}^{N} \frac{[q_{exp}(k) - q_{calc}(k)]^2}{q_{exp}(k)}, \tag{18}$$

where the sum in k is over the N data points. In this analysis, all data points have been weighted equally. Weighting data points according to their intensity did not significantly affect the results of this study. Pauling [5, 7] has undertaken an investigation of the weaker peaks in the X-ray diffraction patterns

Figure 3. Room temperature Cu K_α X-ray diffraction pattern of quasicrystalline $Al_{80}V_{12.5}Fe_{7.5}$. Peaks are indexed according to the scheme of Bancel et al. [13], as given in Table 2.

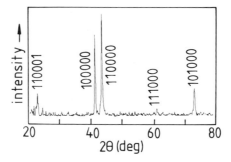

Table 2. Room Temperature Cu K_α X-Ray Diffraction Results for Single-Phase Icosahedral $Al_{80}V_{12.5}Fe_{7.5}$.[a]

Measured			Quasicrystal Model			Cubic (444) Model		Cubic (610) Model	
$2\theta(°)$	d (Å)	$q(Å^{-1})$	I	Indices	$q(Å^{-1})$	Indices	$q(Å^{-1})$	Indices	$q(Å^{-1})$
22.67	3.923	1.602	27	110001	1.608	4 4 4	1.596	6 1 0	1.605
26.30	3.388	1.854	4	11101̄0	1.865	7 4 0	1.860	6 3 2	1.848
35.14	2.554	2.460	8	211001	2.456	10 3 2	2.456	9 2 1	2.449
40.99	2.202	2.854	76	100000	2.857	12 2 2	2.850	10 4 1	2.858
43.19	2.095	2.999	100	110000	3.004	10 8 2	2.997	11 2 2	3.001
50.87	1.795	3.501	8	210001	3.531	14 6 0	3.524	13 3 0	3.526
61.10	1.517	4.143	6	111100	4.145	16 8 0	4.141	15 4 2	4.138
62.97	1.467	4.257	4	111100	4.248	16 8 4	4.244	16 1 1	4.247
73.14	1.294	4.856	25	101000	4.860	20 6 2	4.858	18 3 2	4.854

[a] a_q is found to be 4.677 Å and a_c is found to be 27.06 Å for the (444) model and 23.73 Å for the (610) model. I is the relative intensity.

of these alloys. In the present work, we have concentrated on the strongest peaks.

For the Al-based alloys, Pauling has proposed two different indexing schemes [7, 9]. The peak near $q = 1.62$ Å^{-1} has either been indexed to (444) [9] or to (610) [7]. In the present analysis, we have considered both of these schemes; they are referred to as the (444) cubic model and the (610) cubic model.

Table 2 shows the results of indexing the X-ray diffraction pattern of a typical Al-based quasicrystal $(Al_{80}V_{12.5}Fe_{7.5})$ using the quasicrystalline model and the two different cubic models [(444) and (610)]. Fitted values of the q-vectors for each of the peaks are given.

Table 3 gives the results of the kind of investigation shown in Table 2 for a variety of quasicrystal X-ray diffraction patterns. The findings of this study are represented by the values of the quantity $R = (\chi_q^2 N_q)/(\chi_c^2 N_c)$. For values of $R < 1$, the statistical analysis of the X-ray data favors the quasicrystalline model, while for $R > 1$ the cubic model is favored. Here we have considered data for 14 different Al-based alloys. From a consideration of R for several different alloys, we can define an overall quality of fit measure as the weighted mean of the individual R values, i.e.,

$$\langle R(hkl) \rangle = (N_q/N_c) \left[\sum_i^n m_i \right]^{-1} \left\{ \sum_i^n m_i \chi_{q,i}^2 / \chi_{c,i}^2 (hkl) \right\}, \tag{19}$$

where the weighting factor m_i is the number of peaks in the ith data set. For the Al-based alloys, the above definition along with the data of Table 3 yields $\langle R(444) \rangle = 0.623$ and $\langle R(610) \rangle = 0.492$. In both cases, the average over a large number of data sets indicates that the quasicrystal model gives a better description of the X-ray diffraction patterns of these materials than either of the cubic indexing schemes.

It is important to consider possible difficulties that could affect the validity of the present results. The principal sources of error are as follows:

1. *Incorrect sample geometry.* This can affect the accuracy of the measured lattice parameter values, but will have no effect on the relative line positions and, hence, will not affect the validity of the present analysis.

2. *Goniometer nonlinearity.* This is a small effect, $\leq 0.005°$, and is less than the error quoted on the measurements of 2θ.

3. *Errors in peak locations due to random noise.* This is the major source of error responsible for the number of quoted digits in the experimental results. It amounts to an error of ~ 0.01 in 2θ, but is still smaller than the differences between measured and calculated q-values for all patterns. The fact that we have studied a large number of samples and, in some cases, have repeated measurements on the same or different samples of the same nominal stoichiometry, and all measurements were given consistent results, eliminates this as a major concern.

Table 3. Results of Fits to Cubic and Quasicrystalline Models for X-Ray Diffraction Patterns.[a]

Alloy	No. of Peaks	Quasicrystal Model		Cubic (444)		Cubic (610)		$\dfrac{\chi_q^2 N_q}{\chi_c^2(444)N_c}$	$\dfrac{\chi_q^2 N_q}{\chi^2(610)N_c}$	Ref.
		χ_q^2	a_q (Å)	χ_c^2	a_c (Å)	χ_c^2	a_c (Å)			
$Al_{86}V_{14}$[b]	6	1.1×10^{-4}	4.735	6.3×10^{-6}	27.44	1.1×10^{-5}	24.00	1.50	0.86	c
$Al_{80}V_{12.5}Fe_{7.5}$	9	4.1×10^{-5}	4.677	3.0×10^{-6}	27.06	3.8×10^{-5}	23.73	0.12	0.093	14, c
$Al_{80}V_{11.5}Co_{8.5}$	9	6.6×10^{-6}	4.683	1.1×10^{-5}	27.10	8.6×10^{-6}	23.76	0.052	0.066	14, c
$Al_{80}Cr_{20}$	9	8.8×10^{-5}	4.676	3.0×10^{-5}	27.05	5.6×10^{-5}	23.72	0.25	0.14	c
$Al_{74}Cr_{20}Si_6$	6	5.5×10^{-6}	4.655	3.0×10^{-6}	27.00	4.0×10^{-6}	23.58	0.16	0.12	15, c
$Al_{65}Cr_{20}Ge_{15}$	3	3.5×10^{-5}	4.659	1.2×10^{-6}	27.01	3.8×10^{-6}	23.57	2.51	0.79	c
$Al_{80}Cr_{13.5}Fe_{6.5}$	5	4.8×10^{-5}	4.636	1.2×10^{-6}	26.89	5.1×10^{-6}	23.48	3.44	0.81	14, c
$Al_{65}Cr_{10}Fe_{10}Ge_{15}$	3	8.3×10^{-6}	4.569	1.6×10^{-6}	26.50	4.5×10^{-6}	23.12	0.45	0.16	16, c
$Al_{80}Cr_{13}Co_7$	5	9.5×10^{-5}	4.593	5.2×10^{-5}	26.64	6.5×10^{-4}	23.27	0.16	0.013	14, c
$Al_{86}Mn_{14}$[b]	21	1.7×10^{-5}	4.620	9.5×10^{-6}	26.69	2.2×10^{-6}	23.41	0.15	0.66	13
$Al_{74}Mn_{20}Si_6$	6	2.2×10^{-5}	4.603	7.3×10^{-6}	26.70	8.8×10^{-6}	23.32	0.26	0.22	17
$Al_{50}Mn_{20}Si_{30}$	5	5.8×10^{-3}	4.458	3.2×10^{-4}	25.85	2.9×10^{-4}	22.58	1.42	1.72	18, c
$Al_{55}Mn_{20}Si_{25}$	5	6.0×10^{-3}	4.460	3.8×10^{-4}	25.86	3.5×10^{-4}	22.58	1.36	1.47	18, c
$Al_{65}Mn_{20}Ge_{15}$	4	9.3×10^{-4}	4.602	5.9×10^{-4}	26.62	5.9×10^{-5}	23.23	0.14	0.14	c

[a] N_q/N_c is obtained from Eq. (10).
[b] Indicates an alloy that contains a significant quantity of impurity (nonquasicrystalline).
[c] From this work. In cases where c is indicated in addition to reference, data in addition to that explicitly published before have been used in the present analysis.

93

4. *Statistical errors due to fitting of a.* Again, as a large number of data sets have yielded consistent results, this is, presumably, not a significant effect.

5. *Uncertainty in stoichiometry.* As the exact stoichiometry of the quasicrystal phase is not well defined and a variety of compositions, e.g., in the Al−Mn−Si system, can be made with the same structure, this is not an important factor.

6. *Sample inhomogeneities.* It is commonly believed that significant disorder exists in quasicrystals. Since at least some of the X-ray diffraction peaks are of a line width comparable to that of a well-ordered crystalline material, sample inhomogenity, presumably, does not seriously affect X-ray diffraction measurements.

7. *Presence of impurity phases.* As the present results do not show any clear distinction between single phase quasicrystal and those which are a mixture of quasicrystal and fcc-Al, e.g., Al−V and Al−Mn, the presence of impurity phases are not an important concern.

8. *Phason and phonon stains.* These are an important concern, as it is known that the X-ray diffraction patterns are sensitive to those strains [16]. In fact, the presence of such strains may mean that any long-range description of order in these materials may be questionable.

Despite the above potential difficulties, we feel that the analyses of diffraction patterns of a number of different materials yields statistically significant results, as well as some interesting trends, when viewed in the context of other studies.

In cases where phason strains, for example, cause significant anomalies in the X-ray diffraction peak locations we expect that any model based on the assumption that these effects are small will yield relatively poor agreement between calculated and experimental line positions. As we have incorporated these effects into neither the quasicrystalline nor the cubic models then we expect two features to appear in our results for alloys with extensive phason strains: (1) generally high χ^2 values for both models and (2) values of R that do not differ significantly from unity, i.e., equally poor agreement for both models.

An investigation of the results in Table 3 shows that those alloys in which phason strains and disorder are suspected of playing an important part, e.g., $Al_{86}V_{14}$ [19]. $Al_{65}Cr_{20}Ge_{15}$ [20], $Al_{50}Mn_{20}Si_{30}$, $Al_{55}Mn_{20}Si_{25}$ [21], and $Al_{65}Cr_{10}Fe_{10}Ge_{15}$ [8], generally show the two expected features as given above. On the other hand, those alloys which are believed to be well ordered, e.g., $Al_{80}V_{12.5}Fe_{7.5}$, $Al_{80}V_{11.5}Co_{8.5}$, $Al_{80}Cr_{13}Co_{7}$ [14], $Al_{86}Mn_{14}$ [13], and $Al_{74}Mn_{20}Si_{6}$ [17], generally show somewhat lower χ^2 values and clearly lower values of R. It is in these cases that we expect to be able to clearly differentiate between more and less physically realistic models. As Table 2 indicates, the present results clearly favor an interpretation of the X-ray diffraction patterns based on the true quasicrystalline structure.

The large density of diffraction peaks available in either the quasicrystal model or the large unit cell cubic model makes a definitive conclusion concerning the validity of the two models difficult. The present investigation concludes, however, that a statistical comparison of experimental data on the basis of quasicrystalline and cubic models clearly favors the quasicrystalline picture for the atomic structure of these materials.

Acknowledgments

The authors are most grateful for comments and suggestions of Drs. D. Bahadur, M. E. McHenry, and R. C. O'Handley. This work was funded by grants from the Natural Sciences and Engineering Research Council of Canada.

REFERENCES

1. D. Shechtman, I. Blech, D. Gratias, and J. W. Cahn, "Metallic phase with long-range orientational order and no translational symmetry," *Phys. Rev. Lett.*, **53**, 1951–1953 (1984).

2. D. Levine and P. J. Steinhardt, "Quasicrystals: A new class of ordered structures," *Phys. Rev. Lett.*, **53**, 2477–2480 (1984).

3. R. A. Dunlap, R. C. O'Handley, M. E. McHenry, and R. Chatterjee, "Quasicrystal structure of rapidly solidified Ti-Ni-based alloys," *Phys. Rev. B*, **37**, 8484–8487 (1988).

4. J. W. Cahn, D. Shechtman, and D. Gratias, "Indexing of icosahedral quasiperiodic crystals," *J. Mater. Res.*, **1**, 13–26 (1986).

5. L. Pauling, "Comment on 'Quasicrystal structure of rapidly solidified Ti-Ni-based alloys,'" *Phys. Rev. B*, **39**, 1964–1965 (1989).

6. L. Pauling, "So-called icosahedral and decagonal quasicrystals are twins of an 820-atom cubic crystal," *Phys. Rev. Lett.*, **58**, 365–368 (1987).

7. L. Pauling, "Additional evidence from x-ray powder diffraction patterns that icosahedral quasicrystals of intermetallic compounds are twinned cubic crystals," *Proc. Natl. Acad. Sci. U.S.A.*, **85**, 4587–4590 (1988).

8. L. Pauling, "Icosahedral quasicrystals as twins of cubic crystals containing large icosahedral clusters of atoms: the 1012-atom primitive cubic structure of Al_6CuLi_3, the C-phase $Al_{37}Cu_3Li_{21}Mg_3$, and $GaMg_2Zn_3$," *Proc. Natl. Acad. Sci. U.S.A.*, **85**, 3666–3669 (1988).

9. L. Pauling, "Apparent icosahedral symmetry is due to directed multiple twinning of cubic crystals," *Nature (London)*, **317**, 512–514 (1985).

10. H. B. Elswijk, P. M. Bronsveld, and J. Th.M. De Hosson, "Field ion microscopy contradicts latest model based on twinning of a cubic crystal for quasicrystals," *Phys. Rev. B*, **37**, 4261–4264 (1988).

11. V. Elser, "Indexing problems in quasicrystal diffraction," *Phys. Rev. B*, **32**, 4892–4898 (1985).

12. V. Elser and C. L. Henley, "Crystal and quasicrystal structures in Al-Mn-Si alloys," *Phys. Rev. Lett.*, **55**, 2883–2886 (1985).

13. P. A. Bancel, P. A. Heiney, P. W. Stephens, A. I. Goldman, and P. M. Horn, "Structure of rapidly quenched Al-Mn," *Phys. Rev. Lett.*, **54**, 2422–2425 (1985).

14. D. W. Lawther, R. A. Dunlap, and V. Srinivas, "On the question of stability and disorder in icosahedral aluminum-transition metal alloys," *Can. J. Phys.*, **67**, 463–467 (1989).

15. M. E. McHenry, V. Srinivas, D. Bahadur, R. C. O'Handley, D. J. Lloyd, and R. A. Dunlap, "Structure, thermal and magnetic properties of icosahedral Al-Cr-Mn-Si alloys," *Phys. Rev. B*, **39**, 3611–3615 (1989).

16. D. Bahadur, V. Srinivas, and R. A. Dunlap, "Structural relaxation in icosahedral $Al_{65}Cr_{10}Fe_{10}Ge_{15}$," *J. Phys.: Condens. Matter*, **1**, 2561–2567 (1989).

17. M. E. McHenry, R. A. Dunlap, R. Chatterjee, A. Chow, and R. C. O'Handley, "Magnetic properties of gas atomized powders of $Al_{74}Mn_{20}Si_6$," *J. Appl. Phys.*, **63**, 4255–4257 (1988).

18. R. A. Dunlap, M. E. McHenry, V. Srinivas, D. Bahadur, and R. C. O'Handley, "Ferromagnetism in icosahedral Al-Mn-Si alloys," *Phys. Rev. B*, **39**, 4808–4811 (1989).

19. D. W. Lawther, D. J. Lloyd, and R. A. Dunlap, "Transition metal site distributions in binary aluminum-transition metal quasicrystals: Al-V and Al-Cr," *Mater. Sci. Eng.* (1989), **A123**, 33–39 (1990).

20. V. Srinivas, R. A. Dunlap, D. Bahadur, and E. Dunlap, "Physical properties of icosahedral Al-Cr-Fe-Ge alloys," *Phil. Mag. B* (1989), **61**, 177–188 (1990).

21. V. Srinivas and R. A. Dunlap, "Icosahedral phase formation in high Si and Mn content Al-Mn-Si alloys," *Mater. Sci. Eng.* (1990), in press.

7 *Finite Automata, Quasicrystals, and Robinson Tilings*

Jean-Paul Allouche and Olivier Salon

1. Introduction

The problem of tiling the Euclidean plane or the Euclidean space is related to questions in pure mathematics, crystallography ..., and aesthetics. Moreover religious reasons (such as the impossibility of representing human faces) might have played a role in the systematic study of tilings as ornaments. It is well known that the only way of tiling the plane with regular polygons (a precise mathematical meaning of this expression can be given) is to use equilateral triangles, squares, or regular hexagons (the French "tommettes"). In particular it is impossible to tile the plane with regular pentagons (the reader could try to do so, using cardboard pentagons,... or might wish to read a mathematical proof of this assertion, for instance the elegant proof in [1] derived from the one in [2]).

In 1970, R. Penrose in a paper entitled "The rôle of aesthetics in pure and applied mathematical research" [3] gave inter alia a "kind of tiling" of the plane that uses pieces of a regular pentagon as tiles. These tiles are arranged in a nonperiodic but "attractive" way (hence the title of the paper). Many mathematicians have studied this Penrose tiling and its fivefold symmetry. One of its properties is the occurrence of "worms," which consist of short (S) and long (L) "ties," and the sequence of these ties is generated by what is called a substitution (by mathematicians), amorphism of the free monoid $\{S, L\}^*$ (by theoretical computer scientists), or an inflation rule (by physicists), indeed:

$$\sigma : S \to L, \quad L \to LS,$$

(σ is known as the Fibonacci substitution).

Iterating this rule gives, replacing at each step each letter by its image under σ:

$$L \to LS \to LSL \to LSLLS \to \cdots.$$

In 1984, Shechtman et al. [4] discovered an alloy of aluminum and manganese with a fivefold symmetry; as this violates the classical laws of crystallography, they called this alloy a "quasicrystal." This was the starting point of new interest in what we could call "quasitilings."

On the other hand the notion of substitution of constant length (also called uniform morphism), first introduced by computer scientists, has been developed by mathematicians, and more precisely by number theorists (see the first paper in that direction [5], see also the survey [6]). These substitutions or the associated finite automata have been proposed in physics to generate (or simulate?) a controlled disorder; see for instance the surveys [7] and [8]. One classical example of substitution of constant length is given by $0 \to 01$, $1 \to 10$, (substitution of length 2). Iterating this operation starting from the single symbol 0, we get

$$0 \to 01 \to 0110 \to 01101001 \to \cdots$$

and "finally" an infinite sequence known as the Prouhet–Thue–Morse sequence.

Recently the arithmetic properties of sequences generated by constant length substitutions have been generalized to the multidimensional case ([9]). Such a substitution, instead of replacing a symbol by a string of symbols, replaces a symbol by a square (or a cube, or hypercube . . .) of symbols, for instance:

$$0 \to \begin{matrix} 01 \\ 11 \end{matrix}, \quad 1 \to \begin{matrix} 10 \\ 00 \end{matrix}.$$

Iterating this operation from 0 gives rise to a (double) infinite sequence:

$$0 \to \begin{matrix} 01 \\ 11 \end{matrix} \to \begin{matrix} 0110 \\ 1100 \\ 1010 \\ 0000 \end{matrix} \to \begin{matrix} 01101001 \\ 11000011 \\ 10100101 \\ 00001111 \\ 10011001 \\ 00110011 \\ 01010101 \\ 11111111 \end{matrix} \to \cdots.$$

A natural question then arises: the study of the Penrose tiling, which is a 2-D tiling, involves the Fibonacci substitution, which is a 1-D substitution; is it possible to introduce a 2-D substitution in this study? The answer seems to be no, as there is no underlying square lattice (for a slightly different point of view see [10], see also [11]), but it can be asked whether other quasitilings of the plane can be generated by 2-D substitutions.

In what follows we shall first give the definition and some properties of the 2-D substitutions and finite automata. We shall then recall the construction of the Robinson regular quasitiling of the plane. Finally, we shall outline the proof that this quasitiling can indeed be generated by a 2-D substitution.

2. Sequences Generated by $k \times k$ Substitutions or by Two-Dimensional k-Automata

In this chapter we are going to give an example of a 2×2 substitution and of a 2-D 2-automaton, which, we hope, will replace for the reader a formal definition. Such a definition can be found in [9].

2.1. A 2 × 2 Substitution

Let $\mathfrak{A}$ be the alphabet (finite set) $\mathfrak{A} = \{a, b, c, d\}$, and let us consider the following "2×2 substitution":

$$a \to \begin{matrix} a & d \\ d & b \end{matrix}, \qquad b \to \begin{matrix} b & c \\ c & a \end{matrix}, \qquad c \to \begin{matrix} a & b \\ b & a \end{matrix}, \qquad d \to \begin{matrix} b & a \\ a & b \end{matrix},$$

Iterating this rewriting rule starting from a gives:

$$a \to \begin{matrix} a & d \\ d & b \end{matrix} \to \begin{matrix} a & d & b & a \\ d & b & a & b \\ b & a & b & c \\ a & b & c & a \end{matrix} \to \cdots.$$

After an infinite number of iterations we obtain a (double) infinite sequence which is, by construction, a fixed point of the substitution.

If we now replace, say a and c by 0 and b and d by 1, we obtain a double binary sequence:

$$0110 \cdots$$
$$1101 \cdots$$
$$1010 \cdots$$
$$0100 \cdots$$
$$\vdots$$

We say that this sequence is "the image of a fixed point of a 2×2 substitution".

Note that in the case of a $k \times k$ substitution each letter would be replaced by a $k \times k$ square of letters.

2.2. A Two-Dimensional 2-Automaton

Let us consider the "2-automaton" of Figure 1. It consists of a set of states $S = \{a, b, c, d\}$, where the state a is called the initial state, of four maps from S to S, $\{(0, 0), (0, 1), (1, 0), (1, 1)\}$, and of an output function τ defined by: $\tau(a) = \tau(d) = 0$, $\tau(b) = \tau(c) = 1$.

This machine works as follows: given a pair of integers, say $(3, 5)$, one writes these integers in base 2: $(11, 101)$. Then the shortest integer is completed with leading zeroes to get two strings of binary symbols of equal length: $(011, 101)$. We then feed the machine with the above pairs of digits read from right to left, starting from the initial state a:

$$(011, 101) \cdot a = (0,1)(1,0)(1,1) \cdot a = (0,1)(1,0) \cdot c = (0,1) \cdot c = c.$$

Taking the image of this state c by τ we obtain 1, and say that this is the output generated by the automaton, when fed with the initial pair of integers $(3, 5)$. If this job is done with every pair of integers (m, n), we obtain a (double) sequence $[u(m, n)]$ that is said to be generated by this 2-automaton ("2-automatic sequence"); note that the previous computation gives $u(3, 5) = 1$.

In the case of a k-automaton, the maps from S to S are $\{(i, j); 0 \leqslant i, j \leqslant k - 1\}$ and the integers are written in base k.

Figure 1. Example of a 2-D 2-automaton.

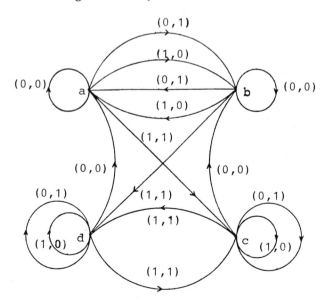

2.3. A Theorem

The attentive reader might have noticed (or proved) that the sequences generated in 2.1 and 2.2 above are actually identical. Moreover if he considers the formal power series

$$F(X, Y) = \sum_{m,n=0}^{+\infty} u(m, n) X^m Y^n,$$

with coefficients in the two element field F_2, he will notice that F satisfies a relation

$$T(X, Y)F^4(X, Y) + U(X, Y)F^2(X, Y) + V(X, Y)F(X, Y) + W(X, Y) = 0,$$

where T, U, V, and W are polynomials with coefficients in F_2. In other words the formal series F is algebraic over the field $F_2(X, Y)$ (the field of rational functions in two variables with coefficients in F_2).

All these facts did not happen by chance and the following theorem due to Salon [9], which is a multidimensional generalization of the theorem of Christol et al. [5] gives a general framework for the preceding observations:

THEOREM [9]. *Let $[u(m, n)]$ be a sequence with values in $\{0, 1\}$. The following properties are equivalent*:

i. the sequence u is 2-automatic,

ii. u is the image of a fixed point of a 2×2-substitution,

iii. the 2-kernel of the sequence u is finite (where the 2-kernel of u is defined as the set of subsequences

$$\{(m, n) \rightarrow u(2^k m + r, 2^k n + s); k \geq 0, 0 \leq r, s \leq 2^k - 1\}),$$

iv. the formal power series $\sum_{m,n=0}^{+\infty} u(m, n) X^m Y^n$ is algebraic over $F_2(X, Y)$.

This theorem gives a link between combinatorial properties of a sequence (the first three properties above) and an arithmetical property of the formal power series associated to this sequence.

2.4. Some Results Concerning Automatic Sequences

The reader is referred to [9] (see also [5] or [12]): 2-automatic sequences are stable under finite modification, addition, termwise or Cauchy multiplication. These sequences have zero entropy (hence they are deterministic); their Fourier spectra can be either discrete or continuous (singular or not): in some sense they are simultaneously "ordered" and "chaotic," or at least they are mimicking a controlled disorder.

Finally note that everything works in the same way for k-automatic sequences; note also that the frequency of each symbol (if it exists) in an automatic 1-D or 2-D sequence is rational (which implies that the fixed point of the Fibonacci substitution is *not* automatic).

3. The Robinson Tiling

The Robinson tiling was introduced in 1971 (see [13], pp. 525–529) and is related to problems of decidability in theoretical computer science. This tiling uses six kinds of tiles, which are represented on Figure 2.

One starts with the 3×3 block of tiles of Figure 3, and one extends it to a 7×7 block, then to a 15×15 block ... then to a $(2^k - 1) \times (2^k - 1)$ block ... by choosing at each step a tile adjacent to a corner of the already constructed block (the details can be found in [13]). In the case where all

Figure 2. The six fundamental Robinson tiles and their symbolic representation.

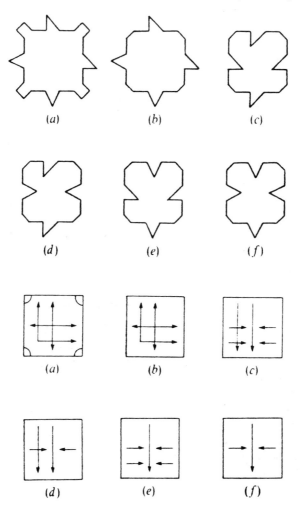

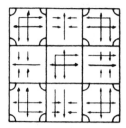

Figure 3. The 3 × 3 regular Robinson tiling.

these choices are identical, we obtain what we call the regular Robinson tiling (see Fig. 4).

As this tiling clearly admits an underlying square lattice structure, it is then natural to ask whether this tiling can be generated by a finite automaton, i.e., by a 2-D substitution.

4. The Main Result

We give here a positive answer to the question above, more precisely:

THEOREM. *The regular Robinson tiling of the plane can be generated by a 2-D 2-automaton with 48 states.*

The proof consists, in view of theorem in Section 2.3, of proving that the 2-kernel of the (double) sequence of tiles in the regular Robinson tiling is finite. We shall skip the technical details which can be found in [14], where some relations for the sequence $[a(m, n)]$ of tiles are proved, e.g.,

$a(4m, 4n) = $ constant,
$a(8m + 1, 8n) = $ constant,
$a(4m + 3, 4n) = a(4m + 3, 4n + 2),$
$a(8m + 7, 8n) = a(4m + 3, 4n),$
. . .

Together, these relations prove that the 2-kernel of our sequence is a set of cardinality 48, and it is not difficult (at least in principle!) to draw a bidimensional 2-automaton with 48 states which generates this sequence.

The reader interested in the relationship between tilings, fractals, and finite automata can read three recent works on this subject: [15], [16], and [17].

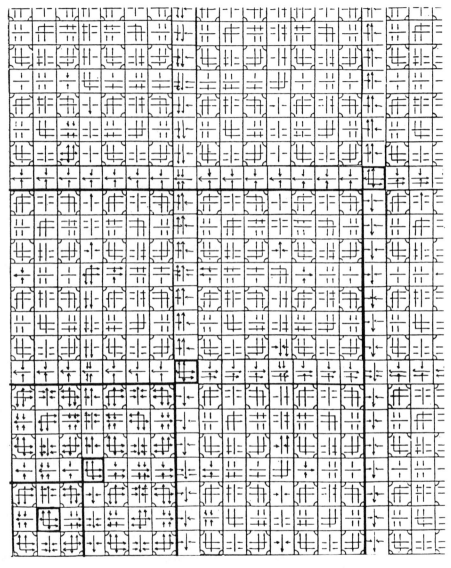

Figure 4. Part of the 31 × 31 Robinson tiling.

Acknowledgments

This work was initiated while the first author was visiting the Institute for Theoretical Physics in Santa Barbara (California) during the Semester on Quasicrystals. This research was supported in part by National Science Foundation Grant PHY82-17853 supplemented by funds from the National Aeronautics and Space Administration.

REFERENCES

1. P. A. B. Pleasants, "Quasicrystallography: Some interesting new patterns," in *Elementary and Analytic Theory of Numbers*, Banach Center Publications, Vol. 17, pp. 439–461. Polish Scientific Publishers, Warsaw, 1985.

2. R. L. E. Schwarzenberger, *N-Dimensional Crystallography*. Pitman, London, 1980.

3. R. Penrose, "The rôle of aesthetics in pure and applied mathematical research," *Bull. Inst. Appl.*, **10**, 266–271 (1974).

4. D. Shechtman, I. A. Blech, D. Gratias, and J. W. Cahn, "Metallic phase with long-range orientational order and no translational symmetry," *Phys. Rev. Lett.*, **53**, 20, 1951–1953 (1984).

5. G. Christol, T. Kamae, M. Mendès France, and G. Rauzy, "Suites algébriques, automates et substitutions," *Bull. Soc. Math. France*, **108**, 401–419 (1980).

6. J.-P. Allouche, "Automates finis en théorie des nombres," *Expo. Math.*, **5**, 239–266 (1987).

7. J.-P. Allouche, *Finite Automata in 1-D and 2-D Physics*. Colloquium Number Theory and Physics, Les Houches, 1989.

8. F. Axel, "Controlled disorder generated by substitutions or finite automata: The state of the art," 3rd international meeting on quasicrystals, incommensurate structures in condensed matter, Mexico, 1989.

9. O. Salon, "Suites automatiques à multi-indices," Séminaire de Théorie des Nombres de Bordeaux, 4-01–4-27 (1986–1987), followed by an appendix of J. Shallit. See also "Suites automatiques à multi-indices et algébricité," *C.R. Acad. Sci. Paris*, t. 305, Sér. I, 501–504 (1987).

10. F. M. Dekking, "Recurrent sets," *Adv. Math.*, **44**(1), 78–104 (1982).

11. J. Peyrière, "Frequency of patterns in certain graphs and in Penrose tilings," *J. Phys.* Colloque C3, Suppl. 7, tome 47 (1986).

12. A. Cobham, "Uniform tag sequences," *Math. Syst. Theory*, **6**, 164–192 (1972).

13. B. Grünbaum and G. C. Shephard, *Tilings and Patterns*. Freeman, New York, 1987.

14. O. Salon, "Quelles tuiles! (pavages apériodiques du plan et automates bidimensionnels)," *Sémi. Théorie Nombr. Bordeaux*, 2ème Sér. 1 (1), 1–26, (1989).

15. S. Mozes, "Tilings, substitution systems and dynamical systems generated by them," *J. An. Math.*, **53**, 139–186 (1989).

16. W. P. Thurston, "Groups, tilings, and finite state automata," Lecture given at the 92nd meeting of the A.M.S., University of Colorado, Boulder (1989).

17. J. Stolfi and J. Shallit, "Two methods for generating fractals," *Comp. Graphics*, **13**(2), (1989).

8
Fivefold Symmetry in Higher Dimensional Spaces and Generalized Penrose Patterns

E. J. W. Whittaker and R. M. Whittaker

1. Introduction

Symmetry may be of any order, and each order of symmetry has some unique qualities that mark it off from all others. Nevertheless it is for low values of n that n-fold symmetry changes most markedly from one value to the next, and fivefold symmetry has, perhaps, more differences both from its even neighbors (fourfold and sixfold) and its odd neighbors (threefold and sevenfold) than symmetry of any other order. In this paper we explore some of the geometrical and mathematical aspects of these differences, with particular reference to n-dimensional spaces with $n > 3$. However, in order to develop the less familiar features of symmetry in higher dimensions, we first review some aspects of fivefold symmetry in two and three dimensions.

2. Symmetry in Two Dimensions

2.1. Fivefold Symmetry in the Plane

Figure 1a shows the effect of successive reflection of a point P in two lines OA, OB intersecting at 36° at O. The effect is to rotate P through 72° about O to the position P'. Four further applications of the operation take P' to P'', P''', P'''', and back to P, giving fivefold rotation symmetry about O. This process is of course entirely general: if the two

107

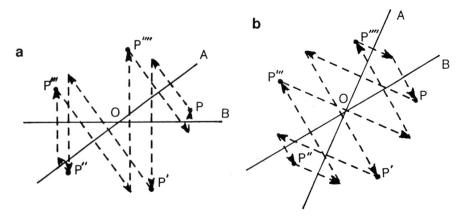

Figure 1. Two lines *OA*, *OB* intersect at 36°. Reflection of *P* in *OA* and then in *OB* takes it to *P′* so that *PÔP′* = 72°. Repetition of the process leads to fivefold rotation symmetry round *O*. The effect is independent of the orientation of *OA*, *OB*.

lines intersect at an angle θ they produce a rotation through 2θ and if 2θ is 360°/*n* then an *n*-fold rotation symmetry is produced. It is emphasized here in order to point out that the two mirror lines have no defined orientation; if they lie in a different orientation as in Figure 1b, the individual reflection operations are quite different but their combined effect is exactly the same. The lines may be described as *cryptomirror* lines. The only point that is invariant under the operation is *O*, the fivefold rotation point. Related phenomena appear in higher dimensions (Section 4.4).

It is readily proved in elementary analytical geometry textbooks that if a point referred to rectangular coordinates (x, y) is rotated through an angle about the origin it is transformed to the point (x', y') given by

$$x' = x \cos \theta + y \sin \theta \text{ and } y' = -x \sin \theta + y \cos \theta$$

This may be expressed in matrix form (with θ replaced by 72°) to give

$$\begin{pmatrix} \cos 72° & \sin 72° \\ -\sin 72° & \cos 72° \end{pmatrix} \begin{pmatrix} x \\ y \end{pmatrix} = \begin{pmatrix} x' \\ y' \end{pmatrix} \tag{1}$$

and multiplication by the 2 × 2 matrix in Eq. (1) provides a convenient representation of a fivefold rotation in the plane.

The simplest regular plane geometric figure with fivefold rotation symmetry is the regular pentagon (Figure 2). In addition to a fivefold rotation point at its center it possesses five mirror lines intersecting in this point, one such line passing through each vertex and the midpoint of the opposite side. This symmetry may be symbolized as **5m**. These are *overt mirrors* which (unlike the cryptomirrors) are not inherent in the fivefold symmetry; more complicated fivefold figures can be drawn that lack overt mirrors.

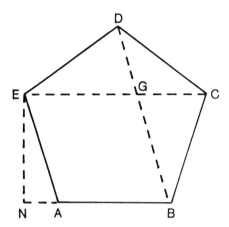

Figure 2. In the regular pentagon a diagonal $EC = AB + 2AN$. Hence if $AB = 1$, $EC = 1 + 2 \cos 72° = \phi$, and $EG : GC = 1 : (\phi - 1) = \phi : 1$.

Inasmuch as plane regular polygons exist with n-fold symmetry for all values of n, fivefold symmetry is less special in two dimensions than in higher dimensions. However even here there is one special property. The broken construction lines in Figure 2 show that a diagonal of the pentagon having unit sides is of length

$$1 + 2 \cos 72° = \phi,$$

where ϕ is the golden number (1.6180339 . . .), and pairs of diagonals cut one another in the ratio ϕ:1.

2.2. *Crystallographic Symmetry in Two Dimensions*

It is well known that the repeating patterns and lattices that characterize crystals can have only two-, three-, four-, and sixfold rotation symmetries, and that fivefold symmetry and orders higher than 6 are forbidden. This is already true in two-dimensional lattices such as that shown in Figure 3, a fact that can be readily demonstrated graphically. However such a demonstration is difficult to generalize to higher dimensions (where in fact it is relaxed in certain ways), and for our purpose it is more useful to discuss the matter analytically. Consider a pattern like that in Figure 3 in terms of the inclined axes Ox_c, Oy_c rather than in terms of the orthogonal axes Ox_o, Oy_o. If all adjacent spots are at unit distances apart, then their coordinates x_c, y_c are integers p, q (whereas x_o, y_o are in general irrational). Thus the matrix corresponding to any symmetry operation in the pattern must also contain only integers, or be transformable to such a form, if it is to convert integral p, q to integral p', q' for all values of p and q.

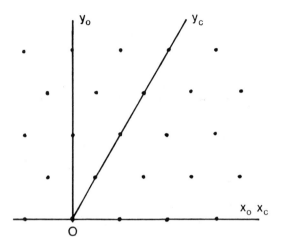

Figure 3. A lattice of points at unit separations may lie at irrational coordinates on orthogonal axes Ox_o, Oy_o, but at integral positions on suitably inclined axes Ox_c, Oy_c.

The form of the entries in a matrix changes radically when we change the axes with respect to which it is expressed, but certain quantities related to the entries are invariant regardless of these axes—in particular, the determinant and also the trace, the sum of the entries on the diagonal from top left to bottom right. Thus the matrix for a sixfold rotation operation in terms of orthogonal axes is

$$\begin{pmatrix} \cos 60° & \sin 60° \\ -\sin 60° & \cos 60° \end{pmatrix} = \begin{pmatrix} \dfrac{1}{2} & \dfrac{\sqrt{3}}{2} \\ -\dfrac{\sqrt{3}}{2} & \dfrac{1}{2} \end{pmatrix}. \tag{2}$$

In terms of axes at 60° it is

$$\begin{pmatrix} 1 & 1 \\ -1 & 0 \end{pmatrix} \tag{3}$$

and the trace in each case is 1. A matrix that does not have an integral trace cannot be converted by a change of axes to an integral form and therefore cannot represent a symmetry operation in a repeating pattern. Since the matrix of an n-fold two-dimensional rotation referred to orthogonal axes is

$$\begin{pmatrix} \cos 360°/n & \sin 360°/n \\ -\sin 360°/n & \cos 360°/n \end{pmatrix}, \tag{4}$$

its trace is $2 \cos 360°/n$, which (excluding the case with $n = 1$) is integral only if $n = 2, 3, 4,$ or 6 (the values of the trace being $-2, -1, 0,$ and 1 in these cases).

3. Fivefold Symmetry in Three Dimensions

Because we live in a three-dimensional world, we tend to regard the way symmetry and rotation occur in three dimensions as normal. Wheels and revolving doors rotate about axles, and so we regard it as normal for a rotation to take place about an axis; in fact we tend to regard the rotation points in the two-dimensional figures of the preceding section as simply points in the paper where "real rotation axes" would pierce it. Equally, we are so used to reflection taking place in the surface of a piece of glass or a pool of water that we tend to regard a mirror line in two dimensions as simply the place where a "real mirror plane" would cut the plane. However this is to allow ourselves to be blinkered by the fact that our minds have been brought up to deal with a three-dimensional world. Reflection is a fundamentally one-dimensional operation: it takes place in every line perpendicular to a mirror line in two dimensions and in every line perpendicular to a mirror plane in three dimensions. In n dimensions it leaves $n - 1$ dimensions invariant, while the reflection occurs in every line perpendicular to such an $(n - 1)$-dimensional mirror hyperplane. Similarly, rotation is a fundamentally two-dimensional operation around a point. If it takes place in n dimensions it leaves $n - 2$ dimensions invariant, which in our three-dimensional world is just our familiar rotation axis, and the rotation occurs in every plane perpendicular to this axis.

The matrix representation of a rotation in three dimensions is a 3×3 matrix, and by a suitable choice of axes it can always be arranged that the two dimensions involved in the rotation relate to the four symbols in the top left-hand corner, and the invariant dimension relates to the bottom right-hand entry. Thus a fivefold rotation with the z axis as rotation axis is represented by

$$\begin{pmatrix} \cos 72° & \sin 72° & 0 \\ -\sin 72° & \cos 72° & 0 \\ 0 & 0 & 1 \end{pmatrix}. \tag{5}$$

Since the trace of this matrix ($2 \cos 72° + 1$) is not integral, it follows that fivefold symmetry is not crystallographic in three dimensions.

The most instructive way to derive a three-dimensional figure with fivefold symmetry is to join the vertices of a regular pentagon to a point displaced along an axis through its center and orthogonal to its plane. This gives a pentagonal pyramid (Fig. 4). It is bounded by five isosceles triangles and one regular pentagon. It possesses the same symmetry as the pentagon, **5m**, but now the rotation is about an axis instead of a point and the reflections are in planes instead of lines.

The fivefold rotational symmetry of the pentagonal pyramid is based on the operation of the matrix (5). The 1 on the diagonal of this can be replaced by -1 without interfering with the fivefold rotation specified by the 2×2 entries at the top left, leading to

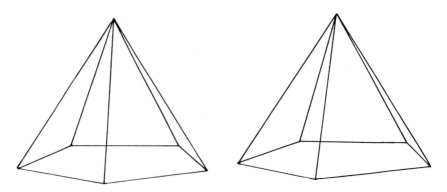

Figure 4. Stereo pair of a regular pentagonal pyramid.

$$\begin{pmatrix} \cos 72° & \sin 72° & 0 \\ -\sin 72° & \cos 72° & 0 \\ 0 & 0 & -1 \end{pmatrix}, \tag{6}$$

which is the matrix of a $\tilde{\mathbf{5}}$ rotation–reflection operation. A simple figure that possesses this symmetry is the pentagonal bipyramid shown in Figure 5, which is bounded by 10 isosceles triangles. The operation of matrix (6) on an upper face turns it by 72° about the vertical axis and simultaneously reflects it in a horizontal plane into the next lower face. Thus the operation must be done 10 times before the face returns to its original position, so that it is strictly a tenfold symmetry operation. However the overall result is the same as that of a combination of a fivefold rotation axis with a perpendicular mirror plane,[1] and the most obvious symmetry in Figure 5 is fivefold.

It is possible for a solid figure to have more than one fivefold symmetry axis only if a **5** axis is set at $\tan^{-1} \phi$ (where ϕ is the golden number) to a **2** axis. Each of them is then repeated round the other to give a set of six **5** axes and fifteen **2** axes, and at the same time ten **3** axes are generated. This is the symmetry of two of the five Platonic regular polyhedra—the dodecahedron and the icosahedron. In the dodecahedron (Figure 6) the **5** axes emerge at the centers of the faces, the **3** axes at the vertices, and the **2** axes at the midpoints of the edges. It is bounded by 12 regular pentagons. The icosahedron is the dual of the dodecahedron, so its equilateral triangular faces correspond to the vertices of the latter and its vertices to the faces.

Symmetry of order higher than 5 does not occur in three-dimensional regular figures, which gives fivefold symmetry in three dimensions a unique position, which it does not hold in two dimensions.

---

[1] A corresponding statement holds good for all $\tilde{\mathbf{n}}$ operations with n odd, but not with n even.

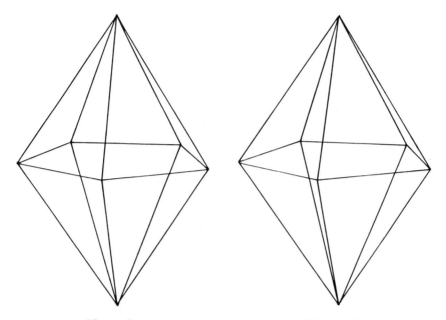

Figure 5. Stereo pair of a regular pentagonal bipyramid.

4. Fivefold Symmetry in Four Dimensions

4.1. General Principles

Symmetry in three dimensions is familiar from everyday life, and symmetry in two dimensions is easily understood because it can be embedded in three dimensions and is more restricted. When we come to generalize to four (or more) dimensions, the problems are greater, but they are illuminated by the changes that take place between two and three dimensions.

The first point to be made is that in four dimensions a rotation does not take place about an axis, but about a plane [1]. We have already seen that in two dimensions it takes place about a point and that any rotation involves movements only within a plane. In three dimensions there is one dimension not involved in the rotation, and this lies along the rotation axis. In four dimensions there are necessarily two not involved in the rotation, and they define a plane that is not affected by it.

Although it is not possible to make truly four-dimensional diagrams, it is possible to use a continuously varying color coordinate to represent the fourth dimension [2]. The idea is illustrated by using color to represent the third dimension in a two-dimensional diagram as in Plate 1. Here the pentagonal pyramid of Figure 4 is viewed from above; the vertex is therefore the nearest point to the viewer and is red, and increasing distance away is

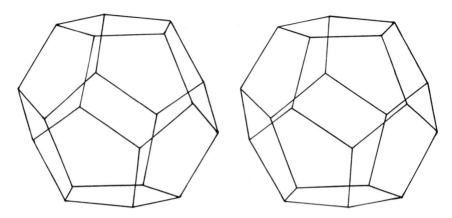

Figure 6. Stereo pair of a regular dodecahedron.

coded by a progressive color change through the spectral sequence to blue. Sets of points of the same color correspond to sections of the pyramid parallel to the base.

The most obvious four-dimensional analogue to this would be to illustrate a dodecahedral pyramid: that is, a regular dodecahedron (Fig. 6) with all its vertices joined to an apex along a fourth axis perpendicular to the three dimensions of the dodecahedron. If this object were viewed down the fourth axis, the apex would project to the center of the dodecahedron and would be colored red; the edges of the dodecahedron would be blue; and the joins of the dodecahedral vertices to the apex would appear as 20 radial spokes colored with the spectral sequences from red to blue. Such a figure would have the same symmetry as the dodecahedron, but each rotation axis combines with the color axis to produce a rotation plane, and each mirror plane combines with the color axis to become a mirror hyperplane.

4.2. Simple Fivefold Rotation

To take the discussion further, we need to take a simple example. Plate 2 shows a regular pentagon with its vertices joined to two points in two dimensions (z and w) perpendicular to it. With one (along z) it forms a pyramid as in Figure 4 and with the other (along w) as in Plate 1, and the two points are also joined. We could regard the figure as a three-dimensional pyramid joined to an apex on the w axis (just as we imagined joining the vertices of a dodecahedron to such a point), and so describe it as a (pentagonal pyramid) pyramid. In Plate 2 the projection is down the w axis, but if we project in a direction in the zw plane making 30° with w, we obtain Plate 3. The apex on the w axis is now displaced by -0.5 from the origin in 3-space, but its color coordinate is reduced by a factor 0.866

(say to orange). The apex on the z axis is displaced by a factor of only 0.866 but has a color coordinate of 0.5 (say a greenish yellow).

4.3. Double Rotations

The matrix of the fivefold rotation of Plates 2 and 3 is clearly

$$
\begin{pmatrix}
\cos 72° & \sin 72° & 0 & 0 \\
-\sin 72° & \cos 72° & 0 & 0 \\
0 & 0 & 1 & 0 \\
0 & 0 & 0 & 1
\end{pmatrix}
\tag{7}
$$

if we write the w coordinate last. We can modify this by analogy with the change from matrix (5) to (6) to give

$$
\begin{pmatrix}
\cos 72° & \sin 72° & 0 & 0 \\
-\sin 72° & \cos 72° & 0 & 0 \\
0 & 0 & -1 & 0 \\
0 & 0 & 0 & -1
\end{pmatrix}
\tag{8}
$$

The figure will clearly by bipyramidal along both z and w. Viewed down the w axis there would be a three-dimensional bipyramid like Figure 5 at the greenish yellow level; the vertices on the positive and negative w axes would be red and blue, but both would be projected to the origin, giving a very confusing figure. The confusion is diminished if we project down a direction in the zw plane as we did in Plate 3. The result is Plate 4 where the projection direction makes 10° with the w axis. The color coordinates of the "central" apices are negligibly reduced by a factor of 0.98 but they are displaced up and down by ±0.17. The top and bottom vertices suffer a correspondingly negligible spatial displacement by a factor of 0.98, but they are displaced from greenish yellow toward red and blue by a factor of 0.17. The vertical joins still overlap one another and were shown side by side in the model, but one is obscured by the other in the photograph.

The nature of the figure is made clearer if the projection direction is changed further to make 30° with w in a plane containing a line in the plane of the pentagon (Plate 5). This displaces the "central" apices off the vertical axis by ±0.5 and foreshortens the pentagon in the same direction by a factor of 0.866. The color of the "central" apices becomes orange-red and blue-green, while the pentagon ranges from orange-yellow to mid-green (though on the model the color range has been renormalized to cover the full range for maximum discrimination). The four apices thus form a rhombus that has equal status with the pentagon as a central feature of the figure. The rhombus has twofold symmetry corresponding to the bottom right-hand corner of matrix (8), which can be written

$$
\begin{pmatrix}
\cos 180° & \sin 180° \\
-\sin 180° & \cos 180°
\end{pmatrix}.
\tag{9}
$$

The symmetry therefore includes a *double rotation* represented by **52**. It would be equally possible to devise a corresponding figure with a **5n** double rotation with any value of n, corresponding to putting

$$\begin{pmatrix} \cos 360°/n & \sin 360°/n \\ -\sin 360°/n & \cos 360°/n \end{pmatrix} \tag{10}$$

in the bottom corner of the 4×4 matrix.

4.4. The 55 Double Rotations

With $n = 5$, the resulting matrix is

$$\begin{pmatrix} \cos 72° & \sin 72° & 0 & 0 \\ -\sin 72° & \cos 72° & 0 & 0 \\ 0 & 0 & \cos 72° & \sin 72° \\ 0 & 0 & -\sin 72° & \cos 72° \end{pmatrix} \tag{11}$$

and a projection of a figure that possesses this symmetry is shown in Plate 6. It may be described as a pentagonal pentapyramid. It is constructed from two regular pentagons, one in the *xy* plane and one in the *zw* plane with all the vertices of the one joined to all those of the other. The projection is similar to that in Plate 5, so that the horizontal pentagon is foreshortened spatially by 0.866 and the vertical one by 0.5, with corresponding converse effects on their color ranges. The plane of each pentagon is individually the fivefold rotation plane of the other (and of the whole figure) when account is taken of both spatial and color coordinates—that is, of full four-dimensional coordinates. However just as Figure 5 had an explicit fivefold axis and an explicit mirror plane but also could be regarded as generated by the $\tilde{5}$ operation, so there is symmetry in Plate 6 that can be regarded as generated by the **55** double rotation itself. Any point on the figure is repeated by this operation to the vertices of a regular pentagon lying in a plane through the origin in 4-space—a result that would not arise from either single rotation plane. An example is shown by the white dots in Plate 6. They lie at equivalent positions on edges that join successive vertices of the two regular pentagons.

There is however a defect in this analogy with the $\tilde{5}$ operation. That is generated by combining a **5** rotation axis with a perpendicular mirror plane, but it also generates these overt symmetries. The **55** double rotation is generated by combining two orthogonal **5** rotation planes, but it does not in turn generate such overt symmetries—they are only cryptorotation planes. If in Plate 6 the two pentagons were connected only by the five joins that bear the white dots, it would be a figure (but not a polytope) having only the **55** double rotation and no other kind of symmetry. The two planes on which the pentagons lie would have no preeminent role in such a figure, there being infinitely many pentagonal plane sections (such as that shown by the white dots), and any orthogonal pair of these could be taken as the cryptorotation planes, whose orientations are incompletely determinate.[2] This phenomenon corresponds to the indeterminate orientations of the mirror lines in Figure 1; it occurs in all symmetry operations that are composed of two (or more) equivalent operations in component subspaces; see, for example, the discussion of the **44** double rotation in Ref. [1].

Starting from the two overt fivefold planes in Plate 6 it is possible to define a second kind of fivefold double rotation if one combines a rotation through 72° about one plane with a rotation through 144° about the other. A set of five points in Plate 6 that is related in this way is shown by the black dots; they lie at equivalent positions on edges that join successive vertices of one regular pentagon to successive alternate vertices of the other. The two kinds of fivefold double rotation may be conveniently distinguished by the symbols $5^1 5^1$ and $5^1 5^2$. Whereas the former repeats any point to the vertices of a plane regular pentagon, the latter does so only if the point lies in one of its two cryptorotation planes; otherwise it repeats it to the vertices of a pentatope. If the point is in an appropriate position with respect to the component rotation planes, the pentatope will be a regular one (see Section 4.5). Again the rotation planes need not be overt as in Plate 6; if the two pentagons were connected only by the joins bearing the black dots, the figure would have only $5^1 5^2$ double rotation and no other kind of symmetry. However, the cryptorotation planes would be determinate in this case because they define the planes in which the generated pentatope degenerates to a regular pentagon.

The double rotation **55** is the lowest order one to give rise to two different operations in this way, because in threefold symmetry a 240° rotation is equivalent to −120°, and in fourfold rotation a 180° rotation is twofold, not fourfold. The $5^1 5^2$ operation is also the first occurrence of a crystallographic symmetry of order 5, since the trace of

$$\begin{pmatrix} \cos 72° & \sin 72° & 0 & 0 \\ -\sin 72° & \cos 72° & 0 & 0 \\ 0 & 0 & \cos 144° & \sin 144° \\ 0 & 0 & -\sin 144° & \cos 144° \end{pmatrix} \tag{12}$$

is −1, an integer.

4.5. Regular Solids in Four Dimensions

The simplest regular solid in four dimensions is the regular pentatope. It may be constructed by making a regular tetrahedron into a "tetrahedron pyramid" keeping all the edges equal. The most symmetrical view of it (Figure 7) is obtained by projecting perpendicular to the hyperplane

---

[2] The orientation of an orthogonal pair of planes in four dimensions is defined by four parameters, but two of these parameters become free in specifying possible cryptorotation planes.

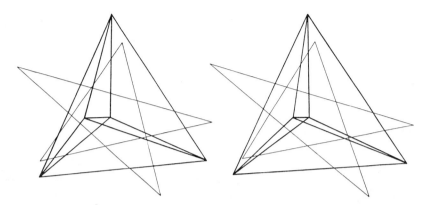

Figure 7. Stereo pair of a three-dimensional projection of a four-dimensional regular pentatope down the fourth (w) axis. One of the two cryptorotation planes of one of the six $5^1 5^2$ symmetry elements intersects five of the triangular faces (internally) at the vertices of a pentagon and the other five (externally) at the vertices of the associated pentagram.

containing the starting tetrahedron so that the fifth vertex projects to the center of the tetrahedron formed by the other four.

The pentatope is most conveniently described in terms of four axes all at $\cos^{-1}(-\frac{1}{4})$ to each other so that the vertices are at $(1, 0, 0, 0)$, $(0, 1, 0, 0)$, $(0, 0, 1, 0)$, $(0, 0, 0, 1)$, and $(-1, -1, -1, -1)$. They are then related by the pentatope operation whose matrix is

$$\begin{pmatrix} -1 & 1 & 0 & 0 \\ -1 & 0 & 1 & 0 \\ -1 & 0 & 0 & 1 \\ -1 & 0 & 0 & 0 \end{pmatrix}. \tag{13}$$

This matrix is equivalent to (12) and takes that form when referred to suitable orthogonal axes [3]. The regular pentatope is bounded by five regular tetrahedra and possesses six differently oriented $5^1 5^2$ operations. A cryptorotation plane of one of these intersects the triangular plane faces in irrational points, which are shown in Figure 7. Five of the intersections are interior to the faces and outline a plane pentagon, while the other five are exterior to the faces and extend the pentagon to a pentagram. The plane of these 10 points goes through the center of the pentatope, but not of course through the vertex that coincides with that point in the three-dimensional projection. Figure 7 provides an interesting demonstration of plane fivefold symmetry associated with the pentatope. The fivefold plane orthogonal to the one shown interchanges the two sets of five triangles that are intersected internally and externally.

Of the other five regular solids in four dimensions [4], three (the hypercube, its inverse, and the related 24-vertex figure) do not have fivefold symmetry. The 120-vertex figure is bounded by regular dodecahedra, and so both this and its inverse (the 600-vertex figure) possess this symmetry. Projections of them are discussed by Coxeter [4] and Shephard [5], and is not discussed further here.

5. Fivefold Symmetry in n Dimensions ($n \geqslant 5$)

With increasing numbers of dimensions, visualization becomes increasingly difficult, and we shall confine our attention to the symmetry of the regular solids. Rather surprisingly, the number of these falls to three when $n = 5$ and remains at this value for all higher values of n [4]. These three are the $(n + 1)$-tope (the analogue of the tetrahedron), the n-dimensional hypercube with 2^n vertices, and its dual with $2n$ vertices (the analogue of the octahedron).

Since the n-dimensional $(n + 1)$-tope is bounded by $(n - 1)$-dimensional n-topes, which are bounded by $(n - 2)$-dimensional $(n - 1)$-topes, etc., it follows that they always possess fivefold symmetries for all values of $n \geqslant 5$.

The hypercube and its dual both necessarily possess the same symmetry. Just as the cube has an axis of threefold symmetry down any one of its diagonals, which relates its three perpendicular edges to one another, so an n-dimensional hypercube must have an axis of n-fold symmetry down any one of its diagonals to relate its n perpendicular edges to one another. Of course this is not an axis of *rotation* symmetry except in the case of $n = 3$, because it is only in three dimensions that simple rotation occurs about an axis. In five dimensions it is double rotation that occurs about an axis, since a double rotation occupies the other four dimensions. Thus in the five dimensional hypercube there is an axis of $5^1 5^2$ double rotation down any diagonal. A 5×5 matrix of the type of matrix (12) with an additional 1 in the bottom right diagonal position will cycle the axes. This means that at any distance along a diagonal from a vertex, the diagonal can be intersected by a perpendicular four-dimensional hyperplane which intersects the five perpendicular edges at the vertices of a regular pentatope that are related by the $5^1 5^2$ operation in that four-dimensional space. This pentatope is the vertex figure of the five-dimensional hypercube, just as the equilateral triangle is the vertex figure of the cube.

Since the n-dimensional hypercube is bounded by $(n - 1)$-dimensional hypercubes, it follows as before that n-dimensional hypercubes possess fivefold symmetries for all values of $n \geqslant 5$. This is therefore true of all regular solids of dimension $n \geqslant 5$.

6. Diagonal Symmetry of Hypercubes and Generalized Penrose Patterns

A cubic lattice consists of a stack of cubes, and a small portion of a diagonal slice of such a lattice is shown in Figure 8. It contains cubes at three levels (marked 1, 2, and 3 at their uppermost vertices) and

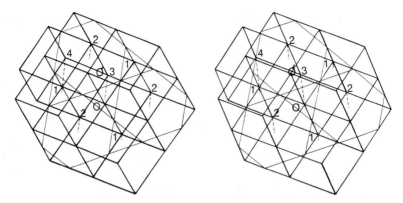

Figure 8. Stereo pair of part of a diagonal slice of a stack of cubes, showing cubes at three levels (1, 2, and 3) that are intersected by a plane perpendicular to the cube diagonal *OO'* at a height $\gamma \cdot OO'$ with $\gamma = 0.18$. One cube at level 4 is also shown.

consists of all cubes that are intersected by a plane perpendicular to the cube diagonal *OO'* at a height $\gamma \cdot OO'$. In Figure 8 $\gamma = 0.18$. If the uppermost vertex of each cube is projected onto the plane, together with its joins to other points that are also so projected, the pattern shown in Figure 9a is obtained. The cube faces all project to 120° rhombuses. If the value of γ were increased to more than ⅓, the plane would no longer intersect cubes at level 1 and would instead intersect cubes at level 4 (one of which is shown in Figure 8). It may be seen that the uppermost vertices of these would project to the same points as those at level 1, but the pattern would change to that in Figure 9b because the vertices at level 4 are joined to the level 3 vertex that projects to the origin, whereas those at level 1 are not. This changes the symmetry round the origin from threefold to sixfold (although the two patterns are congruent). The symmetry changes back to threefold at $\gamma > ⅔$, though with the opposite orientation to Figure 9a.

Five-dimensional hypercubes would lie at five levels in an analogous construction and would be cut by a four-dimensional hyperplane containing $5^1 5^2$ double rotation symmetry. If the upper vertices of these hypercubes are projected onto the hyperplane and then further projected onto a plane perpendicular to the 5^1 cryptorotation plane within it, a generalized Penrose pattern is obtained. The analogy with three dimensions is of course only partial. Most importantly the pattern is nonrepeating because the projection plane is irrational in orientation with respect to the lattice, and it contains rhombuses of two shapes becuase the hypercube faces do not all make the same angle with the projection plane. This derivation of the patterns was devised by de Bruijn [6] and extended by Kramer and Neri [7], Duneau and Katz [8], and Gähler and Rhyner [9].

If the projection plane is arranged to intersect the 5^1 cryptorotation plane on the hypercube diagonal, the patterns obtained have global fivefold sym-

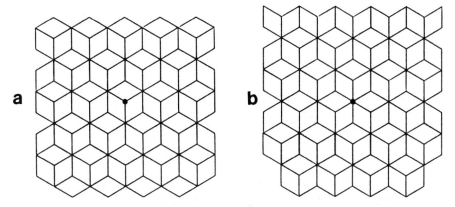

Figure 9. Projection of the uppermost vertices of the cubes (and edges connecting them) in an extended portion of Figure 8. (a) With $0 < \gamma < \frac{1}{3}$. (b) with $\frac{1}{3} < \gamma < \frac{2}{3}$.

metry. Then with $\gamma = \frac{1}{5}$ and $\frac{2}{5}$, the original Penrose patterns (Fig. 10) are obtained, and as shown by Penrose [10], they can be regarded as parts of the same infinite pattern. With $\gamma = \frac{1}{10}$, $\frac{3}{10}$, and $\frac{5}{10}$, one obtains the patterns of Figure 11 [11]. These have more complex properties than Figure 10, but can again be considered as parts of another infinite pattern. Thus there is a cyclical change in the patterns with a period of $\frac{1}{5}$ in γ, just as there is a cyclical change with a period of $\frac{1}{3}$ in the three-dimensional case. Moreover the change from fivefold to tenfold symmetry round the origin when γ is in the neighborhood of 0.5 is exactly analogous to that from threefold to sixfold symmetry in Figure 9. However, as γ changes there is a progressive change from one pattern to another effected by "flips" of units of the pattern that start at large radial distances and become progressively more numerous and closer to the origin. The patterns at intermediate values of γ are therefore much more irregular.

If the projection plane intersects the cryptorotation plane off the diagonal axis, the patterns may lose their global symmetry but retain the same properties as those produced on the axis at the same value of γ.

Generalized Penrose patterns are not uniquely associated with fivefold symmetry [11]. By proceeding in the same way from seven-dimensional and eight-dimensional hypercubic lattices, one can obtain the patterns shown in Figures 12a and b with sevenfold and eightfold symmetry, respectively. They are however much more complex than the fivefold ones. Figure 12a contains three different shapes of rhombuses: in fact the number of shapes is $(n - 1)/2$ when n is prime, and it increases rather less regularly when n is composite. The patterns are also susceptible to more kinds of variation as the position of the projection plane is changed. The possibility of these higher order, generalized Penrose patterns is associated with the fact that as dimensionality increases, so do the possibilities of higher order crystal-

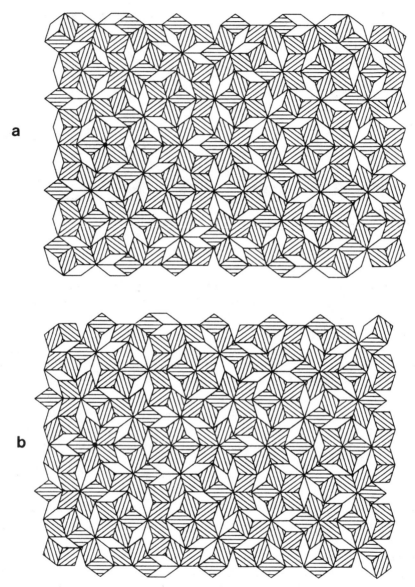

Figure 10. Penrose patterns corresponding to Figure 9 with five-dimensional cubes and (a) $\gamma = \frac{1}{5}$, (b) $\gamma = \frac{2}{5}$. The origin is marked with a dot, and the pattern has global fivefold symmetry about this point. Different kinds of rhombus are distinguished by shading.

122

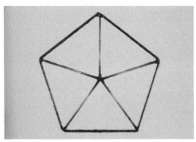

Plate 1. The regular pentagonal pyramid projected from above onto its base. Distance above the projection plane is represented by spectral distance from blue to red.

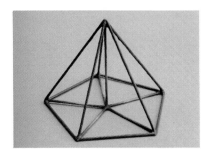 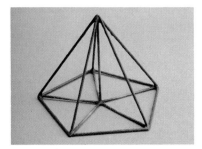

Plate 2. Stereo pair of a projection of a four-dimensional (pentagonal pyramid) pyramid into three dimensions down the fourth (w) axis. Distance "above" the three-dimensional projection is shown by color as in Figure 7.

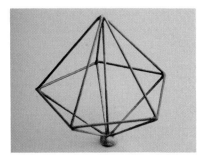 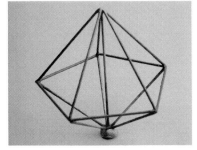

Plate 3. Stereo pair of a three-dimensional projection of the four-dimensional (pentagonal pyramid) pyramid of Figure 8 projected down a direction in the zw plane making $30°$ with the w·axis. Color shows distances in the projection direction.

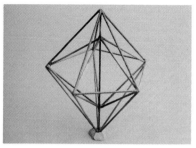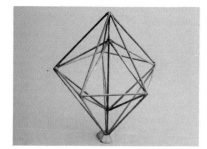

Plate 4. Stereo pair of a three-dimensional projection of the four-dimensional (pentagonal bipyramid) bipyramid projected down a direction in the zw plane making 30° with the w axis, as in Figure 9. Color shows distances in the projection direction.

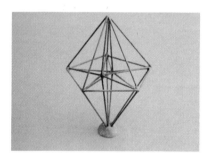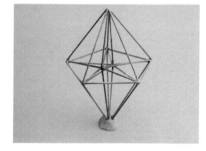

Plate 5. Stereo pair of a three-dimensional projection of the four-dimensional (pentagonal bipyramid) bipyramid of Figure 10 projected down a line (at 30° to the w axis) in a plane containing the w axis and a line in the plane of the pentagon. Color represents distances in the projection direction.

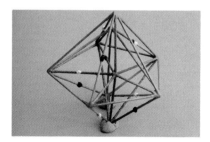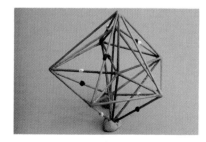

Plate 6. Stereo pair of a three-dimensional projection of a pentagonal pentapyramid with fivefold symmetry in two orthogonal planes projected as in Figure 11. Color represents distance in the projection direction. The white dots indicate the operation of the $5^1 5^1$ symmetry and the black dots the operation of the $5^1 5^2$ symmetry.

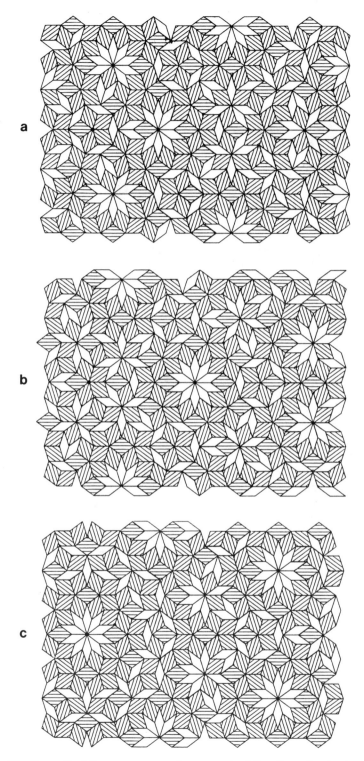

Figure 11. Generalized Penrose patterns from a stack of five-dimensional cubes with (a) $\gamma = \frac{1}{10}$, (b) $\gamma = \frac{3}{10}$, and (c) $\gamma = \frac{5}{10}$.

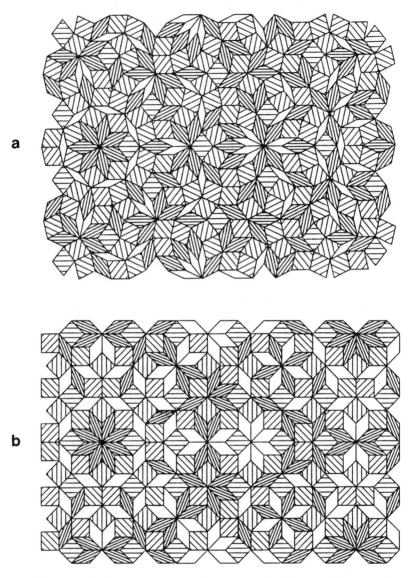

Figure 12. Generalized Penrose patterns from stacks of n-dimensional cubes. (a) $n = 7$, $\gamma = \frac{1}{7}$ and (b) $n = 8$, $\gamma = \frac{1}{4}$.

124

lographic symmetry. Thus in six dimensions, symmetry matrices are of 6×6 order and can accommodate three rotations. If such rotations are through $360°/7$, $720°/7$ and $1080°/7$, the trace is given by

$$2 \cos 360°/7 + 2 \cos 720°/7 + 2 \cos 1080°/7 = -1$$

so that sevenfold symmetry in the form of the triple rotation $7^1 7^2 7^3$ becomes crystallographic and is available in the seven-dimensional cubic lattice.

7. More Exotic Kinds of Fivefold Symmetry

Finally one may note that in higher dimensions not only do ever more exotic orders of symmetry become crystallographic, but this also applies to new combinations of lower orders. Thus in eight dimensions the $5^1 5^1$ operation becomes a cryptocomponent of the crystallographic quadruple rotation $5^1 5^1 5^2 5^2$. Since this can also be regarded as $5^1 5^2 \, 5^1 5^2$, it has $5^1 5^2$ cryptocomponents whose orientation is incompletely determinate.

REFERENCES

1. E. J. W. Whittaker, *An Atlas of Hyperstereograms of the Four-Dimensional Crystal Classes.* Clarendon Press, Oxford, 1985.

2. E. J. W. Whittaker, "A representation of hypercubic symmetry and its projections," *Acta Crystallogr.*, **A29**, 678–684 (1973).

3. E. J. W. Whittaker and R. M. Whittaker, "Graphic representation and nomenclature of the four-dimensional crystal classes: IV," *Acta Crystallogr.*, **A42**, 387–398 (1986).

4. H. S. M. Coxeter, *Regular Polytopes*, 2nd ed. Macmillan, London, 1963.

5. G. C. Shephard, "Plane projections of four-dimensional regular polytopes with five fold symmetry," in *Fivefold Symmetry in a Cultural Context*, I. Hargittai (ed.). To be published.

6. N. G. de Bruijn, "Algebraic theory of Penrose's non-periodic tilings of the plane," *Proc. K. Ned. Akad. Wet. Ser. A.*, **43**, 39–66 (1981).

7. P. Kramer and R. Neri, "On periodic and non-periodic space fillings of E^m obtained by projection," *Acta Crystallogr.*, **A40**, 580–587 (1984).

8. M. Duneau and A. Katz, "Quasiperiodic patterns," *Phys. Rev. Lett.*, **54**, 2688–2691 (1985).

9. F. Gähler and J. Rhyner, "Equivalence of the generalised grid and projection methods for the construction of quasiperiodic tilings," *J. Phys. A*, **19**, 267–277 (1986).

10. R. Penrose, "Pentaplexity," *Eureka*, **39**, 16–22 (1978).

11. E. J. W. Whittaker and R. M. Whittaker, "Some generalised Penrose patterns from projections of *n*-dimensional lattices," *Acta Crystallogr.*, **A44**, 105–112 (1988).

9

Planar Patterns with Fivefold Symmetry as Sections of Periodic Structures in 4-Space

Michael Baake, Peter Kramer,
Martin Schlottmann, and Dieter Zeidler

1. Introduction

The discovery of an icosahedrally symmetric phase in a rapidly solidified Al–Mn alloy [1] has stimulated a large variety of research activities, experimental as well as theoretical ones. Many materials with similar properties, called quasicrystals, have been found so far and equally many explanation models have been discussed (for a survey, the reader is referred to the anthologies [2–5]). From the theoretical point of view, the structure of quasicrystals is deeply related to the existence of nonperiodic space fillings or tilings with long-range orientational order, especially with crystallographically forbidden symmetries. The most popular example is the icosahedral quasilattice [6] often used for the description of the Al–Mn phase. The paradigm in 2-D space is the famous Penrose pattern [7] with fivefold symmetry, which—in a slightly modified version with convex tiles, the well-known rhombi—has been analyzed by de Bruijn in an algebraic fashion [8]. There, the connection to a high-dimensional lattice emerges implicitly. The construction of quasilattices with possible point symmetry from high-dimensional lattices has been worked out thoroughly along different lines [9].

Independently of the idea of tiles, periodic structures in a higher dimensional space underlie the derivation of quasiperiodic functions in the sense of Bohr [10] and illuminate the intriguing structure of the diffraction patterns of the quasicrystalline phases. It is clear then that lattice-derived models

127

should not only provide a transparent geometric frame, generalizing the classical concepts of crystallography as weakly as possible, but also explain the new symmetries found in quasicrystals and yield well-defined tools for the development of their diffraction theory. In a sense, these lattice projection or cut algorithms produce "ideal" quasicrystals, i.e., the "single crystals" of the quasiperiodic game. Real quasicrystals are then defected structures, nontrivially to describe, of course. The latter problem will not be discussed in this chapter.

We will restrict ourselves to a description of the geometric structures, and only for a 2-D example for pedagogical reasons. More precisely, we select the planar paradigms with fivefold symmetry and reformulate them in a simplified and unified fashion, which profits from a certain 4-D lattice and does not suffer from nonuniqueness problems in the standard 5-D description (cf. [11]). The 4-D lattice is nonhypercubic; thus we present the geometric analysis to show that there is nothing mysterious about such lattices. On the contrary, we obtain (originally as a by-product) a second quasiperiodic tiling [12] from a lattice projection, which is built from two triangles that also appear in Robinson's triangulation of the Penrose pattern [13, 14]. (In contrast to [12], we will not call this pattern "Robinson tiling" because its structure is completely different from that of Robinson's decomposition of the Penrose pattern. This point will be discussed in Section 5.)

In contrast to the standard folklore, we would like to emphasize the cell structure also of quasiperiodic tilings—a commonplace in ordinary crystallography. To this end, we reject the ubiquitous cut-and-project prescription and present a simple tile-by-tile construction algorithm instead. We try to explain every step in an intuitive way and with the aid of figures. We hope to convince the reader of the consistent geometric line we follow that culminates in a well-defined Fourier theory from which calculations of various diffraction patterns can be derived. The present chapter is an abridged version of a more comprehensive article [15], which is referred to quite frequently.

The chapter is organized as follows. In Section 2, we introduce the root lattice A_4 and briefly describe its geometric structure. The generalized Wigner–Seitz or Voronoi cell and its 2-D boundaries as well as the corresponding dual polytopes are given explicitly and the 4-D point and space groups are presented. Then, in Section 3, we choose a 2-D cut space—the so-called "physical" space—from the demand that fivefold symmetry survives in this space. A careful analysis shows the representative projection images with respect to the possible point symmetries in this 2-D space and the 4-D translations. Section 4 deals with the triangle pattern. We give the possible vertices together with their relative frequencies, and present an algorithm for the generation of the pattern tile by tile. In Section 5, the same program is briefly outlined for the well-known Penrose pattern, which naturally emerges as a configuration with two types of vertices.

The Fourier theory for all these patterns is then presented in Section 6 in a unified fashion based on the so-called Klotz construction, which is a modification of the dualization method. Several intensity plots are shown that correspond to different arrangements of point scatterers. The calculations stipulate a stable density on the tiles. The typical structures and differences are commented on, followed by the Conclusion in Section 7.

2. The Four-Dimensional Geometry

As was mentioned in the Introduction, we shall present a minimal scenario for the generation of patterns with fivefold symmetry. We start with the determination of a lattice of minimal dimension that produces the Penrose tiling via the dualization scheme. (The common cut-and-project mechanism will be insufficient.) On the one hand, this minimal dimension must at least be $n = 4$ because lattices with dimensions $n < 4$ never show fivefold symmetry [16]. On the other hand, $n = 5$ is sufficient—the 5-D primitive hypercubic lattice does the job [8]. However, the codimension is three and it is apparent from de Bruijn's work that one dimension is superfluous. In fact, the pattern can be obtained from a 4-D lattice plane within the hypercubic lattice (for details, see Section 5). This lattice plane turns out to be the intersection of the 5-D lattice with the 4-D hyperplane $S := \{x \in \mathbb{E}^5 | s \cdot x = 0\} \simeq \mathbb{E}^4$, where $s = e_1 + \cdots + e_5$ and the e_i, $1 \leq i \leq 5$, denote the standard Euclidean basis in $\mathbb{E}^5$. The 4-D space S can also be written as

$$S = \left\{ x \in \mathbb{E}^5 \Big| \quad x = \sum_{i=1}^{5} x_i e_i, \quad x_i \in \mathbb{R}, \quad \sum_{i=1}^{5} x_i = 0 \right\}, \tag{1}$$

while the orthogonal complement reads

$$\hat{S} = \{x \in \mathbb{E}^5 | \quad x = \xi \cdot s, \quad \xi \in \mathbb{R}, \quad s = e_1 + \cdots + e_5\} \simeq \mathbb{E}^1. \tag{2}$$

The 5-D primitive hypercubic lattice $\mathbb{Z}^5$ is generated by the integer linear combinations of the e_i, $1 \leq i \leq 5$. Furthermore, it is unimodular, i.e., it is an integer lattice ($q \cdot q' \in \mathbb{Z}$ for any two lattice points q, q') and reciprocal to itself. This implies the volume of the fundamental domain to be one. The structure of the lattice plane $S \cap \mathbb{Z}^5$ is a bit less trivial. The resulting lattice is generated by the integer linear combinations of the four basis vectors

$$b_1 = e_1 - e_2, \quad b_2 = e_2 - e_3, \quad b_3 = e_3 - e_4, \quad b_4 = e_4 - e_5 \tag{3}$$

and thus turns out to be the simplicial lattice A_4, an example of the so-called root lattices, i.e., those lattices that are generated from the roots of a semisimple Lie algebra [17]. This background (for details, see [15, 17, 18] will be very effective for the description of the geometric structure of the 4-D root lattice A_4 we need.

Before we proceed, a general remark is in order. Although the A_4 lattice is four-dimensional and can be described in the standard 4-D Euclidean space, we will keep the 5-D notation because the description of A_4 is simpler and much more aesthetic this way. Furthermore, the connection to the common 5-D description of the Penrose pattern emerges automatically. As mentioned above, A_4 is a root lattice, and it is obviously [cf. Eq. (3)] generated by the 20 vectors

$$\mathbf{e}_i - \mathbf{e}_j, \quad 1 \leq i, j \leq 5, \quad i \neq j. \tag{4}$$

These are the 20 roots of the simple Lie algebra $su(5)$ in the standard representation [17, 18], which fits well into our picture. The roots are all of the same length, $|\mathbf{e}_i - \mathbf{e}_j| = \sqrt{2}$ $(i \neq j)$, wherefore the root system is called simply laced. The corresponding Dynkin diagram—and hence the lattice—is depicted in Figure 1 (for details see [17]).

The 20 vectors of Eq. (4) turn out to be the *Voronoi vectors* of A_4, which determine the generalized Wigner–Seitz cell or *Voronoi cell* around the origin with respect to the A_4 lattice,

$$V_{A4}(\mathbf{0}): = \{\mathbf{x} \in S| \quad \mathbf{q} \in A_4 \Rightarrow |\mathbf{x} - \mathbf{q}| \geq |\mathbf{x}|\}. \tag{5}$$

(From now on we will suppress the index, which denotes the lattice under consideration if misunderstandings are excluded.)

Before we analyze the Voronoi cell in more detail, let us describe the symmetry of A_4. Its *holohedry H*, i.e., the maximal point group at the origin, is isomorphic with the automorphism group of the generating root system [18]. The latter is the semidirect product of the Weyl group of the root system (i.e., the reflection group generated by the roots) with the group of outer automorphisms. The Weyl group is $W(A_4) = S_5$, the permutation group of five elements (namely the five vectors $\mathbf{e}_i, 1 \leq i \leq 5$), while the group of outer automorphisms can be read from the Dynkin diagram (Fig. 1), and is generated by the operation $\mathbf{e}_i \rightarrow -\mathbf{e}_{6-i}, 1 \leq i \leq 5$. Here, a simplification is possible by rewriting H as the direct product

$$H = S_5 \times Z_2, \quad Z_2 = \{\mathbb{1}, \iota\}, \quad \iota^2 = \mathbb{1}, \tag{6}$$

because the generating outer automorphism is the product of the space inversion ι with the permutation (15)(24)(3) (in cyclic notation), the latter

Figure 1. The Dynkin diagram for the root lattice A_4. The dots represent the basis vectors, their number is the dimension of the lattice. If two dots are directly connected, the scalar product of the corresponding basis vectors is -1, otherwise it vanishes.

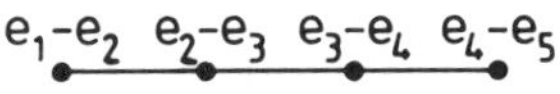

$$\underset{\bullet}{e_1-e_2} \; \underset{\bullet}{e_2-e_3} \; \underset{\bullet}{e_3-e_4} \; \underset{\bullet}{e_4-e_5}$$

already being an element of S_5. The order of H is $|H| = 2 \cdot 5! = 240$ (cf. p. 79 of [16] for an orientation within 4-D point groups). For the representation of H we need, the matrix version with respect to the vectors $\mathbf{e}_i$, $1 \leq i \leq 5$, will be used. Besides the 4-D irreducible representation (irrep) of H that acts on S, an additional 1-D irrep is contained which acts solely on $\hat{S}$. This has the advantage that a permutation $\alpha \in S_5$ is simply represented by the matrix $D(\alpha)$,

$$[D(\alpha)]_{ij} = \delta_{i,\alpha(j)}, \qquad 1 \leq i, j \leq 5. \tag{7}$$

The extension of D to the entire group H is immediate, since $D(\iota) = -\mathbb{1}$. Besides the point group H, we have the translation group T_{A4} generated by the lattice translations of A_4. The semidirect product $T_{A4} \times_s H$ builds the symmorphic *space group* of A_4 (cf. pp. 242ff of [16]), the multiplication rule being

$$(\mathbf{t}_1, h_1) \cdot (\mathbf{t}_2, h_2) = [\mathbf{t}_1 + D(h_1)\mathbf{t}_2, h_1 h_2], \qquad \mathbf{t}_i \in T_{A4}, \qquad h_i \in H. \tag{8}$$

Here, a 4-D translation element $\mathbf{t}$ is written as

$$\mathbf{t} = \sum_{i=1}^{5} n_i \mathbf{e}_i, \qquad \sum_{i=1}^{5} n_i = 0, \qquad n_i \in \mathbb{Z}. \tag{9}$$

The condition $\sum_{i=1}^{5} n_i = 0$ guarantees $\mathbf{t}$ to be a vector in the 4-D space S, cf. Eq. (1).

The reciprocal lattice is spanned by the vectors

$$\mathbf{a}_i = \mathbf{e}_i - \frac{1}{5}\mathbf{s}, \qquad 1 \leq i \leq 5, \qquad \mathbf{s} = \mathbf{e}_1 + \cdots + \mathbf{e}_5, \tag{10}$$

more precisely, by four of them, because

$$\sum_{i=1}^{5} \mathbf{a}_i = 0. \tag{11}$$

From the relation $\mathbf{b}_i = \mathbf{a}_i - \mathbf{a}_{i+1}$, $1 \leq i \leq 4$, one can explicitly recognize A_4 to be a sublattice of A_4^R.

Let us now turn to a more detailed description of the Voronoi cell of A_4 around the origin which can equally well be written as

$$V(\mathbf{0}) = \left\{ \mathbf{x} \in S \middle| \quad \mathbf{x} = \frac{1}{2} \sum_{i=1}^{5} \lambda_i \mathbf{a}_i, \quad |\lambda_i| \leq 1 \right\}. \tag{12}$$

This object is a 4-D convex polytope with the full symmetry $H = S_5 \times Z_2$. The Voronoi cell $V(\mathbf{0})$ possesses bounding polytopes of dimension m, $0 \leq m \leq 4$, which are called m-boundaries from now on. For each m-boundary $P(0 \leq m \leq 4)$ of a Voronoi cell there exists a uniquely defined polytope P^*, called *dual* to P. The general definition of dual polytopes in this context can be found in [19], here we restrict ourselves to the listing of the boundaries and their duals. In general, it turns out that the polytope dual to a Voronoi cell $V(\mathbf{q})$, $\mathbf{q} \in A_4$ [see Eq. (2.20)], is just $\mathbf{q}$ itself (more precisely, the 0-dimensional polytope $\{\mathbf{q}\}$). Furthermore, P^* is always perpendicular to P, and dim $(P^*) = 4 - \dim(P)$. Finally, we have the following relationship between the set of boundaries of Voronoi cells and the set of dual polytopes:

$$P_1 \subseteq P_2 \Leftrightarrow P_1^* \supseteq P_2^*, \tag{13}$$

i.e., if P_1 is a boundary of P_2, then P_1^* is a boundary of P_2^*, and vice versa.

Let us now describe the 2-boundaries of $V(\mathbf{0})$ that play a central role in the generation of the pattern via 2-D tiles. The 2-boundaries form *rhombi*, e.g.,

$$\left\{ \frac{1}{2} (\mathbf{a}_1 + \mathbf{a}_2 - \mathbf{a}_3 + \lambda_4 \mathbf{a}_4 + \lambda_5 \mathbf{a}_5) \mid \quad |\lambda| \leq 1 \right\}. \tag{14}$$

One finds 30 2-boundaries of $V(\mathbf{0})$ obtained from Eq. (14) by permutations of the indexing and 30 others through space inversion. The stability group of these objects is isomorphic with $S_2 \times S_2$, and the 60 rhombi form a single orbit with respect to the holohedry H and split into two orbits with respect to S_5 alone. The dual objects are, e.g.,

$$\{\mu_1(\mathbf{a}_1 - \mathbf{a}_3) + \mu_2(\mathbf{a}_2 - \mathbf{a}_3) \mid \mu_i \geq 0, \ \mu_1 + \mu_2 \leq 1\} \tag{15}$$

and so on. They form regular 2-simplices or *triangles*. Hence, the objects that will yield the tiles of the Penrose pattern (i.e., the rhombi) and of the triangle pattern (i.e., the triangles) are localized. All the dual polytopes will be found as bounding polytopes of the so-called dual complex to be defined in a shortwhile.

Last not least, the Voronoi cell $V(\mathbf{0})$ possesses 30 vertices which are called holes in the geometric treatment of root lattices [17]. They split into two classes, the so-called *deep holes* (i.e., the points with the largest distance to the lattice points of A_4) and the *shallows*. The vertex

$$\frac{1}{2} (\mathbf{a}_1 + \mathbf{a}_2 + \mathbf{a}_3 + \mathbf{a}_4 - \mathbf{a}_5) = \mathbf{a}_1 + \mathbf{a}_2 + \mathbf{a}_3 + \mathbf{a}_4 \tag{16}$$

and its images under H build the 10 shallows with the distance $2/\sqrt{5}$ to the nearest A_4 lattice points. The stability group is S_4, the dual objects are 4-D simplices:

$$\left\{ \sum_{i=1}^{4} \mu_i(\mathbf{a}_i - \mathbf{a}_5) \mid \quad \mu_i \geq 0, \ \mu_1 + \cdots + \mu_4 \leq 1 \right\} \tag{17}$$

and so on. These 10 objects form a single H-orbit and two S_5-orbits. The deep holes have distance $\sqrt{6/5}$ to the closest A_4 lattice points and are of the form

$$\frac{1}{2} (\mathbf{a}_1 + \mathbf{a}_2 + \mathbf{a}_3 - \mathbf{a}_4 - \mathbf{a}_5) = \mathbf{a}_1 + \mathbf{a}_2 + \mathbf{a}_3. \tag{18}$$

Here, one finds 20 pieces under operation of H that split into orbits of length 10 with respect to S_5. The stability group is $S_3 \times S_2$, the dual objects form 4-D Archimedean polytopes, e.g.,

$$\{\mu_1(\mathbf{a}_1 - \mathbf{a}_4) + \mu_2(\mathbf{a}_2 - \mathbf{a}_4) + \mu_3(\mathbf{a}_3 - \mathbf{a}_4)$$

$$+ \mu_4(\mathbf{a}_1 - \mathbf{a}_5) + \mu_5(\mathbf{a}_2 - \mathbf{a}_5) + \mu_6(\mathbf{a}_3 - \mathbf{a}_5)|$$

$$\mu_i \geq 0, \ \mu_1 + \mu_2 + \mu_3 \leq 1, \tag{19}$$

$$\mu_4 + \mu_5 + \mu_6 \leq 1, \ \mu_1 + \mu_4 \leq 1,$$

$$\mu_2 + \mu_5 \leq 1, \ \mu_3 + \mu_6 \leq 1\}.$$

They are bounded by five octahedra and five tetrahedra.

So far, we have described the 4-D geometry of the Voronoi cell $V(\mathbf{0})$ with its boundaries and dual objects. For the construction of the quasiperiodic tilings, we need the extensions to the entire space S. This is easily achieved by translations of the cell and its objects to any other possible position with respect to the lattice. The Voronoi domains are then

$$V(\mathbf{q}) = \{\mathbf{x} \in S \mid \mathbf{q}' \in A_4 \Rightarrow |\mathbf{x} - \mathbf{q}'| \geq |\mathbf{x} - \mathbf{q}|\}. \tag{20}$$

We introduce the notation (cf. [19])

$$\mathcal{V}: = \{P \subseteq S \mid P \text{ is a boundary of some } V(\mathbf{q}), \mathbf{q} \in A_4\}$$

$$\mathcal{V}^{(m)}: = \{P \in \mathcal{V} \mid \dim(P) = m\}, \qquad 0 \leq m \leq 4 \tag{21}$$

where $\mathcal{V}$ is called the *Voronoi complex* of A_4, i.e., the collection of all of its Voronoi domains together with their boundaries, and $\mathcal{V}^{(m)}$ is the associated m-skeleton. If P and P^* denote polytopes that are dual to each other, we can additionally define

$$\mathcal{V}^*: = \{P^* \mid P \in \mathcal{V}\} \tag{22}$$

$$\mathcal{V}^{*(m)}: = \{P^* \in \mathcal{V}^* \mid \dim(P^*) = m\},$$

called the *dual complex* and the dual m-skeleton, respectively. All the dual polytopes mentioned above and their images under the A_4 translations thus form the dual complex $\mathcal{V}^*$. The duality relation $P \Leftrightarrow P^*$ induces a one-to-one correspondence between $\mathcal{V}$ and $\mathcal{V}^*$.

The objects that are dual to the vertices of the Voronoi cells form the so-called dual cells, two types of which exist [cf. Eqs. (17), (19)]. As they will sometimes play a similar role as the Voronoi domains, they will be denoted by $V^*(\mathbf{q}^*)$, where $\mathbf{q}^*$ is a vertex of some Voronoi cell $V(\mathbf{q})$, $\mathbf{q} \in A_4$. Obviously, all dual polytopes P^* of dimension m, $0 \leq m \leq 3$, arise as m-boundaries of certain dual cells. Furthermore, it is an interesting observation that

$$A_4^R = A_4 \cup \mathcal{V}*^{(0)}, \tag{23}$$

i.e., the union of the lattice points of A_4 with all vertices of the Voronoi complex form the lattice points of the reciprocal lattice A_4^R.

To distinguish between the different classes of points, we introduce a modulo function $r(\mathbf{q}^R)$, defined for every reciprocal lattice point $\mathbf{q}^R = \Sigma_{i=1}^4 n_i \mathbf{a}_i$,

$$r(\mathbf{q}^R) := \left[\sum_{i=1}^4 n_i\right]_{\text{mod } 5}. \tag{24}$$

It can take the values $0, 1, \ldots, 4$. From Eq. (24) it is clear that r is defined on A_4 lattice points as well as on vertices of Voronoi cells and one finds

$$A_4 = \{\mathbf{q}^R \in A_4^R \mid r(\mathbf{q}^R) = 0\} \tag{25}$$

and the shallows and the deep holes are characterized as follows:

$$\mathbf{q}^* \text{ is a } \left\{\begin{array}{c} \text{shallow} \\ \text{deep hole} \end{array}\right\} \Leftrightarrow r(\mathbf{q}^*) = \left\{\begin{array}{c} 1 \text{ or } 4 \\ 2 \text{ or } 3 \end{array}\right\}. \tag{26}$$

At this point, we have set up the necessary geometric frame in the 4-D space S, wherefore we now turn to the selection of the 2-D cut space (i.e., the "physical" space) and the construction of the tilings.

3. Two-Dimensional Sections with Fivefold Symmetry

To establish fivefold symmetry, we now turn to the cyclic subgroup C_5 of S_5 generated by the element $g_5 = (12345)$ of order five. This is an abelian group, and it has five well-known inequivalent 1-D complex irreps. Since we consider C_5 as a group of geometric transformations, we need the real irreps. Besides the trivial 1-D representation, there are two inequivalent real 2-D representations, where the generator g_5 is represented by a 2×2 rotation matrix with rotation angle $\phi = 2\pi/5$ or $4\pi/5$, respectively. We denote these three irreps by D^0, $D^{\|}$ and $D^{\perp}$. Now we claim that $\mathbb{E}^5$ has three orthogonal subspaces of dimension 2, 2, and 1 which carry the three irreps $D^{\|}$, $D^{\perp}$, and D^0, respectively. This result may be obtained by explicit construction of an orthogonal 5×5 matrix M with the property

$$MD(g_5)M^{-1} = D^{red}(g_5) = \begin{pmatrix} c & -s & 0 & 0 & 0 \\ s & c & 0 & 0 & 0 \\ 0 & 0 & c' & -s' & 0 \\ 0 & 0 & s' & c' & 0 \\ 0 & 0 & 0 & 0 & 1 \end{pmatrix}, \tag{27}$$

where $D(g_5)$ is defined in Eq. (7), and the orthogonal matrix M turns out to be

$$
M := \sqrt{\frac{2}{5}}
\begin{pmatrix}
1 & c & c' & c' & c \\
0 & s & s' & -s' & -s \\
1 & c' & c & c & c' \\
0 & s' & -s & s & -s' \\
\sqrt{\frac{1}{2}} & \sqrt{\frac{1}{2}} & \sqrt{\frac{1}{2}} & \sqrt{\frac{1}{2}} & \sqrt{\frac{1}{2}}
\end{pmatrix}.
\tag{28}
$$

We have denoted

$$c: = \cos(2\pi/5), \qquad s: = \sin(2\pi/5),$$

$$c': = \cos(4\pi/5), \qquad s': = \sin(4\pi/5).$$

To see the geometric meaning of the matrix M, we consider its columns 1, . . . , 5 as an explicit expression for the components of the five unit vectors $\mathbf{e}_1, \ldots, \mathbf{e}_5$. Then, Eqs. (27) and (28) imply that the irreducible action of g_5 on each one of these five unit vectors yields an integer linear combination of the original vectors, i.e., the action of g_5 transforms the hypercubic lattice spanned by $\mathbf{e}_1, \ldots \mathbf{e}_5$ into itself. Now we specify three orthogonal subspaces $\mathbb{E}_\|$, $\mathbb{E}_\perp$, and $\hat{S}$ by the requirement that they should be transformed into themselves by the three corresponding representations of C_5. Clearly, the components of the vectors $\mathbf{e}_1, \ldots, \mathbf{e}_5$ in these subspaces are given by the column entries in the rows 1 and 2, 3 and 4, and 5 of the matrix M, respectively. When the projections of these vectors are drawn in corresponding planes, they have the form of two stars with fivefold symmetry as shown in Figure 2.

Figure 2. The projections $\pi(\mathbf{e}_i)$, $i = 1, \ldots, 5$ of the orthogonal basis vectors form two stars in the subspaces $\mathbb{E}_\|$, $\mathbb{E}_\perp$. The generator $g_5 = (12345)$ is represented by a rotation with angle $\phi = 2\pi/5$, $4\pi/5$, respectively. The reflection lines corresponding to the generators $s_1 = (1)(25)(34)$ and $s_2 = \iota(1)(25)(34)$ are denoted by $-\bullet-$ and $-\bullet\bullet-$, respectively.

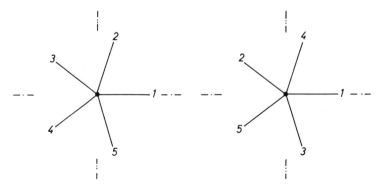

Since its orthogonal complement $\hat{S}$ carries the representation D^0, the subspace S must carry the two representations $D^{\|}$ and $D^{\perp}$. An explicit relation of the representations could easily be found with the help of a matrix that expresses the vectors $\mathbf{b}_1, \ldots, \mathbf{b}_4, \mathbf{s}$ in terms of $\mathbf{e}_1, \ldots, \mathbf{e}_5$. So S decomposes into the two subspaces $\mathbb{E}_{\|}$ and $\mathbb{E}_{\perp}$.

We shall denote the orthogonal projections of geometric objects from S to $\mathbb{E}_{\|}$ and $\mathbb{E}_{\perp}$ by the symbols $\pi_{\|}$ and $\pi_{\perp}$, respectively. The symbol π without index will be used if the expression applies to both $\mathbb{E}_{\|}$ and $\mathbb{E}_{\perp}$. The projections of the basis $\mathbf{b}_1, \ldots, \mathbf{b}_4$ to the representation spaces $\mathbb{E}_{\|}$ and $\mathbb{E}_{\perp}$ are obtained by taking the linear combinations of the projections $\pi(\mathbf{e}_1), \ldots, \pi(\mathbf{e}_5)$ given in the matrix in Eq. (28), and of course we have $\pi_{\|}(\mathbf{s}) = \pi_{\perp}(\mathbf{s}) = 0$.

The translation group T_{A_4} contains no translation vector that is parallel to $\mathbb{E}_{\|}$ or $\mathbb{E}_{\perp}$. This is tantamount to the fact that the projections $\pi(\mathbf{b}_1), \ldots, \pi(\mathbf{b}_4)$ are linearly independent over the integers. Hence, if one chooses one of the 2-D subspaces of S that are invariant under the action of C_5, i.e., equipped with *fivefold point symmetry*, this choice is incompatible with translational symmetry or periodicity in the subspace. A periodic real function which represents a density or potential in S when restricted to one of the subspaces will not be periodic, but only *quasiperiodic*.

Given the subspace S and its decomposition into $\mathbb{E}_{\|}$ and $\mathbb{E}_{\perp}$, there are still reflection elements in the point group H that are compatible with the splitting into these subspaces. These reflections are generated by the permutation $s_1 = (1)(25)(34)$ and by the space inversion ι. We denote the product of s_1 and ι by s_2. The point groups generated by C_5 and one of these reflections will be denoted by $C_{5v} = \langle g_5, s_1 \rangle$, $D_5 = \langle g_5, s_2 \rangle$, and $C_{10} = \langle g_5, \iota \rangle$, and the maximal group generated by any two reflections by $D_{10} = \langle g_5, s_1, \iota \rangle$. The reflection lines which correspond to s_1 and s_2 are indicated in Figure 2. The space inversion ι generates in $\mathbb{E}_{\|}$ and $\mathbb{E}_{\perp}$ the inversion which is equal to a rotation by the angle π.

Now, we analyze the Voronoi domain $V(\mathbf{0})$ of A_4 (Fig. 3) in terms of its projections into the subspaces $\mathbb{E}_{\|}$ and $\mathbb{E}_{\perp}$. In Section 2, we have given the boundaries of the Voronoi domain in terms of the orbits under the point groups H and S_5. For the present purpose, it will be convenient to classify the Voronoi domain, the dual cells, and their 2-boundaries into orbits under the point group C_5 and the translation group T_{A_4}. For a complete classification compare [15]. First assume that two boundaries are related by an element of C_5. Since the action of C_5 is compatible with the subspace splitting, the projections of two such boundaries are related in $\mathbb{E}_{\|}$ or $\mathbb{E}_{\perp}$ by a simple rotation with a multiple of $2\pi/5$. Then, consider boundaries related by translations. A full orbit under the translation group would of course encompass infinitely many elements. In what follows we shall consider only the finite intersections of these orbits with the central Voronoi domain $V(\mathbf{0})$. It may happen that a given boundary is transformed into itself by an element of the space group, which then is called a symmetry of that boundary. The reflection elements s_1, ι, and s_2 when applied to the boundaries can have

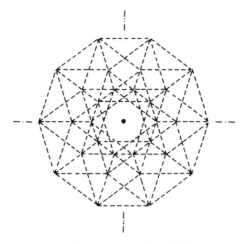

Figure 3. The projection of the Voronoi domain $\pi_{\parallel}[V(0)]$, shown by broken lines, fills a decagon.

two different consequences. Combined with appropriate translations, they may transform a given boundary into itself and therefore generate a symmetry of this boundary, or they may relate boundaries on different orbits. In the latter case, one can always find a space group element that relates the representatives of the orbits.

Of particular interest are the 2-boundaries (see Figs. 4 and 5) since they have the same dimension as the subspaces $\mathbb{E}_{\parallel}$, $\mathbb{E}_{\perp}$ so that there is a one-to-one correspondence between points on these boundaries and points of their projections. The projections of the 2-boundaries have the shape of an acute or thin and an obtuse or thick rhombus known from the Penrose pattern. These shapes occur in pairs related by the space inversion. The projections of their duals are two isosceles triangles with the angle $2\pi/10$ and $6\pi/10$,

Figure 4. Projection into $\mathbb{E}_{\parallel}$ and symmetry of a 2-boundary, "thick rhombus," and dual 2-boundary, "obtuse triangle." The two triangles or the two rhombi of Figures 4 and 5 become the representative tiles of the triangle or the rhombus pattern described in Sections 4 and 5.

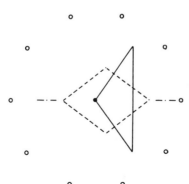

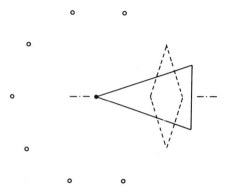

Figure 5. Projection and symmetry of a 2-boundary, "thin rhombus," and dual 2-boundary, "acute triangle."

respectively, along with their inverse images. The projections of the 2-boundaries and of their duals will become the tiles of the 2-D quasiperiodic patterns to be described in Sections 4 and 5. In Section 2, the two types of 0-boundaries were interpreted as the deep holes and shallows in the root lattice A_4. The corresponding dual 4-boundaries (Figs. 6 and 7) project into two pentagons of different size, the fivefold symmetry of which is generated by a space group element conjugate to g_5. As mentioned before, the reflection elements s_1, s_2, and ι may generate symmetries of the boundaries. Since they are compatible with the splitting into subspaces, the reflection symmetries appear in the projections to $\mathbb{E}_{\parallel}$, $\mathbb{E}_{\perp}$ as reflection lines that are indicated in the figures. As these reflections are shifted in part by translation vectors, they do not always pass through the point **0**.

So far we discussed the projection of the central Voronoi domain $V(\mathbf{0})$, its boundaries, and their duals to $\mathbb{E}_{\parallel}$ and $\mathbb{E}_{\perp}$. By applying all elements of

Figure 6. Projection and symmetry of the vertex $\mathbf{e}_1$ and its dual cell. The vertex represents the shallow holes in the lattice A_4. The dual cell has C_{5v} symmetry with respect to its center.

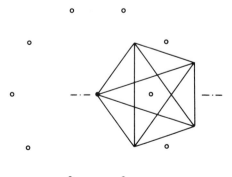

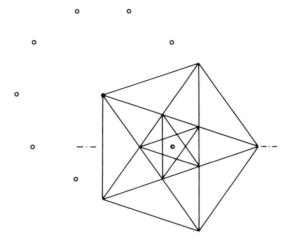

Figure 7. Projection and symmetry of the vertex $\mathbf{e}_1 + \mathbf{e}_5$ and its dual cell. The vertex represents the deep holes in the lattice A_4. The dual cell has C_{5v} symmetry with respect to its center.

T_{A_4} to these geometric objects one obtains the Voronoi complex $\mathcal{V}$ and the dual complex $\mathcal{V}^*$ introduced in Section 2. The projections of geometric objects from the complexes $\mathcal{V}$ and $\mathcal{V}^*$ to $\mathbb{E}_\parallel$ or $\mathbb{E}_\perp$ must then have the same shape as those from the central domain $V(\mathbf{0})$ and its dual, shifted by the projections of appropriate translation vectors. In the following sections we shall construct tilings in the subspace $\mathbb{E}_\parallel$, which we consider as the "physical" subspace, by projection from the "skeletons" $\mathcal{V}^{(2)}$ and $\mathcal{V}^{*(2)}$.

4. The Triangle Pattern

After the detailed discussion of some 4-D geometric structures associated with the A_4 root lattice it is now time to come to the central topic of this article, namely, to the presentation of 2-D tilings with fivefold symmetry that can be derived from that lattice. The method we use is not restricted to our special case but presents a rather general framework for the construction of quasiperiodic patterns.

The space $\mathbb{E}_\parallel$ that will contain the patterns has been chosen in Section 3 by the demand for fivefold symmetry. There are two types of objects in 4-space S which can naturally be projected into $\mathbb{E}_\parallel$: The 2-boundaries of dual cells or of Voronoi cells, i.e., the elements of $\mathcal{V}^{*(2)}_{A_4}$ or $\mathcal{V}^{(2)}_{A_4}$. Choosing the first set, we will obtain the triangle pattern to which the present section is devoted. The analogous method applied to $\mathcal{V}^{(2)}_{A_4}$ leads to a simple way for the construction of the well-known Penrose tiling of the plane to be described in the following section. For the sake of a compact notation, we will drop the index A_4 from now on if no misunderstanding is possible. Therefore, we write just Γ for the set of A_4 lattice points whereas Γ^* will denote the set of vertices of Voronoi cells of Γ.

It is clear that to generate a space filling of $\mathbb{E}_\parallel$ without gaps or overlaps, one has to select carefully those polytopes from $\mathcal{V}^{*(2)}$ that actually are to be projected into $\mathbb{E}_\parallel$. This is done as follows. Each $Q \in \mathcal{V}^{*(2)}$ corresponds to a $P \in \mathcal{V}^{(4-2)} = \mathcal{V}^{(2)}$, which is dual to it, $Q = P^*$. Choose a fixed vector $\mathbf{c}_\perp$, not contained in the projection image $\pi_\perp(P')$ of a 1-boundary P' from $\mathcal{V}^{(1)}$, and consider the set of boundaries $P \in \mathcal{V}^{(2)}$ such that their projection $\pi_\perp(P)$ contains $\mathbf{c}_\perp$. Project exactly the duals to this set from $\mathcal{V}^{*(2)}$ into $\mathbb{E}_\parallel$. We obtain a set of 2-D convex polygons in $\mathbb{E}_\parallel$,

$$\mathcal{T}(\mathbf{c}_\perp) := \{\pi_\parallel(P^*) \mid P \in \mathcal{V}^{(2)}, \mathbf{c}_\perp \in \pi_\perp(P)\}. \tag{29}$$

At this point, we just state that for each $\mathbf{c}_\perp \in \mathbb{E}_\perp$ (with the restriction mentioned above) this set of polygons is in fact a proper space filling of $\mathbb{E}_\parallel$ (i.e., without gaps or overlaps). This construction principle will be called the *dualization method*. A proof of this property can be carried out for very general scenarios and will be published elsewhere. The reader is referred to [19, and refs. therein] for a first orientation.

In order to obtain more explicit expressions for $\mathcal{T}(\mathbf{c}_\perp)$, we make use of the periodicity of the 4-D structure. First of all, $\mathcal{T}(\mathbf{c}_\perp)$ is determined completely if one knows the set of vertex points of its polygons and, for each vertex, the set of polygons that contains this vertex (vertex configuration). But vertices of polygons of $\mathcal{T}(\mathbf{c}_\perp)$ are projection images of vertices of elements of $\mathcal{V}^{*(2)}$, that is, projection images of lattice points of Γ. With the help of Eq. (13) we conclude that the set of vertices in $\mathcal{T}(\mathbf{c}_\perp)$ precisely is

$$\{\pi_\parallel(\mathbf{q}) \mid \mathbf{q} \in \Gamma, \quad \mathbf{c}_\perp \in \pi_\perp[V(\mathbf{q})]\}. \tag{30}$$

Since $V(\mathbf{q}) = V(\mathbf{0}) + \mathbf{q}$, the vertex set takes the form

$$\{\pi_\parallel(\mathbf{q}) \mid \mathbf{q} \in \Gamma, \quad \mathbf{c}_\perp - \pi_\perp(\mathbf{q}) \in \pi_\perp[V(\mathbf{0})]\}. \tag{31}$$

Having determined the vertex set of $\mathcal{T}(\mathbf{c}_\perp)$, we now come to the vertex configurations in $\mathcal{T}(\mathbf{c}_\perp)$. Obviously, the tiles surrounding $\pi_\parallel(\mathbf{q})$ are all those $\pi_\parallel(P^*)$, $P \in \mathcal{V}^{(2)}$, which fulfill $P - \mathbf{q} \subseteq V(\mathbf{0})$, $\mathbf{c}_\perp - \pi_\perp(\mathbf{q}) \in \pi_\perp(P - \mathbf{q})$. The complete vertex configuration around a vertex point $\pi_\parallel(\mathbf{q})$, $\mathbf{q} \in \Gamma$, $\mathbf{c}_\perp - \pi_\perp(\mathbf{q}) \in \pi_\perp[V(\mathbf{0})]$ of $\mathcal{T}(\mathbf{c}_\perp)$ can be written down as

$$\{\pi_\parallel(P^*) + \pi_\parallel(\mathbf{q}) \mid P \in \mathcal{V}^{(2)}, \quad P \subseteq V(\mathbf{0}), \quad \mathbf{c}_\perp - \pi_\perp(\mathbf{q}) \in \pi_\perp(P)\}. \tag{32}$$

We have restricted $\mathbf{c}_\perp$ in such a way that $\mathbf{c}_\perp - \pi_\perp(\mathbf{q}) \in \pi_\perp(P)$ implies $\mathbf{c}_\perp - \pi_\perp(\mathbf{q}) \in \pi_\perp(\overset{\circ}{P})$ [the interior of $\pi_\perp(P)$]. That is, all $\pi_\perp(P)$ for which the corresponding $\pi_\parallel(\mathbf{q}) + \pi_\parallel(P^*)$ belongs to the vertex configuration of $\mathcal{T}(\mathbf{c}_\perp)$ around $\pi_\parallel(\mathbf{q})$ must have a common overlap. Hence, if $\mathcal{W}$ is a subset of 2-boundaries of $V(\mathbf{0})$, then the set $\{\pi_\parallel(P^*) \mid P \in \mathcal{W}\}$ is, up to a translation in $\mathbb{E}_\parallel$, a possible vertex configuration of $\mathcal{T}(\mathbf{c}_\perp)$ if and only if $\mathcal{W}$ is maximal with the property that the interior of $\cap_{P \in \mathcal{W}} \pi_\perp(P)$ is not empty; furthermore, every vertex configuration of $\mathcal{T}(\mathbf{c}_\perp)$ can be described this way. The intersections $\cap_{P\in\mathscr{W}}\pi_\perp(P)$, $\mathscr{W}$ maximal, mentioned are polygons of dimension 2 lying in $\pi_\perp[V(\mathbf{0})]$. We call them elementary polygons of $\pi_\perp[V(\mathbf{0})]$ because every $\pi_\perp(P)$, P a 2-boundary of $V(\mathbf{0})$, is the union of a certain subset of them, but they cannot be subdivided further into smaller 2-D polygons by intersecting them with any $\pi_\perp(P)$ $[P \in \mathcal{V}^{(2)}, P \subseteq V(\mathbf{0})]$. The union of all elementary polygons is $\pi_\perp[V(\mathbf{0})]$, they do not have interior points in common. Now, the above description of the possible vertex configurations is just the establishment of a one-to-one correspondence between elementary polygons in $\pi_\perp[V(\mathbf{0})]$ and classes of vertex configurations of $\mathcal{T}(\mathbf{c}_\perp)$ with respect to translations in $\mathbb{E}_\parallel$. An inspection of $\pi_\perp[V(\mathbf{0})]$ together with the projected boundaries in Figure 8 yields, up to D_{10} operations, just nine different elementary polygons. Therefore, this must also be the number of possible different (up to D_{10} operations) classes of vertex configurations in $\mathcal{T}(\mathbf{c}_\perp)$.

We want to visualize these vertex configurations explicitly. Suppose that $\mathbf{c}_\perp - \pi_\perp(\mathbf{q}) \in \pi_\perp[V(\mathbf{0})]$, $\mathbf{q} \in \Gamma$, so, $\pi_\perp(\mathbf{q})$ is a vertex of $\mathcal{T}(\mathbf{c}_\perp)$. $\mathbf{c}_\perp - \pi_\perp(\mathbf{q})$ is contained in at least one elementary polygon of $\pi_\perp[V(\mathbf{0})]$, and, because of the restriction of $\mathbf{c}_\perp$, even in the interior of this polygon. However, elementary polygons do not overlap in interior points, and that one containing

Figure 8. Projection image of the Voronoi cell $V(\mathbf{0})$ together with its boundaries in $\mathbb{E}_\perp$. The origin of $\mathbb{E}_\perp$ is marked by 0 while the numbers $1, \ldots, 9$ count the elementary polygons that represent the vertices of the triangle pattern (see Fig. 9). The hatching indicates the attaching of boundaries mentioned in the text; for the sake of simplicity we have marked only one set of parallel 1-boundaries. The whole attachment is obtained by C_5-rotations.

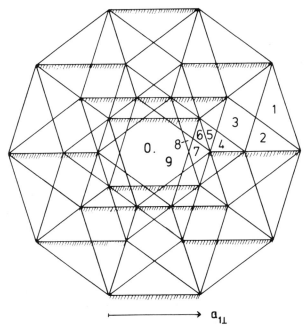

$c_\perp - \pi_\perp(q)$ must be uniquely determined. Let $\mathcal{W} \subseteq \mathcal{V}^{(2)}$ be the set of $P \subseteq V(\mathbf{0})$, $\pi_\perp(P)$ enclosing this elementary polygon. Then, the vertex configuration around $\pi_\parallel(q)$ is $\{\pi_\parallel(P^*) + \pi_\parallel(q) \mid P \in \mathcal{W}\}$. This way, nine different vertex configurations, up to D_{10} operations (and translations), may occur in $\mathcal{T}(c_\perp)$; they are listed in Figure 9.

The space $\mathbb{E}_\parallel$ is not contained in a net plane of the A_4 lattice. As a consequence, the projection images of the lattice points are dense and uniformly distributed in $\mathbb{E}_\perp$. It follows that $\{c_\perp - \pi_\perp(q) \mid q \in \Gamma\} \cap \pi_\perp[V(\mathbf{0})]$ is dense and uniformly distributed in $\pi_\perp[V(\mathbf{0})]$, so exactly the fraction area($\mathcal{E}$)/area $\{\pi_\perp[V(\mathbf{0})]\}$ of all vertices in $\mathcal{T}(c_\perp)$ are from the class represented by the elementary polygon $\mathcal{E}$. The areas of elementary polygons are easy to determine, the resulting frequencies of vertex configuration classes are listed in Figure 9 (we have added up all contributions from the whole D_{10} orbits).

A very simple algorithm for the construction of $\mathcal{T}(c_\perp)$s can be extracted from the analysis so far. Without losing generality, we may assume that one vertex is positioned at $\mathbf{0} \in \mathbb{E}_\parallel$. The following procedure provides arbitrary large parts of triangle patterns as in Figure 10b:

1. Choose $c_\perp \in \pi_\perp[V(\mathbf{0})]$ (restricted as described above), so that indeed the initial vertex configuration surrounds $\mathbf{0}$ in $\mathbb{E}_\parallel$.

2. If $\pi_\perp(q)$ ($q \in \Gamma$) is a vertex of the pattern obtained so far whose configuration is not completely known, one finds the elementary polygon in $\pi_\perp[V(\mathbf{0})]$ that contains $c_\perp - \pi_\perp(q)$. This determines the vertex configuration around $\pi_\perp(q)$ and yields new vertices of the pattern together with their preimages in Γ.

3. Proceed with step 2 with all incompletely known vertices until the tiling reaches the size desired.

The set of $c_\perp$ that must not be chosen [because otherwise some $c_\perp - \pi_\perp(q)$ may be contained in two different elementary polygons] is of measure zero in $\pi_\perp[V(\mathbf{0})]$. But, unfortunately, the only points of $\pi_\perp[V(\mathbf{0})]$ from which a pattern with exact fivefold symmetry can arise, namely, the elements of $\pi_\perp(\Gamma) \cap \pi_\perp[V(\mathbf{0})]$, are forbidden by the restriction. This can be cured by attaching to every $P \in \mathcal{V}^{(2)}$ some of its boundaries and projecting P^* into $\mathbb{E}_\parallel$ as a tile of $\mathcal{T}(c_\perp)$ if and only if $c_\perp$ is an interior point of $\pi_\perp(P)$ (as before), or an "interior" point of the projection image of a boundary attached to P. In Figure 8, one of two possible attachments of 1-boundaries is indicated. Hatching one side of a 1-boundary means that this boundary is attached to each $P \in \mathcal{V}^{(2)}$, $P \subseteq V(\mathbf{0})$, that contains this boundary and whose projection image lies on the marked side. Fivefold symmetry is obviously maintained; the fact that parallel 1-boundaries are hatched on the same side guarantees the translational invariance of the procedure. It is clear from the very construction that no new types of vertex configurations will occur. The construction algorithm is to be modified in an obvious way leading to the exactly symmetric pattern in Figure 10a.

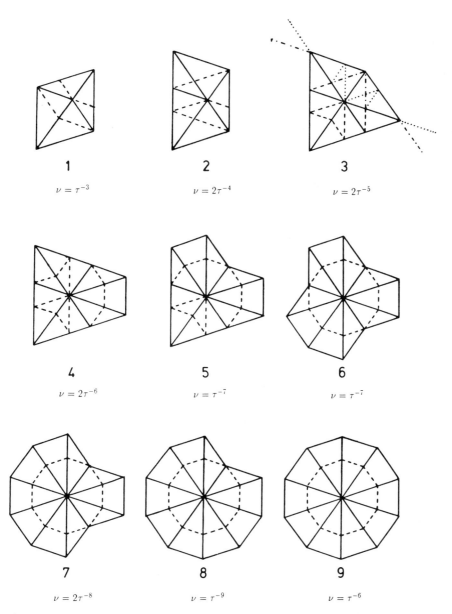

1

$\nu = \tau^{-3}$

2

$\nu = 2\tau^{-4}$

3

$\nu = 2\tau^{-5}$

4

$\nu = 2\tau^{-6}$

5

$\nu = \tau^{-7}$

6

$\nu = \tau^{-7}$

7

$\nu = 2\tau^{-8}$

8

$\nu = \tau^{-9}$

9

$\nu = \tau^{-6}$

Figure 9. The nine vertex configurations of the triangle pattern and their relative frequencies and deflation rules. The numbering refers to Figure 8. Note the additional edges indicated for type 3, which control the deflation of this vertex.

143

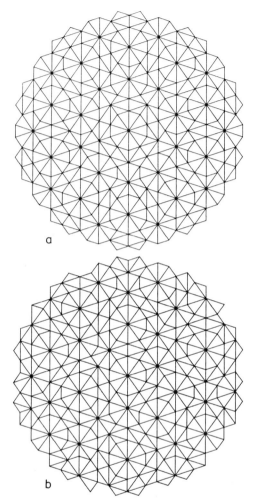

Figure 10. The triangle pattern: (a) symmetric version ($c_\perp = 0$) and (b) generic version $[c_\perp = 0.3\pi_\perp(a_1) + 0.5\pi_\perp(a_2)]$.

Let us now come to a "symmetry" of the A_4 lattice and its consequences for the triangle patterns that has not been mentioned up to this point. By the prescription

$$a_1 \rightarrow a_2 + a_5, \quad a_2 \rightarrow a_3 + a_1, \ldots, a_5 \rightarrow a_1 + a_4 \tag{33}$$

and linear continuation, a linear transformation on S is well-defined. It is easy to see that the A_4 lattice is invariant under this operation. The fact that the transformation, let us denote it by Δ, is not orthogonal prevents one from speaking of a symmetry of the lattice in the usual sense. But the form of the action of Δ in $\mathbb{E}_\parallel$ and $\mathbb{E}_\perp$ makes it an interesting subject of investigation. One verifies easily that $\pi_\parallel[\Delta(\mathbf{a}_i)] = 1/\tau\,\pi_\parallel(\mathbf{a}_i) = \Delta[\pi_\parallel(\mathbf{a}_i)]$ and $\pi_\perp[\Delta(\mathbf{a}_i)] = -\tau\pi_\perp(\mathbf{a}_i) = \Delta[\pi_\perp(\mathbf{a}_i)]$ ($i = 1, \ldots, 5$), therefore, Δ acts in $\mathbb{E}_\parallel$, $\mathbb{E}_\perp$ just as a rescaling by factors $1/\tau$, $-\tau$, respectively. We may perform Δ on the elements of $\mathcal{V}^{(2)}$, $\mathcal{V}^{*(2)}$, which leads to $\mathcal{V}_\Delta^{(2)}: = \{\Delta(P) \mid P \in \mathcal{V}^{(2)}\}$, $\mathcal{V}_\Delta^{*(2)}: = \{\Delta(P^*) \mid P^* \in \mathcal{V}^{*(2)}\}$, and then build the tilings

$$\mathcal{T}_\Delta(\mathbf{c}_\perp): = \{\pi_\parallel(P^*) \mid P \in \mathcal{V}_\Delta^{(2)},\, \mathbf{c}_\perp \in \pi_\perp(P)\}, \tag{34}$$

where we have generalized the duality operator according to $\Delta(P)^*: = \Delta(P^*)$. An easy computation leads to $\mathcal{T}_\Delta(\mathbf{c}_\perp) = 1/\tau\,\mathcal{T}(-1/\tau\,c_\perp)$, which proves $\mathcal{T}_\Delta(\mathbf{c}_\perp)$ to be indeed a tiling of $\mathbb{E}_\parallel$.

The sets $\mathcal{V}^{(2)}$ and $\mathcal{V}_\Delta^{(2)}$ as well as $\mathcal{V}^{*(2)}$ and $\mathcal{V}_\Delta^{*(2)}$ are invariant under the same translation group, namely, the Δ-invariant Γ. Therefore, one expects the tilings $\mathcal{T}_\Delta(\mathbf{c}_\perp) = 1/\tau\,\mathcal{T}(-1/\tau\,\mathbf{c}_\perp)$ and $\mathcal{T}(\mathbf{c}_\perp)$ to be related somehow; this relationship turns out to be very simple. Namely, it is nothing but a *deflation rule* for the triangle patterns. It is not difficult to convince oneself that vertex points in $\mathcal{T}(\mathbf{c}_\perp)$ are also vertex points in $\mathcal{T}_\Delta(\mathbf{c}_\perp)$. Furthermore, it turns out that the complete vertex configuration around $\pi_\parallel(\mathbf{q})$ in $\mathcal{T}_\Delta(\mathbf{c}_\perp)$ consists of those $\pi_\parallel(\mathbf{q}) + 1/\tau\,\pi_\parallel(P^*)$, which fulfill $-1/\tau\,[\mathbf{c}_\perp - \pi_\perp(\mathbf{q})] \in \pi_\perp(P)$, P a 2-boundary of $V(\mathbf{0})$. This can be translated into the prescription for the deflation of a vertex configuration. If $\pi_\parallel(\mathbf{q})$ is a vertex point of the tiling $\mathcal{T}(\mathbf{c}_\perp)$, then it is a vertex point of $\mathcal{T}_\Delta(\mathbf{c}_\perp) = 1/\tau\,\mathcal{T}(-1/\tau\,\mathbf{c}_\perp)$, and the vertex configuration around $\pi_\parallel(\mathbf{q})$ is just the configuration, rescaled by a factor of $1/\tau$, which is represented by the elementary polygon that contains $-1/\tau\,[\mathbf{c}_\perp - \pi_\parallel(\mathbf{q})]$. A look at the decomposition of $\pi_\perp[V(\mathbf{0})]$ into elementary polygons (Fig. 8) leads to the simple rules indicated in Figure 9. Almost every vertex configuration is unambigously transformed under deflation, except for type 3. A more detailed investigation shows that configurations of type 3 must be accompanied by a short edge on at least one of its "wings" (see Fig. 9). The orientation of this edge determines the deflation of vertex type 3 (if it happens that there are short edges on both wings, they will not lead to a contradiction). Figure 11 shows the deflation of a larger part of a $\mathcal{T}$-pattern.

5. The Penrose Pattern

The method used in Section 4 for the construction of the triangle pattern (or shortly, $\mathcal{T}$-pattern) is suitable as well if we want to project objects from $\mathcal{V}^{(2)}$ into $\mathbb{E}_\parallel$ rather than elements of $\mathcal{V}^{*(2)}$. It is almost trivial now to obtain a prescription for the selection of those $P \in \mathcal{V}^{(2)}$, which are to be projected to obtain a proper tiling of $\mathbb{E}_\parallel$: For fixed $\mathbf{c}_\perp \in \mathbb{E}_\perp$, $\pi_\parallel(P)$ ($P \in \mathcal{V}^{(2)}$) is an element of the tiling $\mathcal{P}(\mathbf{c}_\perp)$ if and only if $\pi_\perp(P^*)$ contains $\mathbf{c}_\perp$. This time, we have to restrict the set of possible $\mathbf{c}_\perp \in \mathbb{E}_\perp$ to those that are not contained in the projection images in $\mathbb{E}_\perp$ of 1-boundaries of some

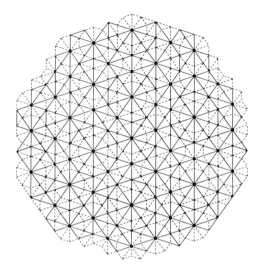

Figure 11. Deflation of a generic triangle pattern.

$P* \in \mathcal{V}*^{(2)}$ (again all $\mathbf{c}_\perp$ except for a set of measure zero in $\mathbb{E}_\perp$ are allowed). The formal expression for $\mathcal{P}(\mathbf{c}_\perp)$ is

$$\mathcal{P}(\mathbf{c}_\perp) = \{\pi_\|(P) \mid P \in \mathcal{V}^{(2)}, \mathbf{c}_\perp \in \pi_\perp(P*)\}. \qquad (35)$$

The vertex points of the pattern are projection images of vertices of some $P \in \mathcal{V}^{(2)}$, i.e., vertices of Voronoi domains of Γ. As we have seen in Section 2, the vertices of Voronoi domains (the elements of $\Gamma*$) split into four orbits under the translation group Γ. The orbits are characterized by the function $r(\mathbf{q}*)$ introduced in Eq. (24). Let us choose the representative elements $\mathbf{q}_1^*: = \mathbf{a}_1$, $\mathbf{q}_2^*: = \mathbf{a}_1 + \mathbf{a}_3$, $\mathbf{q}_3^*: = -\mathbf{a}_1 - \mathbf{a}_3$, and $\mathbf{q}_4^*: = -\mathbf{a}_1$, thus $r(\mathbf{q}_i^*) = i$. Furthermore, for each $\mathbf{q}* \in \Gamma*$, let us write $\mathbf{q}(\mathbf{q}*): = \mathbf{q}* - \mathbf{q}_{r(\mathbf{q}*)}^* \in \Gamma$. In complete analogy to Section 4, we use the periodicity of the 4-D structure to write the vertex set $\mathcal{P}(\mathbf{c}_\perp)$ as

$$\{\pi_\|(\mathbf{q}*) \mid \mathbf{q}* \in \Gamma*, \mathbf{c}_\perp - \pi_\perp[\mathbf{q}(\mathbf{q}*)] \in \pi_\perp[V*(\mathbf{q}_{r(\mathbf{q}*)}^*)]. \qquad (36)$$

Note that now four different dual cells are involved in this expression, as it must be.

The fact that the elements of $\Gamma*$ are arranged in several orbits with respect to the translation group Γ can be used to distinguish between different types of vertices in the patterns $\mathcal{P}(\mathbf{c}_\perp)$. Indeed, we can subdivide $\Gamma*$ into two classes that are not connected by any symmetry operation of the A_4 lattice, namely, the union of the two translation orbits represented by $\mathbf{q}_1^*$ and $\mathbf{q}_4^*$ as well as the union of the orbits of $\mathbf{q}_2^*$ and $\mathbf{q}_3^*$ ($\mathbf{q}_1^*$ and $\mathbf{q}_4^*$ are interchanged by space inversion in S, the same holds for the pair $\mathbf{q}_2^*$, $\mathbf{q}_3^*$). According to this subdivision of $\Gamma*$, we classify the vertices of $\mathcal{P}(\mathbf{c}_\perp)$. Remember that the Voronoi vertices of the class $\mathbf{q}_1^*$ are called shallows and those of class $\mathbf{q}_2^*$ deep holes, so we use this terminology also as names for the vertices themselves.

The projection images of the elements of $\mathcal{V}^{(2)}$ and, consequently, the tiles of $\mathcal{P}(\mathbf{c}_\perp)$ have the shape of the well-known Penrose rhombi [7]. Each has exactly one vertex of the class represented by $\mathbf{q}_1^*$ and three of the $\mathbf{q}_2^*$-class. For the thin rhombi, the $\mathbf{q}_1^*$-type vertex sits at one of the obtuse angles ($4\pi/10$), the $\mathbf{q}_1^*$-type vertex of the thick rhombi is placed at an acute angle ($2\pi/10$). This agrees with the usual decoration of Penrose rhombi in Penrose patterns [8]. Furthermore, the decoration determines the matching rules which define Penrose patterns [8, 7]. If one looks at the vertex configurations of the tilings $\mathcal{P}(\mathbf{c}_\perp)$ (obtained by projection from 4-space S!) studied below, one convinces oneself that the matching rules would be fulfilled if we would decorate the edges of the rhombi according to the original prescription of Penrose. Hence, the tilings $\mathcal{P}(\mathbf{c}_\perp)$ are in fact Penrose patterns.

In order to obtain a complete list of the vertex configurations that are found in $\mathcal{P}(\mathbf{c}_\perp)$'s we proceed exactly as in Section 4 by subdividing the projection images of the representative dual cells $V^*(\mathbf{q}_1^*), \ldots, V^*(\mathbf{q}_4^*)$ in $\mathbb{E}_\perp$ (Fig. 12) into elementary polygons. Because of the close relationship of the A_4 lattice to the 5-D hypercubic lattice it is no surprise that this analysis is essentially equivalent to methods based on pentagrids or on the 5-D lattice [8, 11, 20]. Each elementary polygon represents a class of complete vertex configurations with respect to translations; if we identify vertex configurations related by symmetry operations of the A_4 lattice, we come to the 8 different types of configurations in Figure 13. We distinguish between the two fivefold symmetric vertices 3 and 8 because the vertex points belong to different classes. The relative frequencies of the vertex configurations are again determined by the relative areas of the corresponding elementary polygons.

It is easy to work out an algorithm for the construction of Penrose patterns in complete analogy to that described in the previous section for the $\mathcal{T}$-patterns. The result is shown in Figure 14.

The transformation Δ introduced in Section 4 can be used for the Penrose patterns as well to describe a deflation scenario. The analysis can be carried out in the same way as for the case of $\mathcal{T}$-patterns, we skip the details. Let us just mention that the vertex points will change their class under deflation: "deep hole"-type vertex points become "shallows" and vice versa while the additionally arising vertices will all be "shallows". The deflation of a pattern $\mathcal{P}(\mathbf{c}_\perp)$ is shown in Figure 15, it is immediate that this is in accordance with the well-known deflation of Penrose patterns.

In the literature, one can find an elegant triangulation of Penrose patterns proposed by Robinson [13, 14] that is interesting with respect to our $\mathcal{T}$-pattern. The thin rhombi are halved into two triangles of the shape of the acute triangles occuring in $\mathcal{T}(\mathbf{c}_\perp)$, while the thick rhombi split into two acute and two obtuse triangles (cf. Fig. 13). A superficial look at the resulting pattern in Figure 16 shows that completely different structures arise in comparison with the $\mathcal{T}$-patterns. An analysis of the vertex configurations

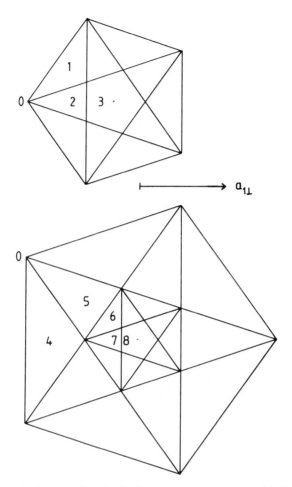

Figure 12. Projection images of the dual cells around $-\mathbf{q}_1^*$ and $-\mathbf{q}_1^* - \mathbf{q}_3^*$ in $\mathbb{E}_\perp$. The numbers 1, ... , 8 refer to the vertex configurations in Figure 13.

of Robinson's decomposition of Penrose patterns is carried out easily with the aid of the vertex list of $\mathscr{P}(\mathbf{c}_\perp)$. There will be one type of vertex configuration that does not arise directly from the decomposition of the Penrose vertices (Fig. 13), namely, the vertex inside a decomposed thick rhombus. Of course this one can also be found in Figure 13. The elaboration of the relative frequencies is an easy exercise with the knowledge of the frequencies of the Penrose vertices, we may skip it here. There are only two vertex configurations that occur also in $\mathscr{T}$-patterns. Even pairings of tiles are found in Robinson's tiling that are never seen in a $\mathscr{T}(\mathbf{c}_\perp)$, namely, the joins of two obtuse triangles along their bases. On the other hand, the pentagon-shaped configuration built from one acute triangle with two obtuse tringles

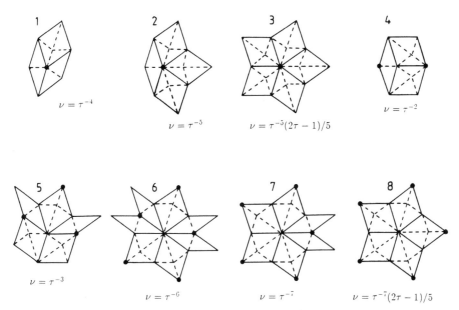

Figure 13. Vertices of the Penrose pattern and their decomposition according to Robinson's triangulation.

at its sides, which is very frequent in the $\mathcal{T}$-patterns, cannot exist in Robinson's tiling. A more detailed comparison between the $\mathcal{T}$-patterns and Robinson's decomposition of the Penrose pattern can be found in [15].

6. Cell Models and Fourier Theory

The method of dualization can be reformulated into a general construction mechanism for quasiperiodic tilings, namely the so-called Klotz construction [19, 21, 22]. Klotz constructions are a certain kind of space fillings of the high-dimensional space (in our case, S) by polytopes, called Klotz polytopes. These space fillings are designed in such a way that cutting through the set of Klotz polytopes by certain affine low-dimensional spaces yields exactly the tilings that are obtained by the dualization method described in the two preceding sections. Besides the necessity of the Klotz construction technique for our proof of the fundamental statement that the dualization method provides in fact space fillings without gaps or overlaps [19], it is also a useful tool for the computation of Fourier transforms of decorations of the tilings by densities.

The present section is devoted to the application of this general method to the calculation of Fourier transforms for the examples of quasiperiodic tilings we have exhibited so far. Let us first set up the Klotz constructions

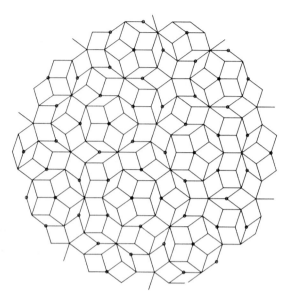

Figure 14. The Penrose pattern.

for the $\mathcal{T}$- and Penrose patterns. For $P \in \mathcal{V}^{(2)}$, we define the Klotz polytopes

$$K_{\mathcal{T}}(P): = \pi_{\parallel}(P^*) + \pi_{\perp}(P)$$
$$= \{\mathbf{x} \in S \mid \pi_{\parallel}(\mathbf{x}) \in \pi_{\parallel}(P^*), \pi_{\perp}(\mathbf{x}) \in \pi_{\perp}(P)\} \tag{37}$$

and

$$K_{\mathcal{P}}(P): = \pi_{\parallel}(P) + \pi_{\perp}(P^*)$$
$$= \{\mathbf{x} \in S \mid \pi_{\parallel}(\mathbf{x}) \in \pi_{\parallel}(P), \pi_{\perp}(\mathbf{x}) \in \pi_{\perp}(P^*)\}. \tag{38}$$

$K_{\mathcal{T}}(P)$ and $K_{\mathcal{P}}(P)$ are a 4-D polytopes in S; the sets $\mathcal{K}_{\mathcal{T}}: = \{K_{\mathcal{T}}(P) \mid P \in$ and $\mathcal{K}_{\mathcal{P}}: = \{K_{\mathcal{P}}(P) \mid P \in \mathcal{V}^{(2)}\}$ are the Klotz constructions for the $\mathcal{T}$- and Penrose patterns, respectively. It follows from the very definition that cutting through $\mathcal{K}_{\mathcal{T}}$ and $\mathcal{K}_{\mathcal{P}}$ with the affine plane $\mathbf{c}_{\perp} + \mathbb{E}_{\parallel}$ and shifting the resulting tilings back to $\mathbb{E}_{\parallel}$ yield exactly the patterns $\mathcal{T}(\mathbf{c}_{\perp})$ and $\mathcal{P}(\mathbf{c}_{\perp})$, respectively. Of course, one has to show that $\mathcal{K}_{\mathcal{T}}$ and $\mathcal{K}_{\mathcal{P}}$ are space fillings of S without gaps or overlaps [19].

One can choose 20 fundamental Klotz polytopes inequivalent with respect to A_4 translations such that $\mathcal{K}_{\mathcal{T}}$ and $\mathcal{K}_{\mathcal{P}}$, respectively, can be generated by shifting these 20 fundamental polytopes by vectors of A_4. This implies that each one of these two sets, which we shall denote by $\mathcal{F}_{\mathcal{T}}$ for the triangles and $\mathcal{F}_{\mathcal{P}}$ for the Penrose case, forms a fundamental domain with respect to the translation group T_{A_4}. The Fourier transform of any density f in S with

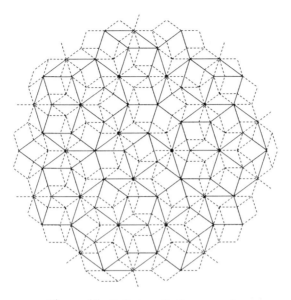

Figure 15. Deflation of a Penrose pattern.

this periodicity is then completely determined by giving its values on one of the two sets.

So far, the Klotz construction yields a nonperiodic tiling in a purely geometric way without any physical input. Let us consider a 4-D density distribution $f(\mathbf{x}_\|, \mathbf{x}_\perp)$, $\mathbf{x}_\| \in \mathbb{E}_\|$, $\mathbf{x}_\perp \in \mathbb{E}_\perp$ which is periodic with respect to the lattice A_4 and restrict this density by the condition that, on each Klotz, it becomes independent of the coordinate $\mathbf{x}_\perp$ perpendicular to $\mathbb{E}_\|$. For fixed $\mathbf{c}_\perp \in \mathbb{E}_\perp$, the cut function $\mathbf{x}_\| \to f(\mathbf{x}_\|, \mathbf{c}_\perp)$, becomes a density in the 2-D space, $\mathbb{E}_\|$, which will be not periodic but quasiperiodic. If we stipulate, furthermore, that part of the point symmetry of A_4 that survives the decomposition of S to be obeyed, we obtain the same density on all tiles of the same shape independently of their orientation. Thus, the tiles get a physical meaning as cells comparable with the unit cell of a periodic crystal. The density may stand for the electron density for X-ray scattering or the potential distribution in the case of electron diffraction. This model with stable densities on the tiles is called quasicrystal model [21, 22]. It is a certain subclass of the general model with unrestricted quasiperiodic densities in $\mathbb{E}_\|$. We have two representative cells with respect to the space group $T \times_s D_{10}$, a thick and a thin rhombus in the Penrose case or an acute and an obtuse triangle in the triangle pattern, respectively, on which a density can be chosen.

Diffraction experiments belong to the most important measurements to acquire physical information on the structure of solids. The simplest method

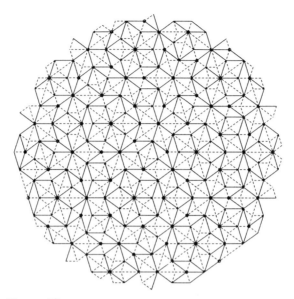

Figure 16. Robinson's decomposition of a Penrose pattern.

to describe diffraction physics is theoretically given by Born approximation, which is a suitable method for the description of elastic s-wave single scattering. It leads to the calculation of the intensities via the squared absolute value of the Fourier transform of the scattering density. Thus, we restrict our attention to the Fourier analysis of quasicrystals henceforth. To describe the Fourier transform, we introduce a 4-D **k**-space and, therein, the reciprocal lattice A_4^R multiplied by 2π. It is spanned by the basis $\mathbf{k}_i^R = 2\pi\mathbf{a}_i = 2\pi(\mathbf{e}_i - 1/5\ \mathbf{s})$, $i = 1, \ldots 4$, the lattice points are $\mathbf{k}_R = \Sigma_{i=1}^4 h_i \mathbf{k}_i^R$, $h_i \in \mathbb{Z}$. As in the **x**-space, we decompose the **k**-vectors in $\mathbf{k} = \mathbf{k}_\parallel + \mathbf{k}_\perp$, $\mathbf{k}_\parallel \in \mathbb{E}_\parallel$, $\mathbf{k}_\perp \in \mathbb{E}_\perp$. Given a density distribution $f(\mathbf{x}_\parallel, \mathbf{x}_\perp)$ of A_4, in the quasicrystal model the Fourier transform $\tilde{f}(\mathbf{k}_\parallel)$ of the cut function $\mathbf{x}_\parallel \rightarrow f(\mathbf{x}_\parallel, \mathbf{c}_\perp)$ is given by

$$\tilde{f}(\mathbf{k}_\parallel) = [\mathrm{vol}[FD(A_4)]]^{-1} \sum_{\kappa \in \mathcal{F}} \tilde{f}_\mathcal{H}(\mathbf{k}_\parallel) Q_\kappa(\mathbf{k}_\parallel), \tag{39}$$

where $\mathcal{F}$ stands for one of the fundamental sets of Klotz polytopes $\mathcal{F}_\mathcal{J}$ or $\mathcal{F}_\mathcal{P}$ in the triangle or the Penrose case, respectively. The structure factor $\tilde{f}_\kappa(\mathbf{k}_\parallel)$ is the (2-D) Fourier transform of the density $f_\kappa(\mathbf{x}_\parallel, \mathbf{c}_\perp)$ on a quasicrystal cell labelled by κ. The quasilattice factor is defined by

$$Q_\kappa(\mathbf{k}_\parallel) = \sum_{\mathbf{k}^R \in 2\pi A_4^R} \delta(\mathbf{k}_\parallel - \mathbf{k}_\parallel^R) \exp(i\mathbf{k}_\perp^R \cdot \mathbf{c}_\perp) K_\kappa(\mathbf{k}_\perp^R), \tag{40}$$

where K_κ is the (2-D) Fourier transform of the Klotz κ projected into $\mathbb{E}_\perp$, i.e., $\pi_\perp(P^*)$ and $\pi_\perp(P)$ in the Penrose and the triangle pattern, respectively.

K_κ will be called the kinematic factor. It is independent of the density and is completely determined by the geometry of the Klotz polytopes. Note that these formulas are correct only for densities defined inside the Klotz polytopes, densities on Klotz boundaries need a special investigation in view of their different symmetry groups [23].

We give some examples to illustrate the properties of Fourier transforms. For the sake of simplicity, we investigate only density distributions with point scatterers defined by δ-functions. The diffraction patterns in Figures 17–20 show the squared absolute value of the Fourier transform for several decorations of the Penrose pattern and the triangle pattern, respectively. For a comparison with the diffraction patterns given by other authors (cf. e.g. [24]) we refer to [15]. The circles represent the diffraction spots, their areas being proportional to the intensities relative to the central peak. Spots with intensity smaller than $\frac{1}{10}$ of the central peak or with coordinates $|\mathbf{k}_\parallel| > 10 \cdot 2\pi$ have been suppressed. The symmetry of the diffraction patterns is D_{10}, i.e., the symmetry group of the decagon with 20 elements. In the small figures right up the dots indicate the sites of the point scatterers on the representative cells. The kinematic diffraction patterns emphasize the difference between the vertex decorations of the Penrose pattern and the triangle pattern as well as the difference between the vertex decorations of the Robinson pattern and that of the triangle pattern, since no scaling and rotation transformation can transport these patterns into one another.

Figure 17. Kinematic diffraction for vertex decoration of the triangle pattern. The circles represent the diffraction spots; their areas are proportional to their intensities relative to that of the central peak. Here, as well as in the following figures, the positions of the point scatterers in the representative cells are schematically depicted right up.

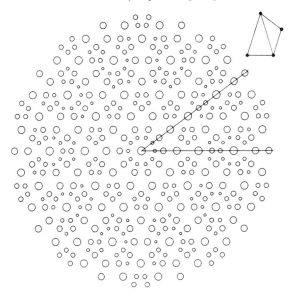

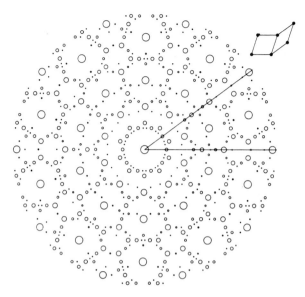

Figure 18. Kinematic diffraction for vertex decoration of the Penrose pattern (minimal intensity is 1% of the central peak).

Figure 19. Kinematic diffraction for the decoration of the "shallows" of the Penrose pattern. The decoration of the "deep holes" yields the same pattern, scaled by a factor τ.

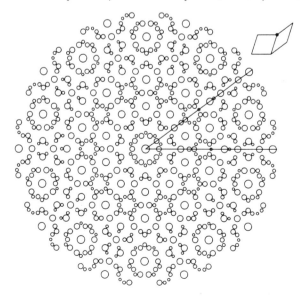

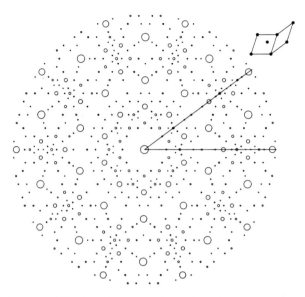

Figure 20. Kinematic diffraction for vertex decoration of the Robinson tiling, seen as a decoration of the Penrose pattern (minimal intensity is 1% of the central peak).

7. Conclusion

In this article we have presented two different quasi-periodic tilings that can be constructed from the 4-D root lattice A_4. We think that these 2-D patterns provide nontrivial but nevertheless easily intelligible examples. In particular, they neither lack the crystallographically forbidden symmetries like the 1-D toy models nor suffer from a structural poorness that could hide some of the noncrystallographic properties. In a sense, the 2-D patterns as derived from the 4-D space build the right pool of models to test physical assumptions with—although real physics generally may call for 3-D patterns. But 2-D structures facilitate their understanding by means of their simple depictability and are without doubt useful for understanding so-called T-phases. The latter are quasiperiodic in two dimensions and periodic in the third. Although several attempts for their description have been made so far [25–28], a rigorous understanding is still missing. We think that especially the triangle pattern is an interesting candidate for such a T-phase. This is supported by the observation that suitable sections through the icosahedral quasilattice [6] yield precisely the vertex arrangement of the triangle pattern.

Acknowledgments
The authors would like to thank the Deutsche Forschungs-gemeinschaft for financial support.

REFERENCES

1. D. Shechtman, I. Blech, D. Gratias, and J. W. Cahn, "Metallic phase with long-ranged orientational order and no translational symmetry," *Phys. Rev. Let.*, **53**, 1951–1953 (1984).

2. *Proc. Int. Worksh. on Aperiodic Crystals*, Les Houches, 1986, D. Gratias and L. Michel (eds.). *J. Phys.* Colloque C3 (1986).

3. P. J. Steinhardt and S. Ostlund (eds.), *The Physics of Quasicrystals*. World Scientific, Singapore, 1987.

4. *Proc. ILL/CODEST Worksh. Quasicrystalline Materials*, Grenoble 1988, Ch. Janot and J. M. Dubois (eds.). World Scientific, Singapore, 1988.

5. M. Jarić (ed.), *Aperiodicity and Order*, vol. 1 *Introduction to Quasicrystals*. Academic Press, San Diego, 1988; Vol. 2. *Introduction to the Mathematics of Quasicrystals*. Academic Press, San Diego, 1989.

6. P. Kramer and R. Neri, "On periodic and non-periodic space fillings of $\mathbb{E}^m$ obtained by projection," *Acta Crystallogr.*, **A40**, 580–587 (1984).

7. R. Penrose, "Pentaplexity—a class of non-periodic tilings of the plane," *Math. Intell.*, **2**, 32–37 (1979).

8. N. G. de Bruijn, "Algebraic theory of Penrose's non-periodic tilings of the plane," *Math. Proc.*, **A84**, 39–66 (1981).

9. F. Gähler and J. Rhyner, "Equivalence of the generalized grid and projection methods for the construction of quasiperiodic tilings," *J. Phys.*, **A19**, 267–277 (1986).

10. H. Bohr, *Fastperiodische Funktionen*. Springer, Berlin, 1932.

11. M. V. Jarić, "Diffraction from quasicrystals: Geometric structure factor," *Phys. Rev.*, **B34**, 4685–4698 (1986).

12. R. Lück, "Penrose sublattices," *Proc. 7th Int. Conf. Liquid Amorphous Metals*, Kyoto 1989, H. Endo (ed.), to be published in *J. Non-cryst. Solids*.

13. R. M. Robinson, *Comments on Penrose Tiles*. University of California, Berkeley, 1975.

14. B. Grünbaum and G. C. Shephard, *Patterns and Tilings*. Freeman, New York, 1987.

15. M. Baake, P. Kramer, M. Schlottmann, and D. Zeidler, "Planar patterns with fivefold symmetry as sections of periodic structures in 4-Space," preprint Tübingen TPT-QC-89-10-1, submitted for publication.

16. H. Brown, R. Bülow, J. Neubüser, H. Wondratschek, and H. Zassenhaus, *Crystallographic Groups of Four-Dimensional Space*. Wiley, New York, 1978.

17. J. H. Conway and N. J. A. Sloane, *Sphere Packings, Lattices and Groups*. Springer, New York, 1988.

18. J. E. Humphreys, *Introduction to Lie Algebras and Representation Theory*. Springer, New York, 1972.

19. P. Kramer and M. Schlottmann, "Dualization of Voronoi domains and Klotz construction: A general method for the generation of proper space fillings," *J. Phys.* **A22**, 1097–1102 (1989).

20. C. L. Henley, "Sphere packings and local environments in Penrose tilings," *Phys. Rev.*, **B34**, 797–816 (1986).

21. P. Kramer, "Space-group theory for a nonperiodic icosahedral quasilattice," *J. Math. Phys.*, **29**, 516–524 (1988).

22. P. Kramer, "On the cell structure for non-periodic long-range order in solids," *Quasicrystals and Incommensurate Structures in Condensed Matter*, M. J. Yacamán et al. (eds.), World Scientific, Singapore, 1990

23. P. Kramer and D. Zeidler, "Structure factors for icosahedral quasicrystals," *Acta Crystallogr.*, **A45**, 524–533 (1989).

24. T. Janssen, "Aperiodic crystals: A contradictio in terminis?," *Phys. Rep.*, **168**, 55–113 (1988).

25. L. Bendersky, "Quasicrystals with 1D translational symmetry and a tenfold rotation axis," *Phys. Rev. Lett.*, **55**, 1461–1463 (1985).

26. J. D. Fitzgerald, R. L. Withers, A. M. Stewart, and A. Calka, "The Al–Mn decagonal phase 1: A re-evaluation of some diffraction effects; 2: Relationship to crystalline phases," *Phil. Mag.*, **B58**, 15–33 (1988).

27. T. C. Choy, J. D. Fitzgerald, and A. C. Kalloniatis, "The Al–Mn decagonal phase 3: Diffraction pattern simulation based on a new index scheme," *Phil. Mag.*, **B58**, 35–46 (1988).

28. Z. Zhang and K. Urban, "Quasicrystalline phases in Al–Cu–Co alloy," preprint, Jülich 1989.

10 *Nondiscrete Regular Honeycombs*

Peter McMullen

1. Introduction

The discrete regular honeycombs or *apeirotopes* in Euclidean space $\mathbb{E}^d$ are well known. Apart from the tilings by cubes for all d (line segments if $d = 1$ and squares if $d = 2$), there are only four more examples: the tilings by equilateral triangles and regular hexagons for $d = 2$, and those by 16-cells (cross-polytopes) and 24-cells for $d = 4$ (see [3]). It is notable that none of these apeirotopes is starry; they all consist of convex cells with convex vertex figures.

It follows that any star-apeirotope which is regular in the traditional sense (that is, its symmetry group is generated by reflections in hyperplanes) is necessarily nondiscrete. Nondiscreteness is usually enough to exclude such apeirotopes from consideration (see, for example, [3]), but nevertheless they can be of interest, as we hope to show.

We shall concentrate on those nondiscrete regular apeirotopes which possess fivefold rotational symmetries. Such apeirotopes are connected with certain quasiperiodic tilings, and then with quasicrystals, which provide physical examples of structures with a basically crystalline nature which still involve local fivefold symmetries.

2. Abstract Regular Polytopes

While we shall concentrate on the concrete aspects of the theory of regular polytopes and honeycombs in this article, it is useful from time to time to bear in mind the more purely combinatorial point of view. We shall therefore briefly discuss that here; we refer to [14, 16, 17] for further details (the theory is also described in other papers, but we shall cite these references again below).

An *abstract n-polytope* or *n-incidence-polytope* is a (finite or infinite) partially ordered set $\mathcal{P}$, whose elements are called *cells* or *faces*, with the following properties (I1)–(I4).

I1. $\mathcal{P}$ has a unique least cell F_{-1} and a unique greatest cell F_n.

I2. The maximal totally ordered subsets or *flags* of $\mathcal{P}$ contain exactly $n + 2$ cells.

Two flags are called *adjacent* if they differ by exactly one cell.

I3. $\mathcal{P}$ is *strongly flag connected*, in that, if Φ, Ψ are two flags of $\mathcal{P}$, then there exists a sequence $\Phi = \Phi_0, \Phi_1, \ldots, \Phi_k = \Psi$, such that $\Phi_j \subseteq \Phi \cap \Psi$ and Φ_{j-1} and Φ_j are adjacent for each j.

I4. If Φ is a flag of $\mathcal{P}$, and if $F \in \Phi \setminus \{F_{-1}, F_n\}$, then there exists exactly one flag Ψ of $\mathcal{P}$ different from Φ which contains $\Phi \setminus \{F\}$.

If $F \leq G$ are two cells of $\mathcal{P}$, then the *section* G/F is defined by

$$G/F = \{H \in \mathcal{P} \mid F \leq H \leq G\}.$$

If the flags of F/F_{-1} contain $j + 2$ cells, then we call F a *j-cell* of $\mathcal{P}$. In addition, 0-, 1-, 2-, and $(n-1)$-cells are called *vertices*, *edges*, *faces*, and *facets*, respectively, while if F is a vertex of $\mathcal{P}$, then the section F_n/F is called the *vertex figure* of $\mathcal{P}$ at F.

It should be remarked that conditions I3 and I4 admit equivalent formulations in terms of sections; we choose the present definitions for brevity.

The *automorphism group* (or, more simply, just *group*) $\Gamma(\mathcal{P})$ of an abstract n-polytope $\mathcal{P}$ consists of the order-preserving permutations, or *automorphisms*, of $\mathcal{P}$. Then $\Gamma(\mathcal{P})$ induces permutations of the family $\mathcal{F}(\mathcal{P})$ of flags of $\mathcal{P}$; if $\Gamma(\mathcal{P})$ is transitive on $\mathcal{F}(\mathcal{P})$, we say that $\mathcal{P}$ is *regular*.

Let $\mathcal{P}$ be an abstract regular n-polytope, and let $\Phi = \{F_{-1}, F_0, \ldots, F_{n-1}, F_n\}$ be a fixed, or *base*, flag of $\mathcal{P}$. Writing Φ^j for the other flag of $\mathcal{P}$ which contains $\Phi \setminus \{F_j\}$ (with $0 \leq j < n$), there is a $\rho_j \in \Gamma(\mathcal{P})$ such that $\Phi^j = \Phi\rho_j$; moreover, $\rho_j \neq \varepsilon$ (the identity), since $\Phi^j \neq \Phi$. It can be seen that if $I \subseteq \{0, \ldots, n-1\}$, $\Omega = \{F_j \mid j \in I\}$ and $\Gamma_\Omega = \{\sigma \in \Gamma(\mathcal{P}) \mid \Omega\sigma = \Omega\}$, then $\Gamma_\Omega = \langle \rho_i \mid i \in I \rangle$, and is simply transitive on the subfamily of flags which contain Ω. Hence, each section of a regular polytope is also a regular polytope. We shall refer to $\rho_0, \ldots, \rho_{n-1}$ as the *distinguished generators* of $\Gamma(\mathcal{P})$.

An automorphism group of a regular polytope can be characterized as a *C-group*, which is a group $\Gamma = \langle \rho_0, \ldots, \rho_{n-1} \rangle$ which satisfies

C1. $(\rho_i\rho_j)^{q_{ij}} = \varepsilon$, with $q_{jj} = 1$ ($j = 0, \ldots, n-1$) and $q_{ij} = 2$ whenever $|i - j| \geq 2$.

C2. For $I, J \subseteq \{0, \ldots, n-1\}$, $\langle \rho_i \mid i \in I \rangle \cap \langle \rho_i \mid i \in J \rangle = \langle \rho_i \mid i \in I \cap J \rangle$ (this is known as the *intersection property*). The *j*-cells of the regular polytope corresponding to the *C*-group Γ are identified with the right cosets of the subgroup $\Gamma_j = \langle \rho_i \mid i \neq j \rangle$.

The *dual* $\mathcal{P}^\delta$ of an n-polytope $\mathcal{P}$ is obtained by reversing the partial order. Thus, if $\mathcal{P}$ is regular with $\Gamma(\mathcal{P}) = \langle \rho_0, \ldots, \rho_{n-1} \rangle$, then $\mathcal{P}^\delta$ is regular with $\Gamma(\mathcal{P}^\delta) = \langle \rho_{n-1}, \ldots, \rho_0 \rangle$; to each j-cell of $\mathcal{P}$ corresponds an $(n-j-1)$-cell of $\mathcal{P}^\delta$.

Particularly in this article, we sometimes wish to emphasize that a polytope is infinite. In that case, we shall refer to it as an *apeirotope*; further, an *apeirohedron* is a 3-apeirotope. (We have already used these terms informally in the previous section. The term *honeycomb* for apeirotope usually only refers to those with convex cells, and so its usage in the title of the paper is, perhaps, inexcusably loose.)

3. Realizations

If two abstract regular n-polytopes are isomorphic (in the obvious sense), then we naturally identify them. However, in a geometric setting, we should wish to distinguish between the isomorphic regular pentagon $\{5\}$ and pentagram $\{\frac{5}{2}\}$ (the notation is the usual Schläfli symbol of [3]). This is most appropriately done in the context of realizations. The following description is largely drawn from [14].

A *realization* of a regular n-polytope $\mathcal{P}$ is a mapping of its vertex-set onto a subset V of some euclidean space E, such that each permutation of V (regarded, if necessary, as a multiply counted set) corresponding to an automorphism of $\mathcal{P}$ is induced by an isometry of E. A realization of $\mathcal{P}$ naturally induces a mapping on all of $\mathcal{P}$; in this context, a j-cell is regarded as the set of all $(j-1)$-cells which it covers (though we must be careful to note that in more general circumstances than we shall need to consider here, a j-cell is not uniquely determined by these $(j-1)$-cells). We shall often tend (particularly in a geometric setting) to think of a realization of $\mathcal{P}$ as composed of these induced images of its cells, which we shall then denote by some symbol such as P.

To a realization of a regular polytope $\mathcal{P}$ corresponds a representation of its automorphism group $\Gamma(\mathcal{P})$ in the isometry group of E. The converse also holds, because if G acting on E (as a group of isometries) is a representation of $\Gamma(\mathcal{P})$, and if G_0 is its subgroup corresponding to Γ_0, let W be the affine subspace of E left pointwise invariant by G_0; we call W the *Wythoff space* of the representation. If $v \in W$ is any point, then $V = vG$ is the vertex set of a realization of $\mathcal{P}$, which we say is obtained by *Wythoff's construction*, with *initial vertex* v.

To each generating involution ρ_j of $\Gamma(\mathcal{P})$ will correspond an involutory symmetry (isometry) or *reflection* R_j in G. We shall often identify a reflection R with its *mirror* $\{x \in E \mid xR = x\}$ of fixed points, and in this spirit we shall use E itself to denote the identity. (In general realizations, one or more of the R_j might equal E, but this will not occur here.) Thus, in view of the intersection property (C2), the Wythoff space W is

$$W = R_1 \cap R_2 \cap \cdots \cap R_{n-1}.$$

There are various ways of combining realizations of $\mathcal{P}$ (and in all that we do, we should observe that it is really only the equivalence class of the realization under isometry which is of interest). First, if $\lambda \in \mathbb{R}$, and V is the vertex-set of a realization P, then λV is the vertex-set of a realization which we shall denote by λP, and call a *scalar multiple* of P. Second, let P and P' be realizations of $\mathcal{P}$, with vertex-sets V in E and V' in E', respectively. Then, with $v \in V$ and $v' \in V'$ corresponding vertices, let $v'' = (v, v') \in V \times V' \subseteq E \times E'$, and write V'' for the set of these v''. Then V'' is the vertex-set of a new realization P'' pf $\mathcal{P}$, which we call the *blend* of P and P', and denote by $P'' = P \# P'$. (A third way of combining realizations, discussed in [14], will not concern us here.)

Observe that if the mirrors of the generating reflections for the groups of P and P' are R_j and R_j', respectively ($j = 0, \ldots, n-1$), then those for λP (for $\lambda \neq 0$) and $P \# P'$ are λR_j and $R_j \times R_j'$, respectively ($j = 0, \ldots, n-1$).

4. Reflection Groups

All the apeirotopes we consider in this article have their origins in three nondiscrete reflection groups. By a *reflection group* we mean a group of isometries of some euclidean space E which is generated by reflections whose mirrors are hyperplanes (affine subspaces of codimension 1). If a reflexion group G acts irreducibly on $\mathbb{E}^d$, then if finite it is generated by d reflexions whose mirrors bound a cone with a simplex as cross section, and if discrete and infinite G is generated by $d + 1$ reflections whose mirrors bound a simplex. It is natural, when considering nondiscrete infinite reflection groups, to impose this latter condition on them as well.

Such a reflexion group G can be represented by its *Coxeter diagram*. This is a graph, with a *node* corresponding to each mirror of a generating reflection, and two nodes are joined by a *branch* marked p if the dihedral angle between the associated mirrors is π/p, with $p > 1$ always a rational number. By convention, a branch is omitted if its mark would be 2, and left unmarked if its mark would be 3. For example, one Coxeter diagram of the group [3, 5] of all symmetries of the regular icosahedron and dodecahedron is

$$\bullet\!\!-\!\!\!-\!\!\bullet\!\!-\!\!\!-\!\!\bullet$$
$$5$$

Wythoff's construction has a nice representation for reflection groups. In fact, we shall describe a much more general construction (see [4]). If G is a reflexion group, we can consider the general polytope with a given initial vertex v relative to the mirrors of its generating reflections (in particular, the specific way of generating G is of importance). This polytope is denoted by ringing just the nodes of the corresponding Coxeter diagram on which

v does not lie. The initial j-cells of P are obtained recursively by deleting nodes (and incident branches) to leave a subdiagram with j nodes, each of whose connected components contains at least one ringed node. We may note here that this cell is a cartesian product of the cells corresponding to the connected components. For example (see also [3]),

$$\odot \!\!-\!\!\!-\!\!\!-\!\!\bullet \overset{5}{\!\!-\!\!\!-\!\!\!-\!\!\!-\!\!} \odot$$

is the (small) rhombicosidodecahedron, with edges $\odot$ of two types, and faces

$$\odot \!\!-\!\!\!-\!\!\bullet \quad , \quad \odot \quad \odot \quad , \quad \bullet \overset{5}{\!\!-\!\!\!-\!\!\!-\!\!} \odot$$

which are triangles, squares, and pentagons, respectively. (It is also the convention in this notation that the initial vertex v is chosen to be equidistant from the mirrors associated with the ringed nodes.)

5. The Regular Honeycombs

We are now ready to discuss certain nondiscrete honey-combs (or apeirotopes) in 2-, 3-, or 4-dimensional euclidean space. We may note that, if $d \geq 5$, there are no appropriate reflection groups from which we might derive such honeycombs. (We shall exclude apeirotopes related to unitary groups, which is a very natural restriction, although one interesting example does arise in this way in $\mathbb{E}^6$; we hope to describe this elsewhere.)

The first thing to note about our honeycombs (and their groups) is that they admit a natural pairing, which is obtained in the following way. Let $\mathbb{T} = \mathbb{Z}[\tau]$ denote the ring of integers of the form $a + b\tau$, with $a, b \in \mathbb{Z}$ and

$$\tau = \frac{1}{2}(1 + \sqrt{5}) = 2 \cos \frac{\pi}{5},$$

the golden section. Then τ is one root of the equation $\lambda^2 - \lambda - 1 = 0$, whose other root is

$$-\tau^{-1} = \frac{1}{2}(1 - \sqrt{5}) = 2 \cos \frac{3\pi}{5},$$

which is obtained from τ by changing the sign of $\sqrt{5}$. There is thus an outer automorphism of $\mathbb{T}$, induced by $\tau \leftrightarrow -\tau^{-1}$; if $\xi = a + b\tau \in \mathbb{T}$, we shall write $\tilde{\xi} = a - b\tau^{-1}$ for its *conjugate*, with the notation extended in the obvious way to vectors $x \in \mathbb{T}^d \subseteq \mathbb{E}^d$. We shall see that we can always choose the vertices of our honeycombs so that the required pairing is induced by this conjugation (compare also [3], pp. 107, 226).

For our planar examples, it is more convenient to pick coordinates for the vertices in certain planes in $\mathbb{E}^3$. (This also relates them more directly to the higher dimensional examples.) The euclidean triangles

$$\frac{5}{2} \quad 10 \qquad , \qquad 5 \quad \frac{10}{3}$$

give rise to the four (nondiscrete) regular planar honeycombs

$$\{5/2, 10\} \overset{\delta}{\leftrightarrow} \{10, 5/2\}$$
$$\sim \updownarrow \qquad \sim \updownarrow$$
$$\{5, 10/3\} \overset{\delta}{\leftrightarrow} \{10/3, 5\}$$

In this diagram, δ and $\sim$ are the duality and conjugacy introduced above, while

$$\{p, q\} = \quad \underset{p \qquad q}{\circ\!\!-\!\!-\!\!-\!\!\bullet\!\!-\!\!-\!\!-\!\!\bullet}$$

is the usual Schläfli symbol (which we shall usually employ instead of the graphical notation; see [2] again).

We can choose the vertex figure $\{10\}$ of $\{5/2, 10\}$ to be one of the diametral decagons of the icosidodecahedron

$$\begin{Bmatrix} 3 \\ 5 \end{Bmatrix} = \quad \underset{5}{\bullet\!\!-\!\!-\!\!-\!\!\circ\!\!-\!\!-\!\!-\!\!\bullet}$$

in $\mathbb{E}^3$, say that with vertices $\pm a_0, \dots, \pm a_4$, where

$$a_0 = (2, 0, 0), \qquad a_{\pm 1} = (\tau^{-1}, \pm\tau, \mp 1), \qquad a_{\pm 2} = (-\tau, \pm 1, \mp\tau^{-1})$$

(all suffixes taken modulo 5), in the plane $\xi_2 + \tau\xi_3 = 0$. The set of all vertices of $\{5/2, 10\}$ is then $\mathbb{Z}\{a_0, \dots, a_4\}$, the family of integer linear combinations of $a_0, \dots, a_4$.

We shall pay little attention to the dual $\{10, 5/2\}$. As far as the conjugate $\{5, 10/3\}$ is concerned, we could take exactly the same set of vertices for the vertex figure, and hence for the entire apeirohedron. The vertex figures are not the same, of course. For $\{5/2, 10\}$, the vertex figure $\{10\}$ has vertices in order $a_0, -a_3, a_1, -a_4, a_2, -a_0, a_3, -a_1, a_4, -a_2$; for $\{5, 10/3\}$, the vertices of its vertex figure $\{10/3\}$ are in order $a_0, -a_4, a_3, -a_2, a_1, -a_0, a_4, -a_3, a_2, -a_1$. Alternatively, we can just apply conjugation, obtaining

$$\tilde{a}_0 = (2, 0, 0), \qquad \tilde{a}_{\pm 1} = (-\tau, \mp\tau^{-1,\mp 1}), \qquad \tilde{a}_{\pm 2} = (\tau^{-1}, \pm 1, \pm\tau)$$

in the plane $\xi_2 - \tau^{-1}\xi_3 = 0$. The quarter-turn

$$\theta: (\xi_1, \xi_2, \xi_3) \mapsto (\xi_1, \xi_3, -\xi_2)$$

then has the effect $\tilde{a}_j \mapsto a_{3j}$.

Of the 15 possible tetrahedra in $\mathbb{E}^3$ which we could consider, none is suitable to be the fundamental region for the group of a nondiscrete regular honeycomb. However, the tetrahedron with graph does give rise to the three quasiregular honeycombs

$$\left\{3, \tfrac{5}{5/2}\right\} = \quad \left\{5, \tfrac{3}{5/2}\right\} = \quad \left\{5/2, \tfrac{3}{5}\right\} =$$

The quasiregularity means that the flags fall into two orbits under the group, corresponding to the two kinds of regular facets. The vertex figures are the three kinds of quasiregular polyhedra in $\mathbb{E}^3$ with group $[3, 5]$, namely the dodecadodecahedron $\{\tfrac{5}{5/2}\}$ the stellated icosidodecahedron $\{\tfrac{3}{5/2}\}$, and the icosidodecahedron $\{\tfrac{3}{5}\}$, all of which have the same vertex-set, which comprise the $\pm a_i$, $\pm b_j$, and $\pm c_k$, with $a_0, \ldots, a_4$ as above, and

$$b_0 = (0, 2, 0), \qquad b_{\pm 1} = (\mp\tau, 1, \tau^{-1}), \qquad b_{\pm 2} = (\mp 1, -\tau^{-1}, \tau),$$

$$c_0 = (0, 0, 2), \qquad c_{\pm 1} = (\pm 1, \tau^{-1}, \tau), \qquad c_{\pm 2} = (\pm\tau^{-1}, \tau, 1),$$

lying in the planes $\xi_2 + \tau\xi_3 = 2$ or 2τ, respectively.

In each case, the two kinds of regular cell, namely $\{3, 5\}$ and $\{3, \tfrac{5}{2}\}$ for $\{3, \tfrac{5}{5/2}\}$, $\{5, 3\}$ and $\{5, \tfrac{5}{2}\}$ for $\{5, \tfrac{3}{5/2}\}$, and $\{\tfrac{5}{2}, 3\}$ and $\{\tfrac{5}{2}, 5\}$ for $\{\tfrac{5}{2}, \tfrac{3}{5}\}$, alternate around each edge of the honeycomb. This explains why each of the three pairs of regular polyhedra with symmetry group the extended icosahedral group $[3, 5]$ which have the same faces ($\{3\}$, $\{5\}$, and $\{\tfrac{5}{2}\}$, respectively) have supplementary dihedral angles.

From our point of view, the first of the three honeycombs is the most interesting, because the vertex figure $\{\tfrac{5}{5/2}\}$ is isomorphic to the abstract regular polytope $\{5, 4\}_6$ (compare [2, 20]; the suffix "6" indicates that the polyhedron, of type $\{5, 4\}$, is specified by the length of its *Petrie polygon*, in this case a hexagon, which traces two successive edges of each face which it meets, but no three). Thus $\{3, \tfrac{5}{5/2}\}$ itself is an abstract regular apeirotope of type $\{\{3, 5\}, \{5, 4\}_6\}$. Though we shall not prove this here, it can actually be shown that it is the universal regular apeirotope of this type; what *universal* means here is that its group $\langle \rho_0, \rho_1, \rho_2, \rho_3 \rangle$ is given by the relations

$$\rho_j^2 = (\rho_0\rho_1)^3 = (\rho_1\rho_2)^5 = (\rho_2\rho_3)^4 = (\rho_0\rho_2)^2 = (\rho_0\rho_3)^2$$
$$= (\rho_1\rho_3)^2 = (\rho_1\rho_2\rho_3)^6 = \varepsilon$$

which are just those determined by the cell and vertex figure.

Among the vertices of the vertex figure are those of 10 regular hexagons

$$\{\pm a_j, \pm b_{j+2}, \pm b_{j+3}\}, \qquad \{\pm a_j, \pm c_{j+1}, \pm c_{j+4}\}$$

(with suffixes modulo 5 as before), so that through each vertex of the honeycomb pass 10 copies of the triangular tessellation $\{3, 6\}$, whose vertices, edges, and faces all belong to the honeycomb. In a similar way, the two other quasiregular honeycombs have six copies of $\{5, \tfrac{10}{3}\}$ (for $\{5, \tfrac{3}{5/2}\}$) or $\{\tfrac{5}{2}, 10\}$ (for $\{\tfrac{5}{2}, \tfrac{3}{5}\}$) through each vertex.

Conjugation by ~, followed by the quarter-turn θ as above, result in

$$b_j \mapsto -c_{3j}, \qquad c_j \mapsto b_{3j}.$$

This interchanges the two kinds of cell {3, 5} and {3, 5/2} of the honeycomb.

We finally come to $\mathbb{E}^4$. Here we have 10 candidates; each of the 10 regular 4-polytopes whose vertex-sets V_4 are those of the 600-cell {3, 3, 5} occurs as a vertex figure. (These candidates are listed in [3], p. 264.) The 120 points of V_4 are obtained from

$$(2, 0, 0, 0), \quad (1, 1, 1, 1), \quad (\tau, 1, \tau^{-1}, 0)$$

by all even permutations and all changes of sign of the coordinates. The 10 resulting honeycombs fall into three pairs, which are conjugate to their duals, and a set of four, as

$$\{5/2, 3, 3, 5\} \overset{\delta, \; \sim}{\leftrightarrow} \{5, 3, 3, 5/2\}$$

$$\{3, 5/2, 5, 3\} \overset{\delta, \; \sim}{\leftrightarrow} \{3, 5, 5/2, 3\}$$

$$\{5/2, 5, 5/2, 5\} \overset{\delta, \; \sim}{\leftrightarrow} \{5, 5/2, 5, 5/2\}$$

$$\{5/2, 3, 5, 5/2\} \overset{\delta}{\leftrightarrow} \{5/2, 5, 3, 5/2\}$$

$$ \sim \updownarrow \sim \updownarrow$$

$$\{5, 3, 5/2, 5\} \overset{\delta}{\leftrightarrow} \{5, 5/2, 3, 5\}$$

The common vertex-set of these 10 honeycombs is thus $\mathbb{Z}V_4$. With each point $(\alpha_0, \alpha_1, \alpha_2, \alpha_3)$ of $\mathbb{Z}V_4$ can be associated the quaternion

$$\tfrac{1}{2}(\alpha_0 + \alpha_1 i + \alpha_2 j + \alpha_3 k);$$

these quaternions then form a ring, called by Conway the *icosians*, whose units are the 120 images of V_4 itself.

The 10 honeycombs are related by the restriction of the method of systematic faceting (see [3]) to the process of vertex figure replacement described in [13]. In this, when the symmetry groups are all generated by reflections in hyperplanes, a vertex figure can be replaced by another with the same vertices, to yield a new regular honeycomb with the same vertices and edges as the original. In particular, the nondiscreteness of one of these 10 honeycombs guarantees the nondiscreteness of them all.

6. Discrete Realizations

In [2], Coxeter allowed the possibility of regular polyhedra or apeirohedra with skew vertex figures, but the restoration of the balance by allowing skew polygons as faces as well had to wait until Grünbaum's paper [9] 40 years later. With the extra freedom the removal of such restrictions permits, we can now, at the cost of doubling the dimension of the space in which we work, find discrete realizations of the regular apeirotopes discussed in Section 5.

The basic idea is very simple. If V is the set of vertices of one of our regular apeirotopes, which, as we have seen, we can always take to be a subset of $\mathbb{T}^k$ for some suitable k, the conjugation $\sim$ produces the set $\tilde{V}$ of vertices of an isomorphic, but nonsimilar, regular apeirotope. These two apeirotopes can then be blended, possibly after scalar multiplication of the terms, to produce a third, of twice the dimension but still isomorphic to the original, which is now discrete.

The discreteness is easy to establish. Each point of V is of the form

$$v = \frac{1}{2}(a + \sqrt{5b}),$$

where $a, b \in \mathbb{Z}^k$ (with the same k as above). Conjugation is then

$$v \mapsto v' = \frac{1}{2}(a - \sqrt{5b}),$$

and the blending (with nonzero scalars λ, μ) is

$$v \mapsto (\lambda v, \mu \tilde{v}) = \frac{1}{2}(\lambda a, \mu a) + \frac{1}{2}\sqrt{5}(\lambda b, -\mu b).$$

It should be clear that the resulting set of points is discrete, but it is even clearer if we apply the further (invertible) linear mapping $(x, y) \mapsto (\mu x + \lambda y, \mu x - \lambda y)$, which takes $(\lambda v, \mu \tilde{v})$ into

$$\lambda\mu(a, \sqrt{5b}) \in \lambda\mu(\mathbb{Z}^k \oplus \sqrt{5}\mathbb{Z}^k).$$

(We shall discuss appropriate choices of λ and μ below.)

We now consider the individual examples. For the blend $\{5/2, 10\}$ # $\{5, 10/3\}$, there is no relative scaling of the vertices (by λ, μ as above) which at once strikes one as natural (but see Section 8).

In $\mathbb{E}^3$, the apeirotope $\{3, \frac{5}{2}\}$ may be abstractly regular, but in its concrete realization it is only quasiregular. However, the blending $v \mapsto (v, \tilde{v}) \in \mathbb{E}^6$ (with v a vertex—and here we *must* choose $\lambda = \mu$ in our relative scaling) admits the additional involutory symmetry

$$(x, y) \to (y\theta, x\theta^{-1}),$$

where θ is the quarter-turn defined in Section 5. The adjunction of this involution (which, with a suitable choice of ρ_0, ρ_1, ρ_2 in Section 5 can be taken to be ρ_3) now gives the blended apeirotope the full symmetry of $\{\{3, 5\}, \{5, 4\}_6\}$.

For the examples in $\mathbb{E}^4$, once again the relative scaling of the components in the blends does not affect the regularity, and so is completely at our disposal. However, there is an additional consideration which may be borne in mind for the three pairs which are conjugate to their duals, because we can now choose the scaling so that the blends of apeirotope and conjugate are geometrically (rather than merely abstractly) self-dual. The appropriately scaled blendings of these pairs are:

$$\tau^2\{5/2, 3, 3, 5\} \ \# \ \tau^{-2}\{5, 3, 3, 5/2\}$$

$$\{3, 5/2, 5, 3\} \ \# \ \{3, 5, 5/2, 3\}$$

$$\tau\{5/2, 5, 5/2, 5\} \ \# \ (-\tau^{-1})\{5, 5/2, 5, 5/2\}$$

(In the last, we scale the second component by $-\tau^{-1}$ rather than by τ^{-1} to accord with conjugation.)

7. Skew Apeirotopes

Coxeter, in his previously mentioned paper [2], described three regular skew apeirohedra in $\mathbb{E}^3$, namely $\{4, 6 \mid 4\}$, $\{6, 4 \mid 4\}$, and $\{6, 3 \mid 3\}$. He showed how each such apeirohedron $\{p, q \mid r\}$ was naturally associated with a regular polyhedron $\{p', q'\}$, with

$$\frac{1}{p} + \frac{1}{p'} = \frac{1}{2}, \qquad \frac{1}{q} + \frac{1}{q'} = \frac{1}{2};$$

moreover, the dihedral angles of $\{p, q \mid r\}$ and $\{p', q'\}$ are supplementary.

There is also a way of obtaining these apeirohedra by applying twisting operations (involutory outer automorphisms) to reflection groups represented by graphs of the form

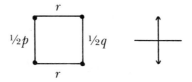

(see [16, 17] for the general theory of such twistings).

It is thus an obvious question to ask whether (nondiscrete) regular skew apeirohedra can be associated, in the same way, with the six regular polyhedra with pentagonal symmetries, and with suitable graphs admitting involutory twists. This is indeed so, because we can also generate the group in $\mathbb{E}^3$ with graph

by different reflections, with corresponding graphs

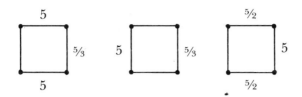

The six resulting regular skew apeirohedra, with their associated regular polyhedra, are as follows:

$\{6, {}^{10}\!/_3 \mid 5\}$	$\{3, 5\}$	$\{5, {}^5\!/_2\}$
$\{{}^{10}\!/_3, 6 \mid 5\}$	$\{5, 3\}$	$\{5, 3\}$
$\{10, {}^{10}\!/_3 \mid 3\}$	$\{{}^5\!/_2, 5\}$	$\{3, {}^5\!/_2\}$
$\{{}^{10}\!/_3, 10 \mid 3\}$	$\{5, {}^5\!/_2\}$	$\{3, 5\}$
$\{6, 10 \mid {}^5\!/_2\}$	$\{3, {}^5\!/_2\}$	$\{{}^5\!/_2, 5\}$
$\{10, 6 \mid {}^5\!/_2\}$	$\{{}^5\!/_2, 3\}$	$\{{}^5\!/_2, 3\}$

In the third column is listed what might be called the *hole polyhedron*, which is the polyhedron $\{r, q/2\}$ (${}^5\!/_3$ and ${}^5\!/_2$ are equivalent marks in this context). Its faces are the holes $\{r\}$ of $\{p, q \mid r\}$, which are the polygonal paths traced out by leaving each vertex by the second edge from that entered by (in a fixed sense); leaving by the first edge gives the faces $\{p\}$. All this is clear when recourse is had to the graphic notation, since the vertices and edges of $\{p, q \mid r\}$ are those of the (uniform, but not regular) honeycomb

while its faces are the *p*-gons

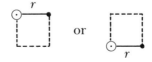

alone; the *r*-gonal faces

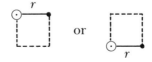

form the holes.

By the way, we should emphasize that our notation $\{p, q \mid r\}$ only indicates the type of the apeirohedron, in that its faces, vertex figures, and holes are polygons $\{p\}$, $\{q\}$, and $\{r\}$, respectively. Thus, while the group $\langle \rho_0, \rho_1, \rho_2 \rangle$ satisfies the relations for $\{p, q \mid r\}$, namely:

$$\rho_j^2 = (\rho_0\rho_1)^p = (\rho_1\rho_2)^q = (\rho_0\rho_2)^2 = (\rho_0\rho_1\rho_2\rho_1)^r = \varepsilon$$

(here, if p is fractional, its numerator should be read instead, and similarly for q and r); other relations may hold as well.

These six regular skew apeirohedra are, indeed, as we should expect, nondiscrete in $\mathbb{E}^3$. But they also occur in conjugate pairs, obtained by interchanging the marks 5 and $\frac{5}{2}$, or 10 and $\frac{10}{3}$:

$$\{6,\ \tfrac{10}{3} \mid 5\} \overset{\delta}{\leftrightarrow} \{\tfrac{10}{3},\ 6 \mid 5\}$$
$$\sim \updownarrow \qquad\qquad \sim \updownarrow$$
$$\{6,\ 10 \mid \tfrac{5}{2}\} \overset{\delta}{\leftrightarrow} \{10,\ 6 \mid \tfrac{5}{2}\}$$
$$\{10,\ \tfrac{10}{3} \mid 3\} \overset{\delta}{\leftrightarrow} \{\tfrac{10}{3},\ 10 \mid 3\}$$

In this diagram, we have indicated dual pairs as well, so that $\{10,\ \frac{10}{3} \mid 3\}$ is conjugate to its dual $\{\frac{10}{3},\ 10 \mid 3\}$. We may thus blend each apeirohedron with its conjugate, to give an isomorphic discrete regular apeirohedron in $\mathbb{E}^6$. In addition, a suitable choice of the scaling will make the last example of the three self-dual.

There are further connections between these six apeirohedra. In the four apeirohedra with decagonal vertex figures ($\{10\}$ or $\{\frac{10}{3}\}$), we can form the polygonal paths by leaving each vertex at the third exit from that entered by (in some fixed sense). In each case, the new polygons happen to be planar (in $\mathbb{E}^3$), and form another regular skew apeirohedron. Denoting this operation by ϕ, the relationships thus obtained are given in the following diagram:

$$\{\tfrac{10}{3},\ 6 \mid 5\} \overset{\delta}{\leftrightarrow} \{6,\ \tfrac{10}{3} \mid 5\}$$
$$\phi \updownarrow$$
$$\{\tfrac{10}{3},\ 10 \mid 3\} \overset{\delta}{\leftrightarrow} \{10,\ \tfrac{10}{3} \mid 3\}$$
$$\phi \updownarrow$$
$$\{6,\ 10 \mid \tfrac{5}{2}\} \overset{\delta}{\leftrightarrow} \{10,\ 6 \mid \tfrac{5}{2}\}$$

(This diagram exactly follows the analogous relationships between the associated regular polyhedra given in [13].)

We are not quite finished with $\mathbb{E}^3$, because the ability to lift into $\mathbb{E}^6$ by blending conjugates admits further possibilities. First, there is one more (self-dial) skew apeirohedron in $\mathbb{E}^3$ which is abstractly regular. It is obtained from one of the graphs already mentioned; it consists of the hexagonal faces

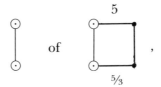

and has hole polyhedra of the two types $\{5, 3\}$ and $\{5/2, 3\}$. Blending it with its conjugate (an identical copy with the marks 5 and $5/3$ interchanged) will yield a regular skew apeirohedron of type $\{6, 6 \mid 5\}$ in $\mathbb{E}^6$, which is (geometrically) self-dual in addition.

We observed above that $\{10, 10/3 \mid 3\}$ and $\{10/3, 10 \mid 3\}$ are dual and conjugate, so that their blend is also self-dual. Now, from a self-dual regular polyhedron or apeirohedron of type $\{p, p\}$ can be obtained a new regular polyhedron or apeirohedron of type $\{p, 4\}$, whose vertices are the mid-points of the edges of the original (this is another example of twisting). The two cases above lead to regular skew apeirohedra of types $\{6, 4\}_{10}$ and $\{10, 4\}_6$. (As before, the suffix denotes the length of the Petrie polygon.)

It is, perhaps, no surprise to find that the faces of one of these apeirohedra are the Petrie polygons of the other, since their common vertices are those of the blend of

with its conjugate (another identical copy). With this *Petrie operation*, which we denote by π, we then obtain a family of six regular skew apeirohedra in $\mathbb{E}^6$, related as in the following diagram:

$$\pi \nearrow \quad \{6, 4\}_{10} \overset{\delta}{\leftrightarrow} \{4, 6\}_{10} \quad \nwarrow \pi$$
$$\{10, 4\}_6 \qquad\qquad\qquad \{10, 6\}_4$$
$$\delta \nwarrow \quad \{4, 10\}_6 \overset{\pi}{\leftrightarrow} \{6, 10\}_4 \quad \nearrow \delta$$

We now consider regular skew apeirotopes derived from the examples in $\mathbb{E}^4$ which we discussed in Section 5. In all cases, what we do is the following. If we have a self-dual regular apeirotope of type $\{p, q, q, p\}$, whose group is, say, $\langle \sigma_0, \ldots, \sigma_4 \rangle$ as in Section 2, then there is an involution, which we call ρ_2, such that $\rho_2 \sigma_j \rho_2 = \sigma_{4-j}$ for each j (so that ρ_2 acts as an outer automorphism of the group). Write $\rho_0 = \sigma_2$, $\rho_1 = \sigma_1$, and $\rho_3 = \sigma_4$. Then $\langle \rho_0, \rho_1, \rho_2, \rho_3 \rangle$ is the group of a regular apeirotope of type $\{q, 4, 4\}$, obtained from the original by *twisting*. In fact, the cells $\{q, 4\}$ are obtained from the central section $\{q, q\}$ by the same twisting just described above, and the vertex figures are tori $\{4, 4\}_{p, 0}$, which are made by identifying opposite sides of a $p \times p$ block of squares.

The three self-dual pairs, with the scalings of the blends with their conjugates, which make them geometrically self-dual in $\mathbb{E}^8$, are listed at the end of Section 5. The resulting apeirotopes and their duals are listed by type:

$$\{\tfrac{5}{2}, 3, 3, 5\} \mapsto \{\{3, 4\}, \{4, 4,\}_{5, 0}\} \overset{\delta}{\leftrightarrow} \{\{4, 4\}_{5, 0}, \{4, 3\}\}$$

$$\{3, \tfrac{5}{2}, 5, 3\} \mapsto \{\{5, 4\}_6, \{4, 4\}_{3, 0}\} \overset{\delta}{\leftrightarrow} \{\{4, 4\}_{3, 0}, \{4, 5\}_6\}$$

$$\{\tfrac{5}{2}, 5, \tfrac{5}{2}, 5\} \mapsto \{\{5, 4\}_6, \{4, 4\}_{5, 0}\} \overset{\delta}{\leftrightarrow} \{\{4, 4\}_{5, 0}, \{4, 5\}_6\}.$$

In this context, it is useful to recall from Section 5 that the quasiregular polyhedron $\{\tfrac{5}{\tfrac{5}{2}}\}$ is the abstract regular polyhedron $\{5, 4\}_6$. Its dual $\{4, 5\}_6$ is represented in $\mathbb{E}^3$ by the small stellated triacontahedron (see [3, 20]).

8. Quasiperiodic Tilings

In this volume, it would be superfluous to spend much time on what constitutes a quasiperiodic tiling. The crucial property of a class $\mathcal{T}$ of quasiperiodic tilings, which we always regard as lying in some fixed euclidean space L (usually a subspace of an $\mathbb{E}^k$) is that its members have a local homogeneity property, which means that if $T, T' \in \mathcal{T}$ and N is a patch of T of diameter δ, then within some fixed multiple of δ from any given point of T' there is a patch of T' which is a translate of N. Local homogeneity is ensured in our tilings by the existence of an *inflation*, which (for our purposes) puts the vertices of a larger copy ρT of a given tiling $T \in \mathcal{T}$ (with $\rho > 1$) among those of some other $T' \in \mathcal{T}$, and its inverse operation *deflation*, which provides a smaller copy $\rho^{-1} T''$ of some $T'' \in \mathcal{T}$ whose vertices include those of T. Since our ρ $(=\tau)$ is irrational, this also ensures that the tilings in $\mathcal{T}$ are *nonperiodic*, in that their symmetry groups contain no translations.

A popular technique for constructing quasiperiodic tilings is the projection method. Rather than use the original method (see, for example, [8, 10, 12, 19] as samples of the huge literature on the topic; these are probably most relevant to the present discussion), we shall employ the generalization of it introduced in [11]. (We are grateful to the referee for pointing out to us the priority of [11]. The general validity of the new method was not established in [11]; we shall do this in [15] in an even more general context.)

Let Λ be a (discrete) lattice in some Euclidean space $\mathbb{E}^n$ (or even, more generally, a set of points on which its symmetry group is transitive, but we shall not consider this case here). The original projection method has $\Lambda = \mathbb{Z}^n$, the integer lattice. The set of points of $\mathbb{E}^n$ no further from some fixed point of Λ, say the origin o, than from any other point of Λ is called a *Voronoĭ region* R of Λ; thus

$$R = \{x \in \mathbb{E}^n \mid \|x\| \leqslant \|x-q\| \text{ for all } q \in \Lambda \setminus \{o\}\}.$$

The translates of R by Λ form a face-to-face tiling $\mathfrak{U}^*$ of $\mathbb{E}^n$. Dual to $\mathfrak{U}^*$ is a tiling $\mathfrak{U}$ of $\mathbb{E}^n$; a j-cell C of $\mathfrak{U}$ is the convex hull of the centers $q \in \Lambda$ of the Voronoĭ regions $R + q$ which contain a given $(n-j)$-cell C^* of $\mathfrak{U}^*$.

Now let L be a d-dimensional subspace of $\mathbb{E}^n$. We call a translate $M = L+t$ $(t \in \mathbb{E}^n)$ of L *general* if M does not meet any k-cell of $\mathfrak{U}^*$ with $k < n-d$. By considering the projections of the $(n-d-1)$-cells of $\mathfrak{U}^*$ along L on to the orthogonal complement $L^{\perp}$ of L, we see that (in an obvious measure-theoretical sense) most translates of L are general. Then such a general M yields a tiling $T = T(M)$ of L in the following way; the d-cells of T are the images under orthogonal projection on to L of the d-cells C of $\mathfrak{U}$ for which $M \cap C^* \neq \varnothing$. In fact, the tiling T is dual, in a natural way, to the tiling induced in M by the intersections $M \cap C^*$ of cells C^* of $\mathfrak{U}^*$ with M.

If the tiling T has a translation among its symmetries, then this translation is necessarily by a lattice point of Λ in L. Thus, if L contains no points of Λ other than o, then T can have no translational symmetries, so that T is (as above) nonperiodic. However, the possibility that T has nontrivial rotational or reflexional symmetries is not excluded.

The quasiperiodic tilings which we consider here are closely related to the regular and quasiregular apeirotopes which we have discussed in Section 5. In each case, we have a nondiscrete lattice Z of vertices of a regular or quasiregular honeycomb in $\mathbb{E}^d$ ($d = 2$, 3, or 4). Blending with the set $\mu \tilde{Z}$ of vertices of the corresponding conjugate (isomorphic) honeycomb, scaled by an appropriate μ, results in a discrete lattice Λ in $\mathbb{E}^{2d}$, to which we apply the technique we have just described. Different choices of μ will yield different families $\mathfrak{T}$ of tilings in $\mathbb{E}^d$, whose vertices all occur in Z.

Before we go on to consider specific examples, we should return to inflation and deflation. For $d = 2$, the vertex figure of the honeycomb which is the starting point is a decagon $\{\pm a_0, \ldots, \pm a_4\}$; these points occur among the vertices of the vertex figure of the honeycomb in $\mathbb{E}^3$, and, in turn, these 30 vertices form the central section of the vertex figure of the honeycomb in $\mathbb{E}^4$. Now we note that $-(a_2 + a_3) = \tau a_0$, and hence easily deduce that $\tau z \in Z$ whenever $z \in Z$, for any of our vertex-sets Z. Under conjugation, to τz corresponds $-\tau^{-1}\tilde{z} \in \tilde{Z}$. Thus each of our blended lattices Λ admits an invertible linear mapping, induced by the mapping $(x, y) \mapsto (\tau x, -\tau^{-1}y)$ on $\mathbb{E}^{2d} = \mathbb{E}^d \times \mathbb{E}^d$. Since this mapping is a contraction in the second component, it is not hard to see that (for a general p) the vertices of $\tau T(L - \tau^{-1}p)$ occur among those of $T(L + p)$, and this provides our inflation.

For simplicity, we shall concentrate here on the case $d = 2$; the higher dimensional cases are more complicated, and space permits them only a cursory mention. But again, the following remarks apply to all the examples.

Some choices of the scaling μ mentioned above may seem more appropriate than others. Note first that changing the sign of μ gives a congruent lattice, and so can be ignored if required. In addition, if we are only interested in the shapes of our lattices Λ, then we can apply a further scaling (to both components together). Bearing in mind the linear mapping $(x, y) \mapsto (\tau x, -\tau^{-1}y)$, which preserves the lattices, we then see that the two scalings by μ and $\tau^{-2}\mu$ yield similar lattices. Thus, if we allow μ to vary over the range $1 \leq \mu < \tau^2$, we shall obtain all possible similarity classes of lattices Λ.

In fact, there are two choices for μ which stand out. One natural choice is $\mu = 1$ (and for regularity of the resulting honeycomb when $d = 3$ this is the only choice). In all cases, the set V of original vertices adjacent to o lies at distance 2 from it; in the blended set, this results in v ($= 10, 30$, or 120, as $d = 2, 3$, or 4, namely the number of points in V) points adjacent to o at distance $2\sqrt{2}$, and (at least) v more at distance $2\sqrt{3}$ (that is, those of the form $(\tau z, -\tau^{-1}z)$, with $z \in V$).

Another natural choice is $\mu = \tau$. Now Λ has $2v$ points at the same minimal (nonzero) distance from o, namely $(z, \tau\bar{z})$ and $(\tau z, -\bar{z})$, with $z \in V$. Note that, in this case, Λ may acquire additional symmetries from those implied by the group of the original honeycomb.

As we have said, we shall concentrate on the case $d = 2$; the method for the remaining cases is very similar, but necessarily more complicated. The main problem is usually to choose a new coordinate system in $\mathbb{E}^{2d}$, in which the lattice Λ, and hence its Voronoĭ region R, can be explicitly and conveniently computed; in fact, for $d = 2$, it turns out to be better still to work in a suitable hyperplane in $\mathbb{E}^5$.

The actual details of calculating the Voronoĭ region R, and the conditions for intersections of the affine subspace M with the faces of its translates $R + t$ ($t \in \Lambda$) are intricate and somewhat tedious (even in the simplest cases $\mu = 1$ or τ), and so we shall omit them. Useful general references for the lattices which arise are [1, 7].

The vertex-set of one of our quasiperiodic tilings in $\mathbb{E}^2$ is going to lie in

$$\mathbb{Z}V = \left\{ \sum_{j=0}^{4} v_j a_j \mid n = (v_0, \ldots, v_4) \in \mathbb{Z}^5 \right\},$$

where $\{a_0, \ldots, a_4\}$ is the set of vertices of a regular pentagon centered at o. In view of the identity $\sum_{j=0}^{4} a_j = o$, a given point in $\mathbb{Z}V$ has many different representations of the form $\sum_{j=0}^{4} v_j a_j$, but, if we wish, for definiteness, we can always suppose that

$$\sum_{j=0}^{4} v_j \in \{-2, -1, 0, 1, 2\}.$$

First, let $\mu = 1$. The lattice Λ then consists of the vertices of the fundamental regions of the group whose Coxeter diagram this is; the Voronoï region R is thus a single cell

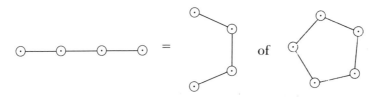

The tilings in T contain four kinds of triangle, with side lengths τ^{-1}, 1, and τ (if we now take $\| a_j \| = 1$), typical examples of which are

$$\text{LA} = \text{conv}\{o, -\tau a_1, \tau a_4\},$$

$$\text{LO} = \text{conv}\{o, -a_2, a_3\},$$

$$\text{SA} = \text{conv}\{o, -a_1, a_4\},$$

$$\text{SO} = \text{conv}\{o, -\tau^{-1}a_2, \tau^{-1}a_3\},$$

where we should recall here relations like $\tau a_j = -a_{j-2} - a_{j+2}$ and $\tau^{-1}a_j = a_{j-1} + a_{j+1}$, with suffixes as before modulo 5. In this notation, L and S stand for *large* and *small*, while A and O stand for *acute* and *obtuse*, all referring to isosceles triangles. The general triangle in a specific tiling $T(L + p)$ is obtained by rotating one of the above by a multiple of $\pi/5$ and translating by an element of $\mathbb{Z}V$ corresponding to a member $n \in \mathbb{Z}^5$; which actually occur in $T(L + p)$ are determined by linear inequalities (all strict) involving n and p. However, for reasons of space, we omit the specific details.

The other choice $\mu = \tau$ yields a lattice Λ whose vertices are those of the tiling

This time, the tilings in T contain only copies of the two triangles LA and LO above, which have edge lengths 1 and τ. Which copies occur in any

given tiling T in $\mathcal{T}$ are again determined by strict linear inequalities in the parameter n (for fixed determining vector p). As before, we do not give the details.

In Figure 1 we give examples of these two kinds of tilings. We have chosen the parameter vectors p for the subspaces $M = L + p$ to correspond, in such a way that every copy of the triangle LA which occurs in the first tiling (which has triangles of all four types) also occurs in the second.

Figure 1. Two tilings derived from the honeycombs $\{5/2, 10\}$ and $\{5, 10/3\}$.

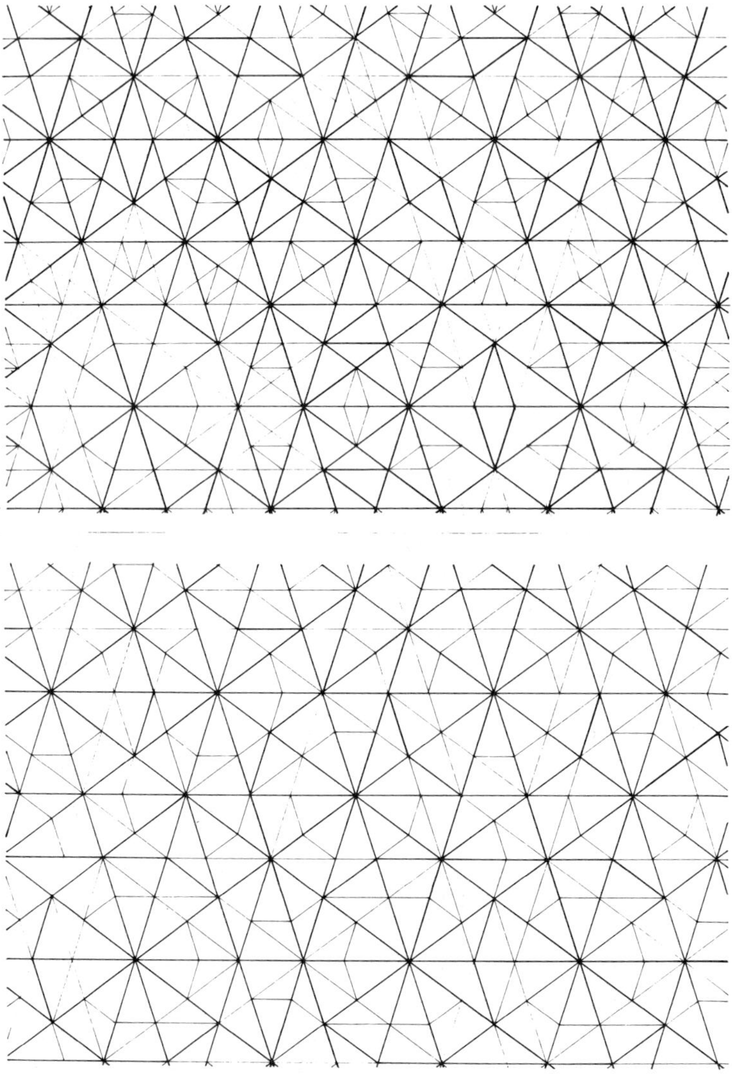

The same kind of pattern is repeated in case $d = 3$ or 4. Of the two best choices $\mu = 1$ and τ, the second yields tetrahedra or 4-simplices with only two edge lengths 1 and τ in the tilings, while the first permits the extra edge length τ^{-1}. (This is, perhaps, a little unfortunate if $d = 3$, since it is the first choice of μ which gives the regular apeirotope, as we remarked above.) Our remaining remarks will concern the case $\mu = \tau$, when fewer types of tile will be involved.

For $d = 3$, the tilings contain six different congruence classes of tetrahedral tiles with edge lengths 1 and τ (and at least one of each). All possible kinds occur, except for the tetragonal disphenoids (see [3], p. 15), which have opposite edges of one length, and the remaining four of the other. It may be interesting to compare these tilings, in general terms, with those recently obtained by Danzer [5]. The latter are composed of four kinds of tetrahedra, whose dihedral angles are multiples of $\pi/2$, $\pi/3$, and $\pi/5$. In contrast, the triangular faces of our tetrahedra have angles $\pi/3$, $\pi/5$, $2\pi/5$, and $3\pi/5$, whereas the dihedral angles are never rational multiples of π. (In informal discussions, Danzer has suggested the possibility of some kind of duality between the two kinds of tilings; if such a duality exists, it is not direct and obvious. A genuine duality of related tilings is discussed in [15].)

In case $d = 4$, we have yet to enumerate the kinds of simplices which can occur in the tilings, even in the simplest case $\mu = \tau$, when again we have just the two edge lengths 1 and τ. This case is very interesting from a number of points of view. The lattice Λ in $\mathbb{E}^8$ consists of the vertices of the semiregular honeycomb

$$5_{21} = $$

composed of regular simplices and cross-polytopes (see [2]). Λ is also the set of centers of the closest lattice packing of $\mathbb{E}^8$ by balls. The $240 = 120 + 120$ points of Λ nearest to o form the vertices of

$$4_{21} = $$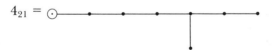

The projections of 4_{21} on to $\mathbb{E}^4$, whose vertices fall on to two homothetic copies of the vertices of $\{3, 3, 5\}$, have recently been discussed in [14].

In this connection, we should also mention [5]. In this paper, the authors consider a four-dimensional quasicrystal, regarded as a point set, which they also derive by projection from 5_{21}. However, we should be careful to observe that there they use the *closed* Voronoĭ region, so that their set of vertices is larger than will occur in any of our tilings; in particular, the minimal distance between points is actually τ^{-1}. Their choice of the linear subspace is inadmissible for our tilings, none of which can be symmetric under the group [3, 3, 5] of the vertex figure of the underlying nondiscrete regular honeycomb. (The same is true of the analogous tilings in $\mathbb{E}^d$ for $d = 2$ and 3; the pictures illustrate the closest approximation to full 10-fold symmetry that can be attained in the plane.)

The same paper prompts our last remark. Danzer has observed (private communication) that the actual tiles in a quasiperiodic tiling may not have much relevance to a corresponding quasicrystal; of most importance are the vertices and edges of the tiles. We have constructed quasiperiodic tilings with just two edge lengths, and with inflation factor τ, in $\mathbb{E}^d$ for $d = 2$, 3, and 4 (and even, as we hinted in Section 5, for $d = 6$ also). Is it possible that, when $d = 3$, our tilings might be relevant to genuine physical quasicrystals?

REFERENCES

1. H. Brown, H. Bulöw, J. Neubüser, H. Wondratschek, and H. Zassenhaus, *Crystallographic Groups of Four-Dimensional Space*. Wiley, New York, 1978.

2. H. S. M. Coxeter, Regular skew polyhedra in three and four dimensions, and their topological analogues. *Proc. London Math. Soc.* (2) **43**, 33–62 (1937). (Reprinted with corrections in *Twelve Geometric Essays*, Southern Illinois University Press, 1968.)

3. H. S. M. Coxeter, *Regular Polytopes*, 3rd ed. Dover, New York, 1973.

4. H. S. M. Coxeter, *Regular Complex Polytopes*. Cambridge University Press, Cambridge, 1974.

5. L. Danzer, "Three dimensional analogs of the planar Penrose tilings and quasicrystals," *Discrete Math.*, to appear.

6. V. Elser and N. J. A. Sloane, "A highly symmetric four-dimensional quasicrystal," *J. Phys. A: Math. Gen.*, **20**, 6161–6168 (1987).

7. P. Engel, *Geometric Crystallography*. Reidel, Dordrecht, 1986.

8. F. Gähler, "Some mathematical problems arising in the study of quasicrystals," *J. Phys.*, **47**, Colloque C3, 115–123 (1986).

9. B. Grünbaum, "Regular polyhedra—old and new," *Aequationes Math.*, **16**, 1–20 (1977).

10. A. Katz and M. Duneau, "Quasiperiodic patterns and icosahedral symmetry," *J. Phys.*, **47**, 181–196 (1986).

11. V. E. Korepin, F. Gähler, and J. Rhyner, "Quasiperiodic tilings: A generalized grid-projection method," *Acta Crystallogr.*, **A44**, 667–672 (1988).

12. P. Kramer and R. Neri, "On periodic and non-periodic space fillings of $\mathbb{E}^m$ obtained by projection," *Acta Crystallogr.*, **A40**, 580–587 (1984).

13. P. McMullen, "Regular star-polytopes, and a theorem of Hess," *Proc. London Math. Soc.* (3), **18**, 577–596 (1968).

14. P. McMullen, "Realizations of regular polytopes," *Aequationes Math.*, **37**, 38–56 (1989).

15. P. McMullen, "Duality, sections and projections of certain euclidean tilings." In preparation.

16. P. McMullen and E. Schulte, "Regular polytopes from twisted Coxeter groups," *Math. Z.*, **201**, 209–226 (1989).

17. P. McMullen and E. Schulte, "Regular polytopes from twisted Coxeter groups and unitary reflexion groups," *Adv. Math.*, **82**, 35–87 (1990).

18. B. R. Monson, "A family of uniform polytopes with symmetric shadows," *Geom. Ded.*, **23**, 355–363 (1987).

19. E. J. W. Whittaker and R. M. Whittaker, "Some generalized Penrose patterns from projections of n-dimensional lattices," *Acta Crystallogr.*, **A44**, 105–112 (1988).

20. J. M. Wills, "The combinatorially regular polyhedra of index 2," *Aequationes Math.*, **34**, 206–220 (1987).

11 A Two-Dimensional Quasiperiodic Dodecagonal Tiling by Two Pentagons

Hans-Ude Nissen

Beauty consists of a certain consonance of different elements.
—THOMAS AQUINAS

1. Introduction

This paper presents a number of new quasiperiodic tilings of the plane including one by only two tiles, two different nonregular pentagons. This tiling, as well as a number of other tilings that can be deduced from it, may prove useful for the crystallographic description of quasicrystal structures with 12-fold symmetric Fourier transform. In addition, these tilings also have potentials of application in graphic art, painting, and design, as well as in the analysis of old (islamic) and new planar ornaments.

The finding of quasicrystals as a third new state of atomic order of solid materials, intermediate between the crystalline state and the amorphous state, [1, 2], has caused worldwide activity by theoreticians to construct quasilattices (which in quasicrystal structures take the role of the Bravais lattices in crystals) as a means of describing especially the translational order of quasicrystal structures. Soon after the first description of two-dimensional dodecagonal quasicrystal structures periodic along one direction, occurring as one of several phases in rapidly cooled small particles of Ni–Cr alloys [3–5], new dodecagonal quasilattices as tilings of the plane were found and described [6–10]. These constructions used two different mathematical techniques, i.e., the grid technique of de Bruijn [11] and the projection method, which was recently discussed for dodecagonal tilings [9]. It was also shown [6, 11] that both techniques render identical results.

The tiling that has proved most adequate for the crystallographic description and the modeling of the real Ni–Cr quasicrystals was found by Gähler [9] and independently by Niizeki and Mitani [8]. It is constructed by the well-known projection method and consists of an aperiodic arrangement of three tiles, i.e., a square, a regular triangle, and a rhombus with a 30° acute angle. It is termed *ship tiling* in this paper and is shown in Figure 1a.

By defining a different acceptance region in applying the projection procedure, both papers cited above also presented a related tiling, termed *shield tiling* for convenience, which contains equilateral nonregular hexagons ("shields") replacing groups of tiles consisting of one 30° rhomb and two equilateral triangles. This tiling is shown in Figure 1b on a scale equivalent to that of Figure 1a. The shield tiling can be regarded as a tiling with holes, in which the shield tiles are regarded as the holes. In this form, the tiling consists of five tiles (supertiles) with zigzag outlines resembling islamic tilings, and the number of corners (touching points with other tiles) for these types of tiles is 4, 6, 8, 8, or 12, respectively. The vertex configurations are identical at all the corners up to rotation; there is an association of two half rhombs and three surrounding triangles forming an irregular heptagon, which can be used as a marker tile in building up the pattern.

Both the ship tiling and especially the shield tiling will be the basis for the construction as well as the description of different tilings, all of which are related to the former two tilings: they can be superimposed on either of them in such a way that equivalent regions are mapped onto each other. It should be pointed out that the Ni–Cr quasicrystals, judging from the presently available electron microscopic evidence, are "random tilings" [12] lacking strict quasiperiodicity as well as inflationary properties. In spite of these differences, the two tilings mentioned above have been successfully used as quasilattices of model structures needed to perform dynamic calculations of the electron diffraction intensities as well as the high-resolution electron microscopic contrast pattern normal to the 12-fold axis [13].

The purpose of this study originally was to provide new tilings that may be used to find an aperiodic set of tiles forcing aperiodicity if the tiles are added in accordance with specific "matching rules" [14] such as exist for the decagonal-symmetric Penrose patterns [15] as well as for the octagonal square-rhombus tiling [16, 17] and have also recently been proposed for the butterfly tiling (see below) [18]. In the course of the present work it appeared desirable, for reasons of application to crystallography and to art, to consistently describe the geometry of these patterns and to show the geometric relationship between them.

2. Construction and Description of the Tilings

To create new dodecagonal quasiperiodic tilings from a given one, we connect certain equivalent points within the tiles or certain selected equivalent vertices. A first step, which may serve as an example of

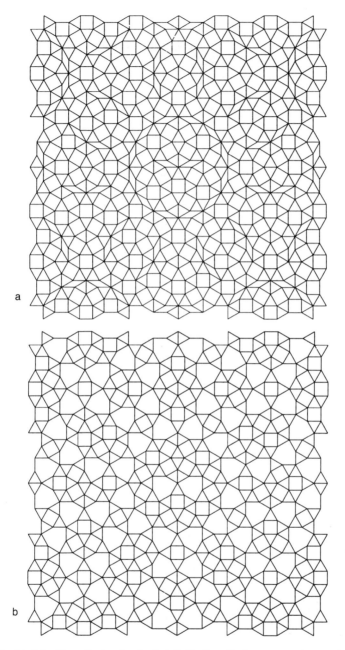

Figure 1. (a) Ship tiling. Notice the single infinite line of mirror symmetry passing through the center of the figure. (b) Shield tiling.

this procedure, is presented in Figure 2a. It shows, in bold lines, a network of connecting lines between the centers of gravity (called center points for convenience) in all regular triangles of the shield tiling shown in dotted lines in the figure. By omitting all connections between the center points of adjacent triangles forming a 60° rhombus, a network of connection lines with constant length is obtained. The resulting tiling, termed *butterfly tiling*, consists of regular hexagons, 30° rhombs and squares, all of equal side lengths. It has been obtained by both techniques: the de Bruijn grid method [18, 19], where the dual of the superposition of two periodic nets of regular triangles is constructed, and the projection method, where a cube honeycomb in six dimensions is projected into two dimensions [18, 19].

This tiling consists of regions having the shape of a regular dodecagon, with an interior mirror symmetric arrangement of two hexagons, three squares, and six 30° rhombs, suggesting the shape of a butterfly. Five of the six rhombs are shared with adjacent dodecagonal regions or with a "void" between them consisting of one hexagon and three squares. Other voids between the dodecagons are formed by a single square. Figure 2a shows the tile-to-tile relation between the butterfly tiling and the shield tiling. The butterfly tiling, without the underlying lines of the shield tiling, is presented in Figure 2b. Compared to the dodecagons of the underlying shield tiling, the radius of the dodecagonal areas in the butterfly tiling is larger, so that there is an overlap of dodecagons in the butterfly tiling: five 30° rhombs are shared by adjacent dodecagonal regions. By contrast, the corresponding adjacent dodecagons in the shield tiling as well as in the ship tiling (Fig. 1a) share only one of their sides.

The next step in constructing new dodecagonal tilings from the shield tiling is to connect the center points of all squares in the shield tiling. To obtain a pattern as simple as the butterfly tiling in which all edges have the same length, it proves useful to interpret the shields as a partial superposition of three squares having the same edge length as the isolated squares in the shield tiling. In this interpretation, each of the three squares touches the outline of the shield in two edges including a right angle, and from the position of these edges the position of the center points of the three overlapping squares can be constructed.

The tiling resulting when the center points of all isolated squares as well as the three "square center points" within each shield are joined by straight lines is shown in Figure 3a superimposed on the shield tiling. The same tiling, drawn without the underlying shield tiling, is shown in Figure 3b. This tiling has been named *three-star tiling* because of the occurrence of a tile having the shape of a star with three rays, i.e., a nonconvex hexagon. As seen in Figure 3b, the square center points are connected in such a way that only one edge length results in the tiling, as is also the case in the ship tiling and the butterfly tiling. Frequently four three-stars have one vertex, a tip of a ray, in common so that they form a shape known as the Maltesian Cross, the coat of arms of the Maltesian Knights. The three-star tiling may

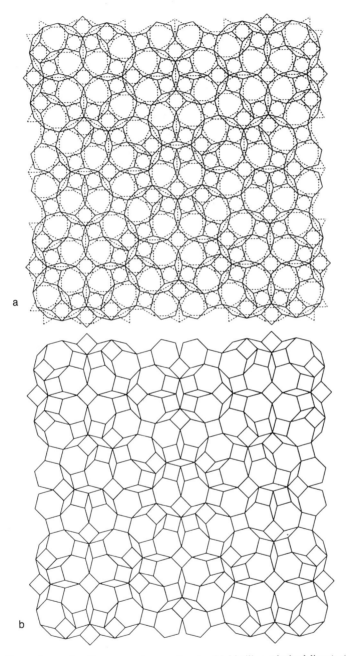

Figure 2. (a) Butterfly tiling, superimposed onto shield tiling (dashed lines). (b) Butterfly tiling.

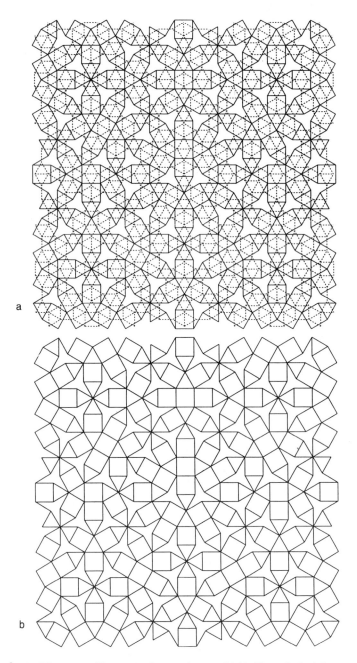

Figure 3. (a) Three-star tiling, superimposed onto shield tiling (dashed lines). (b) Three-star tiling.

186

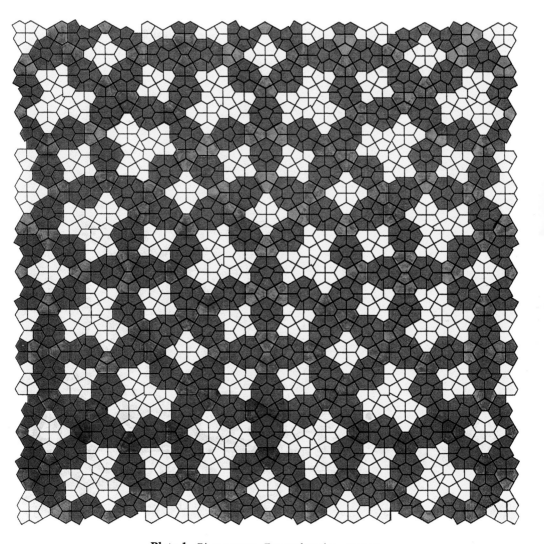

Plate 1. Ring pattern. For explanation, see text.

be useful for modeling dodecagonal quasicrystal structures in which parallel arrangements of two and three squares occur. Comparing it with octagonal quasilattices, the octagonal tiling composed of squares and 45° rhombs [16–18] also contains groups of two parallel squares. Another aspect of the three-star tiling, i.e., its relation to tilings by pentagons, will be discussed below.

A third, new dodecagonal tiling, which can be constructed by a procedure very similar to that used for generating the butterfly pattern and the three-star tiling, is the *wheel tiling*, shown in Figure 4a and b. It is composed of only two tiles and has been termed wheel tiling because of a wheel-like arrangement of one of the two tiles, as can be seen, for example, in the center and near the four corners of Figure 4a and b, respectively. The wheel tiling is essentially a connection of the center points of all triangles and all squares in the shield tiling by straight lines. Applying again the interpretation of the shields as being composed of three partly overlapping squares, each shield contributes three center points of these assumed incomplete squares as vertices in the wheel tiling. These groups of three vertices are identical with the three nonconvex vertices in each three-star of the three-star tiling, as a transparent overlay of Figure 3b on Figure 4a shows.

An inspection of Figure 4a and 4b shows that the more frequent tile in the wheel tiling is a unit of two nonregular pentagons, each having four edges of unit length with the fifth edge shorter by a factor of $(\sqrt{3} + 1)/2$. This pentagon is called a σ-pentagon, because it occurs in the dual of one of the semiregular Archimedean tilings by squares and equilateral triangles (cf. Stampfli, Chapter 12 of this volume) which is used for the description of the crystalline σ-phase [20, 21]. The occurrence of the σ-pentagon in the shield tiling results from the fact that this tiling contains domains of the Archimedean $\{3^2\,4\,3\,4\}$ tiling separated by pairs of shields or interrupted by isolated triangles. To obtain a tiling with only one edge length, as desired, the lines separating pairs of σ-pentagons have been omitted in the wheel tiling. The resulting tile has a dumbbell shape.

To construct a tiling consisting only of σ-pentagons and of any remaining gaps between these, we decorate each vertex of the shield tiling by one σ-pentagon. There are 12 possible orientations for any σ-pentagon, and to discuss the orientation chosen for building up the new pattern, we have to refer to the projection method. The pertinent details of this technique have recently been discussed by Gähler [10]. Each vertex of the shield tiling is the projection of a lattice point of a four-dimentional lattice. The projected positions onto the space orthogonal to the tiling space of those lattice points, which give rise to a vertex of the tiling, fall into a dodecagon-shaped acceptance region. This acceptance region is divided into 12 sectors; each sector corresponds to one of 12 possible orientations of any σ-pentagon, so that the sector into which a particular lattice point falls determines the orientation of that σ-pentagon.

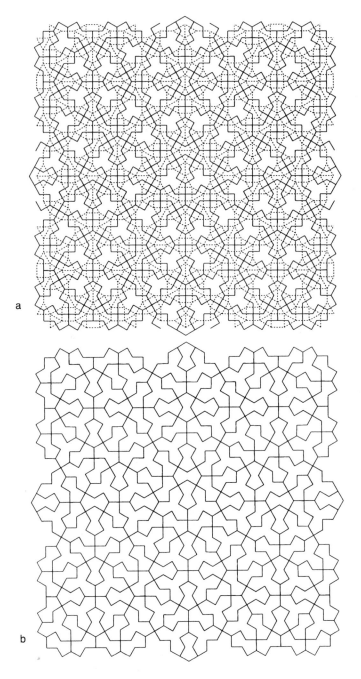

Figure 4. (a) Wheel tiling, superimposed onto shield tiling (dashed lines). (b) Wheel tiling.

If this procedure is carried out for the vertices of a shield tile, a certain amount of space remains uncovered by σ-pentagons and can be defined as the second tile of the new tiling. It also has a pentagonal shape like the first tile, with one nonconvex vertex, suggesting the shape of a rubber fin used by swimmers. Different from the σ-pentagon, which has a line of mirror symmetry, the nonconvex pentagon has no symmetry element. Since the tiling thus consists of two pentagon-shaped tiles only, it is called the *two-pentagon tiling*. It is shown in Figure 5a together with the underlying shield tiling from which it is derived, and in Figure 5b without the shield tiling. Notice that for each shield tile there is one nonconvex pentagon tile in the two-pentagon tiling. The nonconvex pentagon appears in two variants, which are enantiomorphs of each other, and, apparently, occur in equal amounts in the tiling.

Two other procedures that also determine uniquely the orientation of all σ-pentagons and thus also the shape and orientation of the nonconvex pentagons have been applied. In one of them, the ship tiling (Fig. 1a), is converted into a tiling of nonoverlapping dodecagons. Each σ-pentagon is then chosen so that its orientation with regard to the two 30° rhombs in each dodecagon is constant. The second procedure is similar and makes use of the butterfly tiling (Fig. 2b). In this tiling the five 30° rhombs in each complete dodecagonal region occurring in the tiling are cut along their long diagonals, and the cutting lines are prolonged until they meet another cutting line through the adjacent 30° rhomb. These lines are then the outlines of the σ-pentagons in the two pentagon tiling of corresponding size.

Like all tilings presented in Figures 1–4, the two-pentagon tiling has a single infinite line of mirror symmetry (starting from the origin), while all other symmetry lines (which are the only symmetry elements in that tiling) are local. The tiling is quasiperiodic and can be inflated and deflated. (For definition of these terms, see [15].)

The two-pentagon tiling has been used as the basis of another dodecagonal tiling. It was obtained by merging the pentagons into groups forming supertiles. The resulting tiling is called *ring tiling*. In its construction, shown in Figure 6a–d, a drawing of the shield tiling, inflated with respect to the scale in Figure 3b, is superimposed onto the two-pentagon tiling, as shown in Figure 6a. We first consider the surroundings of these σ-pentagons situated on a vertex of the inflated shield tiling. It is evident from Figure 6a, that the seven pentagonal tiles that touch these special σ-pentagons always form the same configuration and have the same orientation relative to each other for all vertices of the inflated shield tiling. They can therefore be merged into one type of the supertiles that form the new tiling (see Fig. 6b). The next step is to eliminate the lines of the superimposed shield tiling inside the new house-shaped tiles. Outside the "houses" these lines are then used to delimit three other supertiles composed of 3, 8, and 22 pentagons, respectively. They are called hexagon, bird, and tribird, respectively, and are

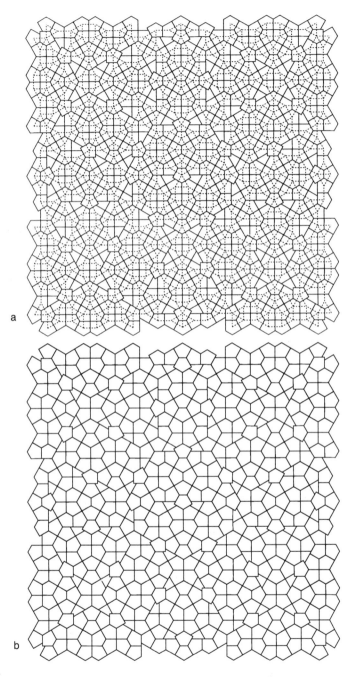

Figure 5. (a) Two-pentagon tiling, superimposed onto shield tiling (dashed lines). (b) Two-pentagon tiling.

groups (identical for each type of tile) of pentagons around the center points of triangles, squares, and shields, respectively, in the superimposed shield tiling (Fig. 6c). If we additionally omit the separation line between two of the hexagon tiles wherever they are in contact along one of their edges, we obtain an additional fifth tile called hourglass because of its shape. This means that all edges in the shield tiling are omitted that divide its 60° rhombs into two triangles. The advantage of introducing the hourglass type of tiles is that in the tiling with these five types of tiles, shown by the thick lines in Figure 6c, no tile belonging to one of the five types is in contact along an edge with any tile of the same type. The ring pattern, together with the underlying two-pentagon tiling, is shown in Figures 6c and Plate 1, and, without the lines of the two-pentagon tiling, in Figure 6d. The latter figure shows a larger region of the tiling in a smaller scale, so that a pattern of interlocking rings can be seen. The size of all five tiles in the ring tiling is intermediate between the size of the two types of pentagons in the corresponding two-pentagon tiling and their size in the inflated two-pentagon tiling.

A remarkable property of the ring tiling is that the shield tiling, the butterfly tiling, and the three-star tiling can be recovered from it as will now be explained. By construction, the centers of the house tiles form an (inflated) shield tiling. If we connect, in an analogous way, the center points of all hexagons and bi-hexagons (blue color in Plate 1) we obtain an inflated butterfly tiling. Similarly, the center points of the bird tiles and the center points of the three birds inside the tribird tiles can be connected to form an inflated three-star tiling (see Plate 1, and compare Fig 3b). Thus the center points of the tiles in the ring pattern (Figs. 6a and Plate 1) can be divided into three subsets, each of which constitutes a dodecagonal tiling in itself. Moreover, these three tilings are put into a fixed orientation and scale relation to each other and to the ring tiling itself by the fact that their relative position to each other is fixed by the geometry of the ring tiling. Viewing the colored tiling (Plate 1) from a distance, all three tilings can be recognized simultaneously by regarding each type of tiles as a set of patches of a specific color forming the vertices of a tiling.

In comparing the ring tiling, Figure 6a, with the two-pentagon tiling, it is noticed that not all lines of the two-pentagon tiling are used as outlines of the tiles in the ring tiling. The remaining lines form a complicated network seen in Plate 1 as black lines superimposed on the colored ring tiling. After omission of certain lines in this network, a further dodecagonal tiling, the *four-star tiling*, results which is shown in Fig. 7a and 7b. This tiling has only one edge length and three types of tiles, a 60° rhomb, a star-shaped octagon with four arms always surrounded by four 60° rhombs and a tile having the shape of a flying seabird (a tern or a slender seagull) seen from above. The four-stars and the tern-shaped tiles are connected by part of their vertices so that they form a continuous network. In this construction

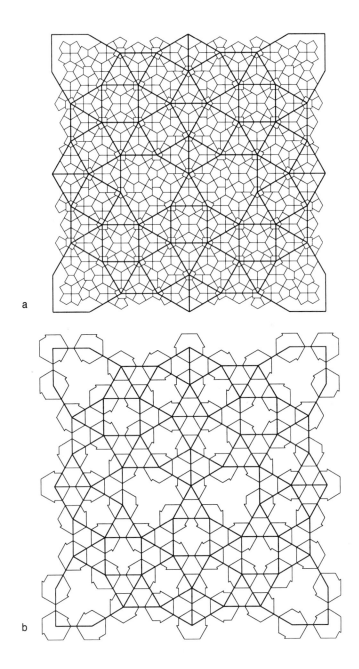

a

b

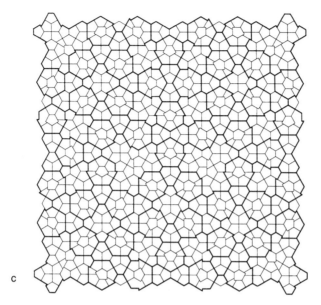

c

Figure 6. (a) Shield tiling (thick lines), inflated relative to Figure 1b, and superimposed onto two-pentagon tiling (thin lines) drawn in the same scale as Figure 5b. (b) Shield tiling (thick lines) with house-shaped regions from the two-pentagon tiling from 6a indicated in thin lines. (c) Ring tiles outlined by thick lines on a two-pentagon tiling (thin lines) in the same size as Figures 5b and 6a. (d) Ring tiling presented in a size smaller than 6c so as to show a wider region of the tiling.

of the four-star tiling, three types of lines are omitted from the network of black lines superimposed on the colored ring tiling in Plate 1:

1. The σ-pentagons in the center of the "house"-shaped tiles (marking the positions of the vertices of the ship tiling as discussed above); the lines pointing toward the center of this pentagon have to be continued up to this center for constructing the four-star tiling.

2. The lines inside the four-star tiles.

3. Certain lines inside the tern-shaped tiles. If these latter lines are left in the tiling, the tern-shaped tile loses its line of mirror symmetry and forms two different enantiomorphic tiles.

The four-star tiling can also be constructed in another way: We use as basis a version of the ship tiling (Fig. 1a) in which two of the edges bounding the 30° rhombs are omitted; the rhombs are then joined with the two adjacent equilateral triangles into a larger tile, a nonconvex hexagon. This verion of the ship tiling is shown in Fig. 7a as dotted lines. If we now replace each edge of the tiling in dotted lines by a 60° rhombus having its long axis parallel to that edge we obtain the *four-star tiling*, shown in bold lines in Fig. 7a.

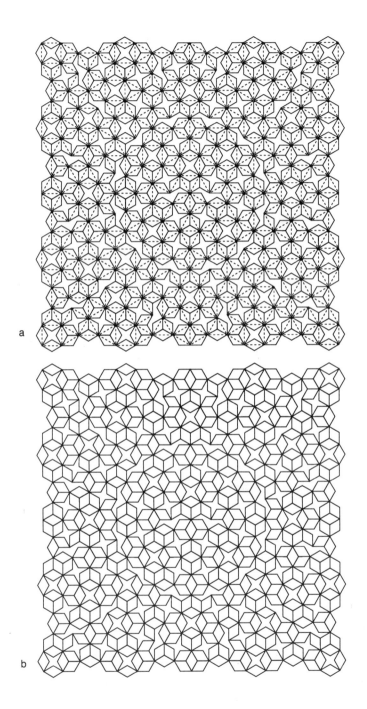

a

b

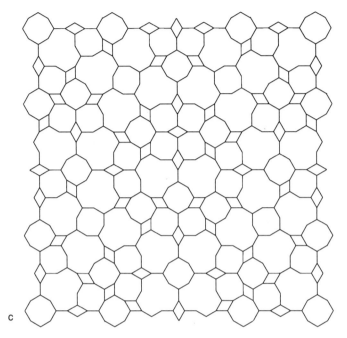

c

Figure 7. (a) Four-star tiling, superimposed onto shield tiling (dashed lines). (b) Four-star tiling. (c) Plate tiling.

By omitting all edges bounding the four-stars and the tern-shaped tiles, the resulting subset of edges forms the *plate tiling*, presented in Fig. 7c. This name was chosen because it can conveniently be built of stone plates or ceramic tiles. It also consists of three tiles, a 60° rhomb (i.e., a tetragon), a nonregular equilateral octagon, and a nonregular nonconvex, equilateral dodecagon, forming a three-connected network of edges of constant length. The motif of a butterfly, similar to that of the butterfly tiling (Fig. 2b), also occurs in this tiling. If the dual of the plate tiling is constructed without considering the 60° rhomb tiles, it is found to be identical to the three-star tiling (Fig. 3b). By topological deformation of the irregular dodecagons into regular ones and introduction of 30° rhombs inside the octagons, the plate pattern can be converted into a different tiling with a single edge length consisting of four different types of tiles.

The shield tiling allows construction of not only a dodecagonal tiling by two pentagon-shaped tiles, but also of a tiling by two quadrangles. This *two-quadrangle tiling*, presented in Fig. 8, is constructed by replacing each vertex of the shield tiling by a house-shaped tile, as in the construction of the ring tiling, except that each house is rotated by 180° with respect to its orientation in the ring tiling. Subsequently, all houses are enlarged continuously until they are in contact along their edges. The small areas not covered by the house-shaped tiles can then be covered either by a second

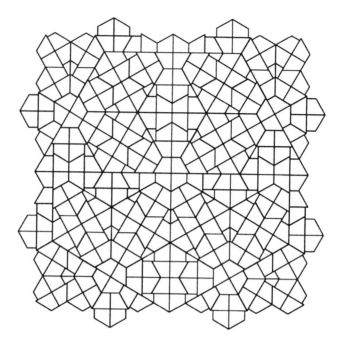

Figure 8. Two-quadrangle tiling.

tile, a nonregular hexagon having a wedge shape, or by combinations of one-half of a house tile and one-half of a wedge tile, where both tiles are halved along their line of mirror symmetry. After dissecting each house tile into two pairs of enantiomorphic quadrangles and each wedge-shaped tile into two quadrangles of the second type, the two-quadrangle tiling as shown in Figure 8 results.

3. Discussion

In the preceding section we have presented a number of dodecagonal tilings, all of which can be superimposed onto the shield tiling in such a way that equivalent regions are mapped onto each other. The local surroundings of any vertex in the new tiling always correspond to one and the same local geometry in the shield tiling. This relation is a consequence of the construction of the new tilings as subsets [22] of the vertices, the edges or a set of centers of one or more types of tiles in the shield tiling. In the case of the construction of the four-star tiling, a few additional edges, occurring in the ship tiling (Fig. 1a) but not in the shield tiling, had to be added.

The tilings generated in this way are the butterfly-, three-star-, wheel-, two-pentagon-, and four-star tiling, using the terms proposed in this paper.

For the four-star tiling and the related plate tiling we have found two different ways for their construction. These tilings all have a strictly dodecagonal Fourier transform and a single line of mirror symmetry as the only infinite symmetry element in real space. They differ in this respect from other dodecagonal tilings, which may have an exact 12-fold symmetry axis. Among the group of tilings discussed above, the butterfly tiling has been described previously (see introduction), while the other five tilings are described here for the first time.

The construction of the ring tiling differs from the method employed for the other tilings in that here tiles of the basic patterns have been welded together into supertiles rather than selecting a subset of points from the initial tiling. Due to the way it has been constructed, the ring tiling allows at least four other tilings (i.e., those described at the beginning of the paper) to be generated as subsets of center points of different types of tiles.

The quasiperiodic dodecagonal tilings presented in this study offer themselves as a basis for artistic designs such as ceramic walls or stone surface covers in a similar way as Penrose tilings have been used, for example, in an attractive tiling made of two types of stone slabs as tiles at the entrance of the EUREKA science center in Helsinki, Finland. It is important, not only for geometry and for the crystallography of quasicrystals, but also for the suggested applications in art, that the minimum number of different tiles needed to construct a dodecagonal tiling has been reduced by the present study from three to two.

Many of the tilings presented in Figures 1–7 and Plate 1 are well suited to be executed in color, e.g., in the form of color prints, colored drawings, or paintings. The basic designs can be varied by systematic changes in the outlines of tiles as in the famous Penrose chickens [15, 23] or in paintings and lithographs by M. C. Escher, or the basic color patterns can be complicated by introducing different colors for sets of tiles belonging to the same type. Such images offer interesting opportunities for the study of color symmetry in quasiperiodic tilings.

Several tilings presented in this paper can be converted, after minor changes, into dodecagonal quasiperiodic infinite layered band patterns [14], with bands as interlocking rings or with extended bands. In past periods such patterns might have been a major novelty for the artisans in the islamic regions, but even today it may, together with such patterns made from decagonal and octagonal quasiperiodic tilings, widen the spectrum of possible designs in the islamic art tradition.

Acknowledgments

This work would not have been possible without the active participation of Dr. F. Gähler, who has made the laser printer drawings presented in this paper and, through discussion and criticism, has improved its content. I am very obliged to him for his inspiration and help. I have also had much benefit from discussions with Professor A. Dress as well as Dr. P. Stampfli, Mr. C. Beeli, and Dr. J. Rhyner. Details were discussed at a private workshop on quasilattices sponsored by Mr. J. and Mrs. B. Stampfli at Locarno in August 1989, and I am very thankful for their invitation and hospitality.

REFERENCES

1. D. Shechtman, I. Blech, D. Gratias, and J. W. Cahn, "Metallic phase with long-range orientational order and no translational symmetry," *Phys. Rev. Lett.*, **53**, 1951–1953 (1984).

2. T. Ishimasa, H.-U. Nissen, and Y. Fukano, "A new ordering state between crystalline and amorphous," Poster and abstract, Workshop on Physics of small particles, Gwatt, Switzerland, 18–20 Oct. 1984.

3. T. Ishimasa, H.-U. Nissen, and Y. Fukano, "New ordered state between crystalline and amorphous in Ni–Cr particles," *Phys. Rev. Lett.*, **55**, 511–513 (1985).

4. T. Ishimasa, Y. Fukano, and H.-U. Nissen, "The crystalloid structure in Ni–Cr small particles," in *Quasicrystalline Materials. Proc. I.L.L./CODEST Workshop, Grenoble*, Ch. Janot and J. M. Dubois (eds.), pp. 168–177. World Scientific, Singapore, 1988.

5. T. Ishimasa, H.-U. Nissen, and Y. Fukano, "Electron microscopy of crystalloid structure in Ni–Cr small particles," *Phil. Mag.*, **A58**, 835–863 (1988).

6. F. Gähler and J. Rhyner, "Equivalence of the generalized grid and projection methods for the construction of quasiperiodic tilings," *Phys. A: Math. Gen.*, **19**, 267–277 (1986).

7. P. Stampfli, "A dodecagonal quasiperiodic lattice in two dimensions," *Helv. Phys. Acta*, **59**, 1260 (1986).

8. P. Stampfli, "New quasiperiodic lattices from the grid method," This volume.

9. N. Niizeki and H. Mitani, "Two-dimensional dodecagonal quasilattices," *J. Phys. A: Math. Gen.*, **20** L405–410 (1987).

10. F. Gähler, "Crystallography of dodecagonal quasicrystals," *Quasicrystalline Materials, Proc. I.L.L./CODEST Workshop, Grenoble*, Ch. Janot and J. M. Dubois (eds.), pp. 272–284. World Scientific, Singapore, 1988.

11. N. G. de Bruijn, "Algebraic theory of Penrose's non-periodic tilings of the plane. I and II," *Ned. Akad. Wetensch. Proc. Ser. A*, **43**, 39–66 (1981).

12. P. W. Leung, C. L. Henley, and G. V. Chester, "Dodecagonal order in a two-dimensional Lennard-Jones system," *Phys. Rev.*, **B39**, 446–458 (1989).

13. C. Beeli, F. Gähler, H.-U. Nissen, and P. Stadelmann, "Comparison of HREM images and contrast simulations for dodecagonal Ni–Cr quasicrystals," *J. Phys. France*, **51**, 661–674 (1990).

14. B. Grünbaum and G. C. Shephard, *Tilings and Patterns*. Freeman, San Francisco, 1986.

15. R. Penrose, "The role of aesthetics in pure and applied mathematical research," *Bull. Inst. Math. Appl.*, **10**, 266–271 (1974). Reprinted: *Pentaplexity. Math. Intelligencer*, **2**, 32–37 (1979).

16. F. P. M. Beenker, "Algebraic theory of non-periodic tilings of the plane by two simple building blocks: A square and a rhombus," Eindhoven University of Technology, Technical Report No 82-WSK-04, 1982 (unpublished).

17. R. Ammann, B. Grünbaum, and G. C. Shephard, "Aperiodic tiles," preprint, 1989.

18. J. E. S. Socolar, "Simple octagonal and dodecagonal quasicrystals," *Phys. Rev.* **B39** (15), 10519–10551 (1989).

19. N. J. Niizeki, "Dodecahedral quasiperiodic tilings of a plane by a projection of a 6d simple hypercubic lattice," *J. Phys. A: Math. Gen.*, **21**, 2167 (1988).

20. F. C. Frank and J. S. Kasper, "Complex alloy structures regarded as sphere packings. I and II," *Acta Crystallogr.*, **12**, 184–190 and 483–499 (1959).

21. M. O'Keeffe and B. G. Hyde, "Plane nets in crystal chemistry," *Phil. Trans. R. Soc. London Ser. A*, **295**, 553–623 (1980).

22. R. Lück, "Penrose sublattices," *Proc. 7th Int. Conf. Liquid Amorphous Metals*, Kyoto, Japan, Sept 4–8, 1989. Preprint. To appear in *J. Non-Crystalline Solids* (1990).

23. M. Gardner, "Extraordinary nonperiodic tiling that enriches the theory of tiles," *Sci. Am.*, Jan. 1977, 110–121 (1977).

12 New Quasiperiodic Lattices from the Grid Method

Peter Stampfli

1. Introduction

The possibilities for obtaining truly three-dimensional quasiperiodic lattices of maximal isotropy are rather limited, because their symmetry has to represent a regular tesselation of the unit sphere and thus they are related to one of the five Platonic solids [1]. Of these, only the symmetry of the icosahedron (and the pentagondodecahedron) is not compatible with spatial periodicity and actually all three-dimensional quasiperiodic lattices observed so far have icosahedral symmetry. By contrast, there are many more possibilities to construct quasiperiodic lattices or tilings in the two-dimensional plane, since the symmetry of any regular polygon could be used to obtain planar tilings. Only the symmetries of the equilateral triangle, the square and the hexagon can generate periodic tilings in the plane.

Two-dimensional quasiperiodic crystal structures, which are simply periodic in the third space direction, have actually been observed using selected area electron diffraction and high-resolution electron microscopy. Examples are the two-dimensional quasiperiodic structures with 10-fold rotational symmetry [2], which are related to the three-dimensional icosahedral structures. Other quasiperiodic crystal structures with 12-fold (dodecagonal) [3] and 8-fold (octagonal) [4] rotational symmetry have also been observed.

Ishimasa et al. have obtained images of the atomic structure of small Ni–Cr particles using high-resolution electron microscopy [3]. The corresponding electron diffraction patterns have twelvefold rotational symmetry. The spatial structure consists of basic units having the shape of equilateral triangles, squares, and rhombs with an acute angle of $2\pi/12$ radians as tiles. A large fraction of these triangles is connected through their sides to squares and not to other triangles. Thus the isolated triangle is a distinct tiling unit of this structure. At the time of this discovery, a quasiperiodic tiling corresponding to this structure could not be constructed using the grid method or the projection method of de Bruijn [5], because these methods did not provide any possibility to use distinct triangles as tiles. Moreover, these two methods are equivalent to each other [6].

At the beginning of the present work has been the observation, that a well-known periodic tesselation with sixfold rotational symmetry is made of equilateral triangles and that a new quasiperiodic tiling would be obtained if the rotational symmetry of this periodic tesselation could somehow be "doubled." The resulting tiling would have dodecagonal symmetry and it would include triangles as distinct tiles. I have been able to construct this quasiperiodic tiling by a suitable modification of the grid method [7]. However, the resulting quasiperiodic tiling contains too many narrow rhombs with an acute angle of $2\pi/12$, such that it is quite different from the structure of the Ni–Cr particles. Later, Gähler has produced the same quasiperiodic tiling with the projection method (published as an example in [8]). Both, Gähler [9] and Niizeki and Mitani [10] have independently discovered a suitable modification of this tiling, which reduces the number of rhombs and results in a close similarity with the Ni–Cr structure.

The content of this article is due to these ideas and due to the important suggestion of H.-U. Nissen that new quasiperiodic tilings could be obtained by applying the modified grid method to Archimedian periodic tesselations and to other nonregular tesselations involving irregular polygons. In this paper the regular and semiregular tesselations of the plane are first presented and discussed. Then it is shown how the grid method can be used to obtain quasiperiodic tilings from these tesselations, and the results obtained in this way are presented. It should be noted that the local rotational symmetry is often partially broken in these quasiperiodic tilings (as discussed in [11]) and only realized in an approximate way. Finally, irregular polygons are introduced as tiles in the grid method. Thus a continous transformation between different periodic tesselations and the corresponding quasiperiodic tilings becomes possible.

2. Periodic Tilings and Their Dual Lattices

Some regular polygons, namely the triangle, the square, the hexagon, the octagon, and the dodecagon, can be joined together to form many different periodic tesselations. Since there are many excellent books that discuss them in detail [12–14], I will present only those properties that are important for the grid method.

The so-called regular tesselations are shown in the first row of Figure 1 (bold lines). They are built of equal polygons (which have to be either squares, triangles, or hexagons). Additional periodic tesselations are obtained if several different polygons (which can also be dodecagons and octagons) are joined together. If always the same set of polygons is joined together in the same sequence at every vertex of the tesselation, then all vertices are equal and one of the semiregular tesselation shown in the rest of Figure 1 (bold lines) is obtained. Many more periodic tesselations could be produced,

Figure 1. Regular and semiregular tesselations (bold lines) and their dual lattices (thin lines).

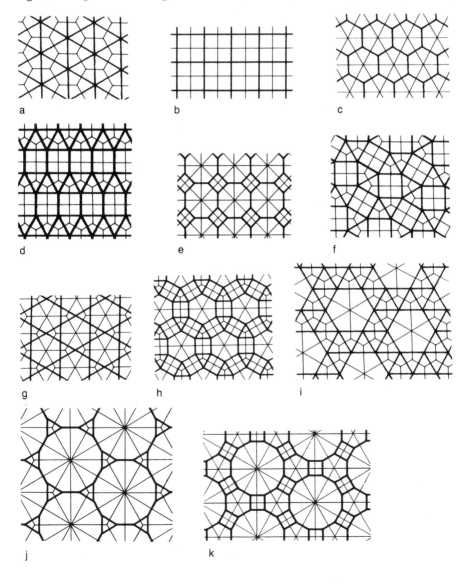

all of which would contain at least two different vertices [13]. They are not considered here, because there are too many of them.

The key to the grid method is the dual lattice of a given tesselation [12, 13] shown in thin lines in Figure 1. It can be obtained by joining the centers of the polygons around each vertex by straight lines. This results in a new tiling by unique polygons, which are images of the vertices of the original tesselation. It is easy to see that the dual of a regular tesselation is itself a regular tesselation. Note that the square lattice is the dual of itself and that the tesselation of triangles (Fig. 1a) and the tesselation of hexagons (Fig. 1c) are dual to each other. Further, the dual of a semiregular tesselation consists of equal irregular polygons, because the vertices of the tesselation are all equal. The dual in turn contains several different vertices that correspond to the unequal polygons in the tesselation.

It is easy to reconstruct the original tesselation from its dual lattice. Each vertex in the dual lattice corresponds to a uniquely defined tile in the tesselation. Three lines meeting at a vertex at angles of $2\pi/3$ give an equilateral triangle, two lines crossing at right angles give a square, and n lines crossing at equal angles give a regular $2n$-gon. If two vertices of the dual lattice are connected by a line, then the two corresponding tiles of the tesselation share a common side that is perpendicular to the connecting line. Thus one can build up a tesselation using its dual lattice and applying simple local rules. The same principle is used to construct quasiperiodic tilings, as discussed in the next section.

3. Discussion of the Grid Method

We will explain the grid method using a typical example. A more analytical presentation of the grid method can be found in [11], and the equivalence with the projection method is discussed in [6, 8]. The intermediate steps required for obtaining a quasiperiodic tiling of dodecagonal symmetry from the periodic tesselation of Figure 1a are illustrated in Figure 2. In Figure 2a, two copies of the original tesselation are shown, rotated by $2\pi/12$ with respect to each other. The corresponding dual lattices with the respective orientations are shown in Figure 2b. These two lattices can be superimposed (see Fig. 2c) to obtain a so-called grid. This grid has a 12-fold rotational symmetry around the center indicated by a dot. Obviously, the grid cannot be interpreted as a tiling, because it is composed of infinitely many different tiles.

The quasiperiod tiling of Figure 2d is obtained from the vertices of the grid (Fig. 2c). Only the arrangement of grid lines at each vertex and the connections between vertices by grid lines are used. Note that the actual length of the lines between the vertices is disregarded; thus only the topology of the grid is considered. The grid is then essentially quasiperiodic, as any local network of vertices in any finite region is repeated infinitely many

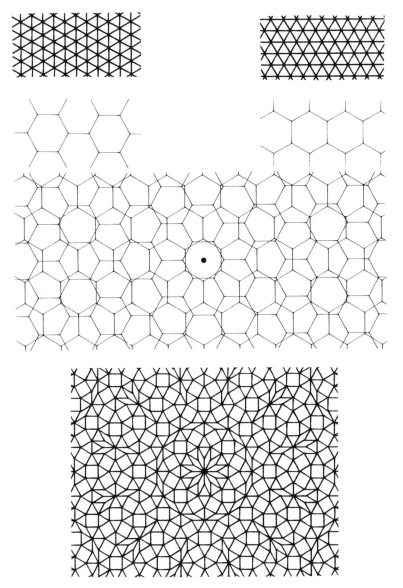

Figure 2. Illustration of the intermediate steps of the grid method as applied on the tesselation of Figure 1a.

times throughout the grid. The average distance between such similar regions increases with their size. This is simply due to two incommensurate periodicities in different space directions in the simple dual lattice, which are superimposed in the same space direction in the grid. The ratio of the period lengths is $1:\sqrt{3}$ in the case of a tesselation with hexagonal symmetry, or $1:\sqrt{2}$ for tesselations with fourfold symmetry. Thus the grid can be used to construct a quasiperiodic tiling in the same way as a periodic tesselation can be reconstructed from its dual lattice. We obtain new vertices in the grid whenever two lines of different copies of the dual lattice cross each other. These vertices generate rhombs or squares as new tiles. Their sides are perpendicular to the lines in the grid, and thus the angle between the crossing lines reappears at the corners of the rhomb. Two tiles are joined together at their sides if the corresponding vertices in the grid are connected by lines. The result is shown in Figure 2d, where the dot indicates the center of global 12-fold symmetry corresponding to the 12-fold symmetry of the grid (Fig 2c). A larger part of the same tiling is shown in Figure 3.

Figure 3. Quasiperiodic tiling with dodecagonal symmetry obtained from the periodic tesselation of Figure 1a (note that an arbitrary section is shown).

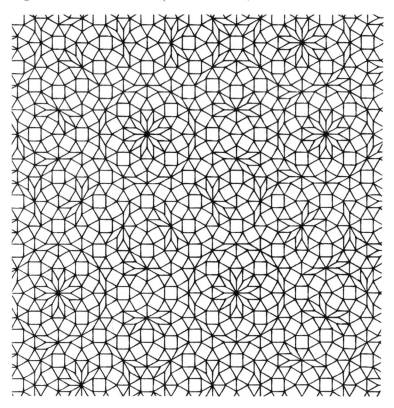

It is interesting to compare Figures 2c and d. There are two hexagons superimposed around the center of the grid, and their respective sides cross at angles of $2\pi/12$. These crossings gives rise to the 12 rhombs joined in a dodecagonal star. The 12 vertices corresponding to the corners of the two hexagons give rise to 12 equilateral triangles that fill in the gap between the rhombs. Each of these vertices is connected by a line in radial direction to a similar vertex. This results again in triangles that are joined through their basis to the triangles generated previously. Two lines go out from each of these vertices, crossing the lines of adjacent vertices at right angles. This results in squares fitting in between the triangles. It would not be difficult to continue in this way, using visual inspection of the grid to determine the local arrangement of tiles. Actually a computer has been used to construct the tilings and to make the drawings.

This method can be applied to any periodic tesselation. It can even contain irregular polygons, as will be shown later. The two copies of the dual lattice of a periodic tesselation with n-fold rotational symmetry have to be rotated by an angle of $2\pi/(2n)$ with respect to each other in order to form a grid. A quasiperiodic lattice with a $2n$-fold rotational symmetry is obtained from this grid. Notice that this is a generalization of the original definition given by de Bruijn [5]. It may be surprising that the tiles can always be joined together without overlap and without gaps resulting in a perfect tiling. But actually it is easy to show that each possible mesh in a grid gives a small cluster of tiles fitting perfectly together around a common vertex in the tiling if the grid is made of superimposed lattices that, taken separately, do result in perfect tilings. Thus no gap and no overlap of tiles arises at any vertex of the tiling, and thus the whole tiling has to be without any gap and without any overlap.

4. Results Obtained from the Regular Tesselations

A large section of the quasiperiodic tiling derived from the periodic tesselation of Figure 1a is shown in Figure 3. This section contains more information than the region around the point with global dodecagonal symmetry. The visual appearance of Figure 3 is dominated by groups of 12 rhombs, having a flower-like shape. Scattered between these groups are many more "flowers," but some of their "petals" (rhombs) are always missing.

In Figure 4 the tiling obtained from the square tesselation (Fig. 1b) is shown. It has been discovered independently by Ammann and by Beenker and it has since been extensively discussed (see e.g. [15]). It is shown here for comparison with the other tilings. This tiling has a rather uniform appearence dominated by right angles and flower-like aggregates of eight rhombs.

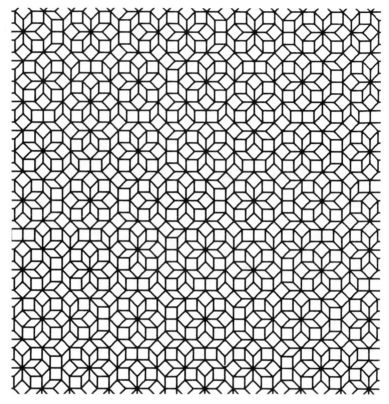

Figure 4. Quasiperiodic tiling with octagonal symmetry obtained from the periodic tesselation of Figure 1b.

The tesselation by hexagons (Fig. 1c) results in the tiling shown in Figure 5. This tiling has been discussed previously by Niizeki [16] and by Socolar [15]. Here the hexagons are predominant and distributed rather uniformly. It is difficult to recognize any remarkable pattern, and only small fragments of the dodecagonal symmetry can be seen.

5. Results Obtained from the Semiregular Tesselations

The tesselation of Figure 1d has only a twofold rotational symmetry but the grid method can be applied in this case too. The grid is made of two copies of the dual lattice shown in Figure 1d (thin lines), rotated by an angle of $2\pi/4$ with respect to each other. The period lengths of this lattice in horizontal and vertical direction have an incommensurate ratio of $1:(2 + \sqrt{3})$. They are superimposed onto each other in

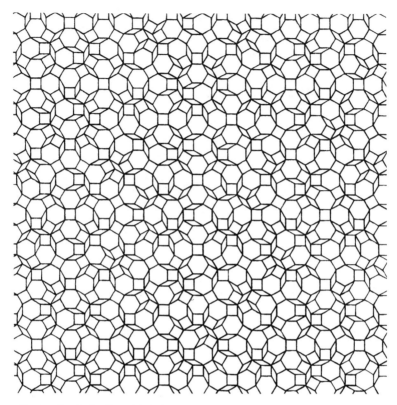

Figure 5. Quasiperiodic tiling with dodecagonal symmetry obtained from the periodic tesselation of Figure 1c.

the grid and thus force the quasiperiodicity. The resulting quasiperiodic tiling has a fourfold rotational symmetry and is shown in Figure 6. There appear many small fragments of the periodic tesselation by squares (Fig. 4) that resemble very small crystals of a regular crystalline phase imbedded in a matrix of triangles and rhombs. Similar periodic inclusions in a quasiperiodic tiling are discussed in [6]. The long range order of this quasiperiodic tiling should result in a diffraction pattern with sharp peaks independently of the small size of the periodic domains.

The tiling of Figure 7 is obtained from the tesselation by octagons and squares (Fig. 1e). It has the same octagonal symmetry as Figure 4, but now octagons appear as new tiles. They form circles and long chains, which wind through the tiling like snakes. It is easy to obtain the tiling of Figure 4 from the new tiling of Figure 7 and thus to demonstrate their similarity: We simply have to draw a new grid of continuous straight lines into the tiling of Figure 7. These lines go through the centers of the rhombs and are perpendicular to the sides of the rhombs. They are called Ammann

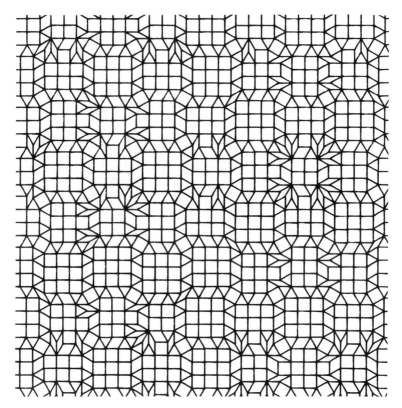

Figure 6. Quasiperiodic tiling with fourfold rotational symmetry obtained from the periodic tesselation of Figure 1d.

lines [15] and result in always the same decoration of the square tiles, the rhombs, and the octagons. The vertices inside each octagon define a unique subdivision of the octagon into two squares and four rhombs. Thus the octagons disappear and the tiling of Figure 4 is obtained, including its characteristic aggregates of eight rhombs.

The tesselation of Figure 1f contains points of fourfold rotational symmetry at the centers of the squares that are not part of a mirror plane. Note that there are separate mirror planes in Figure 1f, going through the triangles. Thus this tesselation is enantiomorphic to itself and contains two different patterns of tiles with a local cyclic fourfold rotational C_4 symmetry that are mirror images of each other. The corresponding tiling of eightfold symmetry (Fig. 8) contains large swirling groups of tiles, which turn clockwise and counterclockwise. They are mirror-images of each other and have a local cyclic C_8 symmetry, if the subdivision of the octagon at their center is disregarded. Interlocking fragments of these swirls appear throughout this

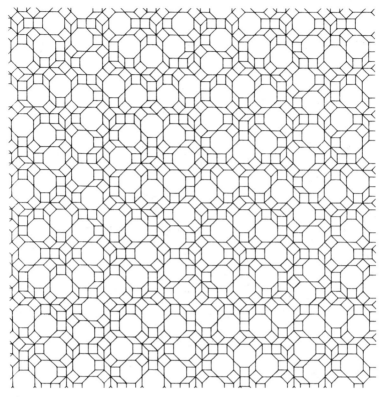

Figure 7. Quasiperiodic tiling with octagonal symmetry obtained from the periodic tesselation of Figure 1e.

tiling and give it a similarity to ice flowers on window panes (shown in [17] as so-called windowpane frost).

The quasiperiodic tiling of dodecagonal symmetry derived from Figure 1g is shown in Figure 9. This tesselation has previously been discussed by Niizeki [11], but no picture has yet been published. It contains both equilateral triangles and hexagons. The number of rhombs and hexagons is reduced in comparision to Figures 3 and 5, respectively, and thus it appears to be somehow intermediate. Very large regions with rather symmetric patterns can be recognized. However, slight perturbations of the local dodecagonal symmetry are always present.

The semiregular tesselation of Figure 1h is made of hexagons, squares, and triangles, but it can also be seen as made of overlapping dodecagons. The corresponding quasiperiodic tiling (Fig. 10) contains mostly squares. Large dodecagonal shapes appear having sides of two times the length of a single tile. Even larger overlapping dodecagonal structures can be recognized

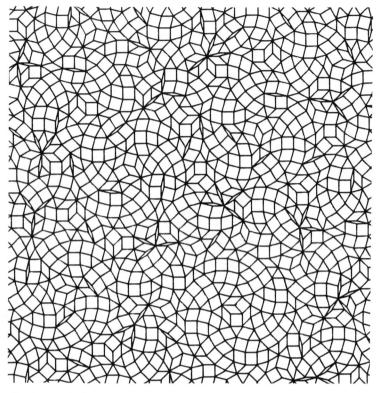

Figure 8. Quasiperiodic tiling with octagonal symmetry obtained from the periodic tesselation of Figure 1f.

if Figure 10 is viewed from a larger distance. This figure is vaguely reminiscent of a microscopic picture of biological tissue, with the large hexagonal tiles as vessels and the other smaller tiles as cells.

The tesselation of Figure 1i has only a cyclic C_6 symmetry and it does not contain a mirror plane. Thus the corresponding quasiperiodic tiling (Fig. 11) has a C_{12} symmetry. Note that the mirror images of these tilings are separate distinct tilings, in contrast to the tilings of Figures 1f and 8, which are identical to their mirror images. The section shown in Figure 11 is too small to make the quasiperiodicity evident and at first sight the structure appears to be a random network similar to a glass. However, it can be seen that there is a definite "handedness" and that this structure is different from its mirror image. Considering pairs of rhombs, which form small "arrows," it is evident that these arrows now form fragments of circular chains, which mainly turn around clockwise, if the direction indicated by these arrows is followed.

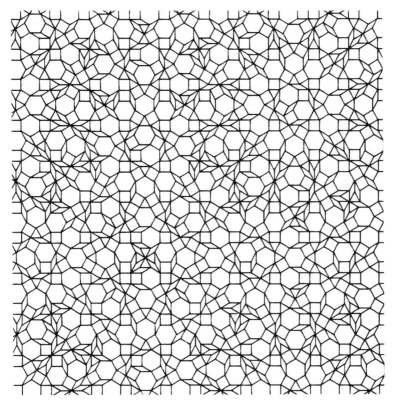

Figure 9. Quasiperiodic tiling with dodecagonal symmetry obtained from the periodic tesselation of Figure 1g.

The tesselation of Figure 1j results in the quasiperiodic tiling of Figure 12 with approximately dodecagonal symmetry. A typical pattern consists of two dodecagons, rotated by $2\pi/12$ with respect to each other and connected by a common corner. Rhombs and squares are filling in the space between these dodecagons, such that a large dodecagonal shape is formed. The dodecagonal symmetry is broken inside these shapes, and the tiling shows the dodecagonal symmetry only in an incomplete way. Many roughly concentric rings can be seen between the dodecagons, resembling waves on the surface of a pond, whereby the dodecagons become small "islands."

Finally, the tesselation of Figure 1k is faintly similar to Figure 1h, and the corresponding quasiperiodic tiling (Fig. 13) resembles roughly Figure 10. The group of two dodecagons and rhombs inside a large dodecagon seen in Figure 12 appears again, but the overall impression of Figure 13 is much more irregular.

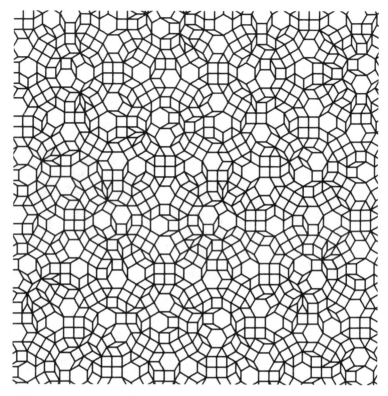

Figure 10. Quasiperiodic tiling with dodecagonal symmetry obtained from the periodic tesselation of Figure 1h.

6. Similarities between Different Tilings

Although the tilings of Figures 3 to 13 have a rather diverse visual appearance, there is a close similarity between the different tilings at a large scale of length, which concerns only large groups of tiles. This can be seen by connecting "special points" of high local symmetry by straight lines. Choosing, for example, the vertices in Figure 3, which are at the center of a complete star of 12 rhombs, or choosing in Figures 12 and 13 the vertices where two dodecagons meet, one obtains in each case a large lattice made of squares, equilateral triangles, and rhombs with an acute angle of $2\pi/12$. The ratio of the size of these new tiles to the original tiles is an integer power of the basic self-similarity ratio $1:(1 + \sqrt{2})$ for tilings with octagonal symmetry and $1:(2 + \sqrt{3})$ for dodecagonal symmetry. This suggests that quasiperiodic lattices of the same symmetry have the same underlying "basic geometry."

The similarity between these quasiperiodic tilings is a consequence of the similarity between the corresponding regular tesselations in Figure 1.

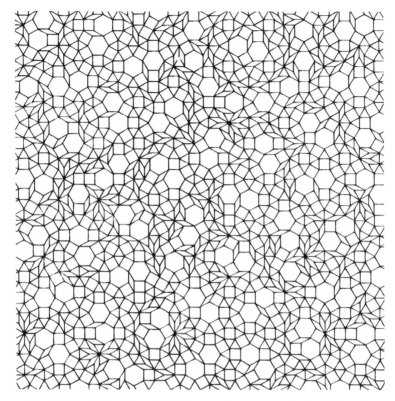

Figure 11. Quasiperiodic tiling with dodecagonal symmetry obtained from the periodic tesselation of Figure 1i.

The common basis of the periodic tesselations is simply their rotational symmetry, mirror symmetry, and periodicity. Similarly, the common basis of the quasiperiodic tilings is their rotational symmetry (which may be broken) and their quasiperiodicity (due to the combination of two incommensurate periodicities, as discussed in Section 3).

7. A Transformation between Different Tilings

We can make the similarity between quasiperiodic tilings explicit by a continuous deformation of the tiles, which transforms these tilings one into another. First, we present the transformation between periodic tesselations.

An example is given in Figure 14. Note that Figure 14c is identical to the usual regular tesselation by hexagons (Fig. 1c). If some of the hexagons are reduced in their size and the surrounding six hexagons are deformed

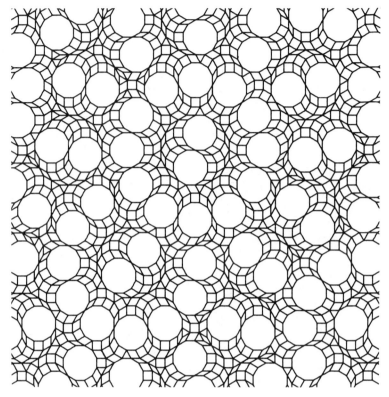

Figure 12. Quasiperiodic tiling with dodecagonal symmetry obtained from the periodic tesselation of Figure 1j.

into irregular hexagons, then Figure 14b is obtained. This procedure can be continued (see Fig. 14a) until the small hexagons disappear and the irregular hexagons are transformed into equilateral triangles. Thus we obtain the tesselation by triangles (Fig. 1a). On the other hand, increasing the size of the hexagon, we obtain Figure 14d and e until the small sides of the irregular hexagons disappear and the semiregular tiling of Figure 1g is reached. Another example is Figure 1b, which can be obtained from Figure 1e by shrinking the squares into points and distorting the octagons into squares. Such transformations from the other semiregular tesselations to regular tesselations of the same symmetry are easily found. Note that Figures 1f and c are exceptions that might pose a problem. In these cases, one has to distort some of the equilateral tringles into isosceles triangles until their basis disappears and they are thus reduced to a single line. Examples of such intermediate tesselations are found as "plane isogons" in [14].

The corresponding transformation between quasiperiodic tilings is obtained by applying the grid method to intermediate periodic tesselations. The dual lattice to our example of Figure 14 is shown in Figure 15 in thin lines, while

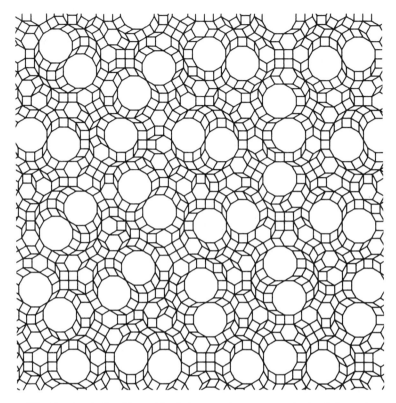

Figure 13. Quasiperiodic tiling with dodecagonal symmetry obtained from the periodic tesselation of Figure 1k.

the bold lines show the tesselation of Figure 14b. Note that the dual lattice consists of two different sublattices. The first sublattice, drawn as three parallel lines, corresponds to the set of long lines in the particular tesselation and is identical to the dual lattice of Figure 1a. The second sublattice, drawn as a single line, corresponds to the short lines and is identical to the dual of Figure 1g. As discussed in Section 3, two dual lattices are rotated with respect to each other and superimposed to make a grid, from which the quasiperiodic tiling is produced. A parallelogram or a rectangle with sides of different lengths is obtained at each vertex where two lines of inequivalent sublattices cross. If lines of two equivalent sublattices cross, one obtains small or large rhombs and squares. These tiles, together with the polygons of the irregular periodic tesselation, form a new quasiperiodic tiling. Note that an arbitrary ratio between the different lengths can be used.

The quasiperiodic tiling shown in Figure 16 is obtained in this way from the tesselation shown in Figure 14b or 15 (bold lines). Its striking visual appearance is due to the varying density of black lines that results in large interconnected circular shapes or "wheels." A comparison shows that indeed

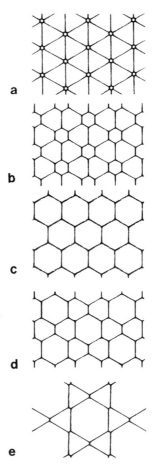

Figure 14. Illustration of the continuous deformation (top to bottom) from the periodic tesselation by triangles (Fig. 1a) to the tesselation by hexagons (Figs. 1c or 14c) to the semiregular tesselation of Figure 1g.

the tiling of Figure 3 is obtained, if the length of the short lines is reduced to zero. This moves the large rhombs together to form circular clusters. The gaps between the rhombs fill in with irregular hexagons that have changed into equilateral triangles. The resulting local arrangement is characteristic for Figure 3, but note that not exactly the same part of the tiling is shown. On the other hand, the tiling of Figure 5 is obtained if all lines have equal length.

Thus, as discussed earlier, we can find continuous transformations or interpolations between all tilings of the same symmetry. The tiles simply grow or shrink like the cells in a soap froth. Another analogy is the varying shapes of snow crystals [17], which are all based on hexagonal symmetry and fractal structure.

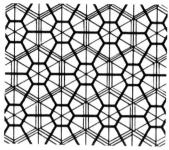

Figure 15. Dual lattice to the irregular tesselations of Figure 14. The single and triple thin lines indicate the two distinct sublattices corresponding to the two sets of lines with different length in Figure 14. The bold lines reproduce Figure 14b.

Figure 16. Quasiperiodic tiling with dodecagonal symmetry obtained from the irregular periodic tesselation of Figure 15 (or Fig. 14b).

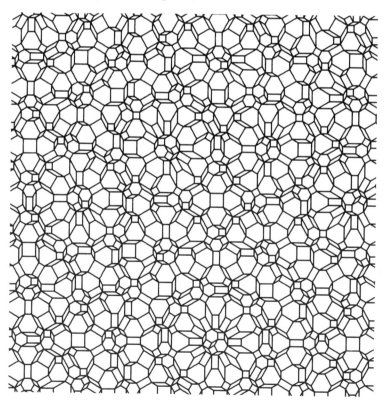

219

Note that, as suggested by Gähler, the deformable irregular tesselation of Figure 14 could also be used to compose a grid instead of the usual dual lattices. Thus an additional transformation between different quasiperiodic tilings is obtained. The single tesselation always has the regular tesselation of triangles (Fig. 1a) as its dual. But if two copies of the tesselation are combined in a grid, they result in quasiperiodic tilings that depend on the distortion. These tilings are always made of equilateral triangles, squares, and rhombs. But note that the probability that six triangles are combined to form a hexagon in the tiling increases for a decreasing size of the regular hexagons in the grid (this corresponds to going from Fig. 14c to Fig. 14a). In the end all triangles are part of a hexagon, such that essentially the tiling of Figure 5 is obtained. Similar transformations are possible between the quasiperiodic tilings corresponding to the periodic tesselations of Figures 1a, c, g, and i, because the tiles of these tesselations can all be divided into equilateral triangles.

8. Summary

In this article we have shown how the grid method can be used to obtain a quasiperiodic tiling from any periodic tesselation if the tesselation can be reproduced from an appropriate dual lattice. Two copies of this lattice are then used to make a grid, which serves as a plan for the quasiperiodic tiling. A complete account of the results obtained from the regular and semiregular periodic tesselations is given. The underlying similarity between different tilings is discussed and transformations between these tilings are presented. An explicit example, involving deformable irregular tiles, is shown.

Acknowledgments

I am very grateful to Dr. Hans-Ude Nissen and Dr. Franz Gähler for inspiring discussions and for generously sharing new ideas. The present work would never have been done without their help and encouragement.

REFERENCES

1. H. S. M. Coxeter, *Regular Polytopes*, 3rd ed. Dover Publications, New York, 1973.

2. L. Bendersky, "Quasicrystal with one-dimensional translational symmetry and a ten-fold rotation axis," *Phys. Rev. Lett.*, **55**, 1461–1463 (1985).

3. T. Ishimasa, H.-U. Nissen, and Y. Fukano, "New ordered state between crystalline and amorphous in Ni–Cr particles," *Phys. Rev. Lett.*, **55**, 511–514 (1985).

4. N. Wang, H. Chen, and K. H. Kuo, "Two-dimensional quasicrystals with eightfold rotational symmetry," *Phys. Rev. Lett.*, **59**, 1010–1013 (1987).

5. N. G. de Bruijn, "Algebraic theory of Penrose's non-periodic tilings of the plane. I and II," *Kon. Nederl. Akad. Wetensch. Proc. Ser A*, **84**, 39–66 (1981).

6. F. Gähler and J. Rhyner, "Equivalence of the generalized grid and projection methods for the construction of quasiperiodic tiling," *J. Phys. A: Math. Gen.*, **19**, 267–277 (1986).

7. P. Stampfli, "A dodecagonal quasiperiodic lattice in two dimensions," *Helv. Phys. Acta*, **59**, 1260–1263 (1986).

8. V. E. Korepin, F. Gähler, and J. Rhyner, "Quasiperiodic tilings: A generalized grid-projection method," *Acta Crystallogr.*, **A44**, 667–672 (1988).

9. F. Gähler, "Crystallography of dodecagonal quasicrystals," in *Quasicrystalline Materials*, Ch. Janot and J. M. Dubois (eds.), pp. 272–284. World Scientific, Singapore, 1988.

10. K. Niizeki and H. Mitani, "Two-dimensional dodecagonal quasilattices," *J. Phys. A: Math. Gen.*, **20**, L405–L410 (1987).

11. K. Niizeki, "A classification of two-dimensional quasi-periodic tilings obtained with the grid method," *J. Phys. A: Math. Gen.* **21**, 3333–3345 (1988); K. Niizeki, "An *n*-gonal quasilattice in two dimensions as a dual lattice to a multiple periodic grid," *J. Phys. A: Math. Gen.*, **22**, 1871–1881 (1989).

12. P. Pierce and S. Pierce, *Polyhedra Primer*. Van Nostrand, New York, 1978.

13. K. Critchlow, *Order in Space*. Thames and Hudson, London, 1969.

14. A. V. Shubiknov and I. Koptsik, *Symmetry in Science and Art*. Plenum, New York, 1974.

15. J. E. S. Socolar, "Simple octagonal and dodecagonal quasicrystals," *Phys. Rev.*, **B39**, 10519–10531 (1989).

16. K. Niizeki, "A self-similar dodecagonal quasiperiodic tiling of the plane in terms of squares, regular hexagons and thin rhombs," *J. Phys A: Math. Gen.*, **21**, 2167–2175 (1988).

17. W. A. Bentley and W. J. Humphreys, *Snow Crystals*. Dover, New York, 1962.

13

Fivefold Symmetry in the Context of Potential Surfaces, Molecular Conformations, and Chemical Reactions

Paul G. Mezey

1. Introduction

Symmetry plays an important role in the analysis of molecular conformations and chemical reactions [1–14]. The concept of potential energy surface provides a unified basis for the study of both conformational rearrangements and chemical reactions, and one may formulate a symmetry analysis of conformers and reactions within the framework of potential surfaces [15–19]. Such an analysis has several advantages, most important of which is the possibility of a global description, considering, at least in principle, all possible nuclear configurations. A global analysis lends itself to a clear recognition of the essential interrelations between symmetry properties and energetic stability, as well as to a complete accounting of restrictions imposed by symmetry on a given stoichiometric family of chemical species.

A stoichiometric family of molecules is specified if a fixed family of atoms is given; the family contains all possible chemical species composed from these atoms. One may generalize this concept by allowing variations in the electronic state, as well as in the total number of electrons, that is, in the total charge. A nuclear configuration space M (a so-called *metric* space) is defined by the collection of N atomic nuclei, where the space contains all possible mutual (relative) arrangements of these nuclei [15, 20]. If only internal (relative) configurations are considered, then rigid translations and rigid rotations of the nuclear arrangements are disregarded, consequently, for a system of N nuclei 6 degrees of freedom are eliminated (if $N>2$). In such cases $3N-6$ internal coordinates are sufficient, consequently, the dimension of the internal configuration space M is $3N-6$. Each point of this space M represents a unique nuclear configuration, hence a conformational change or a chemical reaction can be modeled by paths, or in general, by displacements within M. Such a configuration space (internal configuration space) M can be provided with a *metric*, that is, with a global distance function $d=d(K_1, K_2)$, that may be used to express the dissimilarity of any two internal configurations K_1 and K_2 of the space M [21]. This metric d turns M into a *metric space*.

If in addition to the N nuclei, the number of electrons and the electronic state are also specified, then an energy value can be assigned to each nuclear configuration (within the Born–Oppenheimer or similar approximation). This assignment defines an energy function, also called an *energy hypersurface*. The energy hypersurface $E(K)$ for each electronic state contains information on all stable chemical species and all possible conformational changes and chemical reactions confined to the given electronic state [15]. Consequently, it is natural to analyze symmetry properties along potential energy hypersurfaces [22].

In what follows, a short review of the methods and some of the main results of the global approach to symmetry analysis will be given. Fivefold symmetry is rather special within the context of potential energy surfaces, with relevance to transition structures, instabilities, fivefold degeneracies, and multidimensional tiling problems within a configuration space. After the survey of some of the general results, we shall place special emphasis on some problems with fivefold symmetry.

2. Symmetry Properties along Potential Energy Surfaces

The point symmetry and point symmetry groups of the three-dimensional, geometric arrangements of atomic nuclei of molecules are used extensively in many areas of chemistry, including the computation of molecular orbitals, spectroscopic studies, conformational analysis, and synthesis design [1–15].

The nuclear configuration space approach is especially suitable to investigate the following question: which are those nuclear arrangements of a stoichiometric family of atoms that have a prescribed symmetry? These nuclear arrangements form various subsets of the nuclear configuration space M. The entire space M can be subdivided into domains according to the various point symmetry groups of the nuclear arrangements in the three-dimensional space. There are several theorems that interrelate these symmetry domains with topological and geometrical properties of potential energy surfaces $E(K)$, where the variable K symbolizes the nuclear configuration in three-dimensional space, as well as the point representing this configuration in space M. In particular, various symmetry conditions have been derived, interrelating the energy minima, transition structures, and other stationary nuclear arrangements of neutral molecules and ions of electronic ground states and excited states, excimers, and exciplexes. The existence of stationary nuclear configurations within domains of a series of ground and excited state potential energy surfaces can be predicted by testing the point symmetry of a family of nuclear arrangements. The *catchment region point symmetry theorem* [16–18, 22] and various *vertical symmetry theorems* [17–19] can be used for the search of stationary points on potential surfaces.

Here we shall review four of these symmetry theorems.

The first one of these theorems concerns the *catchment regions* of potential surfaces [23]. Catchment regions reflect the deformability of chemical species. Most minor distortions of a nuclear configuration do not change the chemical identity of the corresponding molecular species. *Stable molecules* of the given stoichiometry as well as the *transition structures* of all interconversion processes of these species, all in a specified electonic state, for example, in the electronic ground state, can be represented by *subsets* within the configuration space M. With reference to the potential energy surface of the electronic state, these subsets can be chosen as the *catchment regions* of the nuclear configuration space [15, 23]. The catchment regions are defined by the following condition: infinitely slow, vibrationless relaxation of each distorted nuclear arrangement within a catchment region leads to the same equilibrium nuclear configuration. Evidently, there is precisely one equilibrium configuration within each catchment region. In the above definition, the equilibrium configuration is not necessarily an energy minimum; the catchment region concept applies for *any* stationary point, in particular, for saddle points of transition structures. In a formal, classical sense, a catchment region is the range of distorted configurations that preserves chemical identity. If domains of the nuclear configuration space are provided with mass-weighted coordinates, then the definition of catchment regions can be given in terms of steepest descent paths: from each point of the catchment region the steepest descent path leads to a common stationary point of energy [15, 23]. The stationary point can be used as a "chemical identity label" for each point of a given catchment region.

The catchment region point symmetry theorem [16–18, 22] states the following:

Within each catchment region the nuclear configuration corresponding to the equilibrium point has the highest point symmetry.

One can easily show that it is meaningful to refer to a unique, highest point symmetry within a catchment region.

This theorem describes an important relation between two very different molecular properties. The location of catchment regions and their equilibrium points are the properties of the potential energy surface, that is, they are determined by energy relations. By contrast, the point symmetries of nuclear configurations are purely geometric properties, hence they do not directly depend on energy. In spite of this, there exists a rather general relation between these two molecular properties.

The stable chemical species of different electronically excited states or different overall electronic charges are generally associated with a different partitioning of the nuclear configuration space M into a different set of catchment regions. By contrast, the point symmetry of the nuclei in any given nuclear configuration is fixed and is not affected by the electronic state or net charge. Consequently, point symmetry is a link among all electronic states of neutral and ionic species of a fixed stoichiometry in a global sense, that is, for all possible nuclear configurations. The catchment region point symmetry theorem is general for the potential energy surfaces and catchment regions of all electronic states of all neutral and all ionic species of the given stoichiometry. That is, the theorem gives a general symmetry condition for the catchment regions of various electronic states of neutral and ionic species.

The *catchment region minimum theorem* [18] may be regarded as a partial converse of the catchment region point symmetry theorem. This result, stated below, is also valid for all electronic states of neutral and charged species of the given stoichiometric family:

If within a catchment region there is a point K with a symmetry element R not present anywhere else in the catchment region, then this point K must have the lowest energy value within the catchment region, and point K is the unique critical point of the catchment region.

Point symmetry implies further useful relations for the relative nuclear configurations of neutral and ionic molecular species of ground and various excited electronic states, excimers, and exciplexes, and transition structures. (For sake of simplicity in the terminology, in the following section we shall consider the net charge as being part of the electronic state specification.) Energy may be regarded as a formal "vertical" dimension over a nuclear configuration space M and interrelations among various electronically excited state potential surfaces may be regarded as formal vertical relations over M. The result we shall describe below is referred to as the vertical point symmetry theorem.

Take any surface B that divides the nuclear configuration space M into two parts, M_1 and M_2. We assume that the set M_1 of configurations contains surface B as its boundary. There exist some symmetry elements that are present for all nuclear configurations K' along B. A family of such symmetry elements, $R_1', R_2', \ldots, R_p'$, is denoted by $\boldsymbol{R'}$,

$$\boldsymbol{R'} = \{R_1', R_2', \ldots, R_p'\}. \tag{1}$$

Select a nuclear configuration K from set M_1. A family of symmetry elements, $R_1, R_2, \ldots, R_q$, that is present at point K is denoted by $\boldsymbol{R}$:

$$\boldsymbol{R} = \{R_1, R_2, \ldots, R_q\}. \tag{2}$$

The *vertical point symmetry theorem* [17, 18] states the following:

If (i) *no configuration along B possesses the entire family $\boldsymbol{R}$ of symmetry elements, or, if* (ii) *configuration K does not have all the symmetry elements of family $\boldsymbol{R'}$, then the family M_1 of configurations must contain at least one critical point for the potential energy surface of each electronic state.*

(It is clear that if either one of conditions (i) and (ii) is fulfilled, then K must be an interior point of M_1, that is, K cannot fall on the boundary B.)

The above theorem indicates only the presence of a critical point within a given domain M_1, and the exact location, type, and the number of the critical points are not specified by the theorem. Consequently, some of the critical points may remain undetected by the theorem. For a different electronic state the critical points may have different locations within M_1, and they may also be of different types, for example, minima, or saddle points. The actual test point K itself does not have to be a critical point for any one of the potential surfaces.

The theorem describes some aspects of the behavior of the *potential energy surface* (the presence of a critical point within some region), based on a property of the *configuration space M* (symmetry conditions on boundary B and a test point K). One may obtain conclusions *without any quantum chemical calculation*, simply by checking the symmetry elements for a family of configurations.

The *missing critical point theorem* [18] may be regarded as a partial converse of the above vertical point symmetry therorem. If one knows that there is an electronic state for which no critical point exists within a given domain M_1 of the configuration space M, then this information can be used to obtain global conclusions concerning symmetry within M_1:

If the point symmetry group g_i contains the collection of all the symmetry operators of symmetry elements occurring at various points of the boundary surface B, and if exists an electronic state of any net charge with a potential surface that has no critical point within set M_1, then no point K of set M_1 can have a point symmetry group g that contains g_i as a subgroup (where $g \neq g_i$).

That is, the lack of a critical point within M_1 for any electronic state of any overall charge implies that no interior point of M_1 can have any symmetry element not present at the boundary B.

One may restrict the symmetry analysis to a few of the most important internal coordinates, called active coordinates, using *relaxed cross-sections* of potential surfaces. In such models only the selected active internal coordinates are regarded as variables and all other coordinates (the passive coordinates) are taken at their local optimum values for each choice of the selected coordinates. These optima are defined by the following condition: along a relaxed cross section C of the configuration space M all forces (as defined with respect to a specified potential energy surface) act tangentially to C. Symmetry theorems analogous to those described above can be derived for relaxed cross sections of potential surfaces of ground and excited electronic states [17, 18].

3. An Example of Fivefold Symmetry: The Spider Monkey Saddle Problem of Protonated Cyclopentane

Degenerate critical points of potential energy surfaces are characterized by a singular Hessian matrix $\mathbb{H}(K)$ of second energy derivatives, that is, by the occurrence of at least one zero eigenvalue of $\mathbb{H}(K)$. The so-called monkey saddle point [5, 7] is a degenerate critical point where three mountain ridges and three valleys meet. Such degenerate critical points may exist on actual potential surfaces [5b, 7, 24]. In Figure 1 a contour diagram of a monkey sadle surface is shown. The term "monkey saddle" indicates that the three joining valleys provide space for the two legs and the tail of the monkey sitting on the monkey saddle point.

Figure 1. A contour representation of a monkey saddle surface. Heavy lines are equipotential contours above the energy level of the central critical point; thin lines represent energy contours below the critical point value.

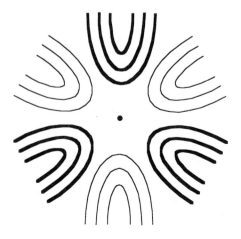

An interesting degenerate critical point is one where five mountain ridges and five valleys meet. If these mountain ridges (as well as the valleys) are equivalent, then the potential has fivefold symmetry. By analogy with the ordinary monkey saddle problem, we shall use the term "spider monkey saddle" for the potential around such a critical point. A spider monkey has all four of its limbs as well as its tail approximately of the same length, and one may imagine a saddle where the joining five valleys may provide comfort for all four limbs and the tail of a spider monkey. A contour diagram of a spider monkey saddle is shown in Figure 2.

We shall use the example of protonated cyclopentane derivatives for illustrating some aspects of the fivefold symmetry of the spider monkey saddle problem on potential surfaces. Earlier *ab initio* studies of the parent compound, protonated cyclopentane, has indicated high mobility of the proton over the ring structure and the presence of a critical point with fivefold symmetry (see, e.g., ref. [25] and references quoted therein).

Take a protonated 1,2,3,4,5-penta-substituted cyclopentane, $C_5H_5X_5H^+$, where all X substituents are on the same side of the formal "molecular plane" P, taken as the plane of optimum fit to the five carbon nuclei. A schematic model of such a protonated cyclopentane derivative is shown in Figure 3. Consider a two-dimensional relaxed cross section C of the potential energy surface, where the two active coordinates are two of the cartesian coordinates of the proton. Assume that all five carbon nuclei are within the plane P. Then the two active coordinates span a plane parallel to the plane P. Within the corresponding relaxed cross section, the potential surface of the proton has a fivefold symmetry, as a consequence of the fivefold symmetry of a three-dimensional arrangement where the proton is located above the

Figure 2. A contour representation of a spider monkey saddle surface of fivefold symmetry. The notation is the same as in Figure 1. At the degenerate critical point in the central location five valleys and five mountain ridges meet.

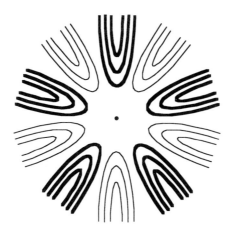

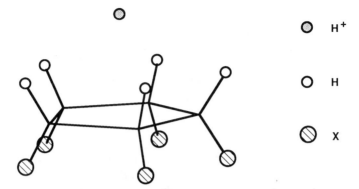

Figure 3. A molecular problem with a two-dimensional relaxed potential surface having a fivefold symmetry at a spider monkey saddle point. The two active coordinates of the proton H^+ in the protonated cyclopentane derivative are spanning a plane parallel to the plane P defined by the five carbon atoms. Depending on the choice of substituent X, one may obtain various two-dimensional relaxed potential surfaces, all having fivefold symmetry. As implied by the vertical point symmetry theorem, a simple inspection of symmetry elements may be used to obtain conclusions on the presence of critical points on the potential surfaces (see text for details).

center of mass of the five carbon nuclei. By varying the substituent X, one may deform the two-dimensional relaxed cross section of the corresponding potential surface, however, in the course of these deformations the fivefold symmetry of the potential surface is retained. For the various choices of substituent X the following possibilities may occur:

1. The nuclear configuration corresponding to placing the proton above the center of the five-membered ring (central location for the proton) is a local maximum of the potential surface.

2. The central location is a local minimum.

3. The central location is a spider monkey saddle.

4. The central location is some other degenerate critical point, e.g., the surface is locally planar.

It is intuitively evident that as long as the fivefold symmetry is preserved, no other possibility may arise, that is, the central location must be a critical point of the relaxed potential energy surface. In the first two cases, (1) and (2), the critical point is nondegenerate, whereas in cases (3) and (4) one finds degenerate critical points.

One may give a formal proof of the above, and the methods we shall apply will also prove some additional, intuitively plausible results. If by an appropriate choice of substituent X the structure with the proton at the central location has a fivefold symmetry in the three-dimensional space, then a fivefold symmetry also follows for the two-dimensional potential energy surface.

As long as a five-membered carbon ring is maintained, there may exist at most one point K of fivefold symmetry within a two-dimensional relaxed cross section C. Cross section C is defined in terms of the active coordinates taken as the two protonic coordinates spanning a plane parallel to the (approximate) molecular plane P. One may use the above information to conclude that K must be a critical point of the potential, *a conclusion based on the presence of fivefold symmetry, taking into account no direct information at all on the energy function.*

Take any closed loop B within C that encircles the configuration K of fivefold symmetry. A family $\boldsymbol{R} = \{R_1, R_2, \ldots, R_q\}$ of symmetry elements at point K may be chosen to contain the symmetry element C_5. By contrast, no family $\boldsymbol{R}' = \{R_1', R_2', \ldots, R_p'\}$ of symmetry elements occurring at every point K' of loop B may contain the symmetry element C_5, since within C, point K is the only point having the symmetry element C_5. By applying the vertical point symmetry theorem, one concludes that point K must be a critical point of the relaxed potential energy surface. Consequently, only the cases 1–4 may occur.

Some of the above, formal conclusions are illustrated by earlier potential surface calculations [25]. If the potential energy surface over a relaxed cross section C is calculated using the *ab initio* Hartree–Fock LCAO-SCF method and the standard STO 4-31G basis set, then for the special case of protonated cyclopentane (case X=H) the critical point above the central location is a spider monkey saddle, similar to the one modeled by the contour surface diagram of Figure 2.

The missing critical point theorem can be used to rule out the presence of a suspected fivefold symmetry configuration within a relaxed cross section C of a potential energy surface. If for some choice of substituent X of the protonated cyclopentane derivative, a domain M_1 of a relaxed cross section C contains no critical point at all, and if along the boundary B of M_1 there is no point with a C_5 point symmetry element, then no configuration K within M_1 can have a C_5 point symmetry element either. Such a problem arises, for example, if the relaxed cross section C shows a slight deviation from a planar arrangement of the five carbon atoms. The theorem is suitable for detecting such deviations: the lack of a critical point within the relaxed cross section C gives indication that no configuration of precise C_5 point symmetry element exists within the chosen domain M_1.

4. A Constraint on Reaction Networks of Fivefold Symmetry

In what follows, we shall review briefly some of the fundamental properties of reaction networks on potential surfaces, representing families of reaction mechanisms [26]. After the clarification of the basic concepts, we shall describe a limitation on certain reaction networks of fivefold symmetry.

Chemical species can be represented by catchment regions of potential surfaces. The notation $C(\lambda, i)$ has been used for such catchment regions, where $\lambda = 0$ for a stable chemical species and $\lambda = 1$ for a transition structure ("transition state"). The index i is a simple index of ordering. The pairwise interrelations of various chemical species that exist along the hypersurface can be represented by relations between catchment regions, described by various graphs and networks. Quantum chemical reaction graphs and reaction networks may be defined in terms of *intersection graphs* of catchment regions $C(\lambda, i)$ and their closures, denoted by *closure* $[C(\lambda, i)]$, where the closure of a catchment region is regarded as the collection of all of its points and all of its boundary points in the configuration space M.

Some of the most essential features of a topological model of energy hypersurfaces can be analyzed using network theory and graph theory. In this section we describe the elements of the reaction network model. An excellent introduction to graph theory can be found in [27]. The reaction network model of potential energy hypersurfaces has been introduced in [26].

The basis of this model is a neighbor relation among catchment regions. A chemical species represented by a catchment region $C(\lambda, i)$ is a strong neighbor of the chemical species represented by $C(\lambda', j)$ if

$$closure\ [C(\lambda, i)] \cap C(\lambda', j) \neq \varnothing, \tag{3}$$

where the symbol $\cap$ stands for intersection (common part), and $\varnothing$ stands for the empty set. The above relation reads: catchment region $C(\lambda, i)$, together with its boundary points, has some common part with catchment region $C(\lambda', j)$. This fact is expressed by a statement that their intersection is not the empty set.

Accordingly, a strong neighbor relation for the catchment regions is defined by

$$N^s[C(\lambda, i), C(\lambda', j)]$$

$$= \begin{cases} 1 & \text{if } closure\ [C(\lambda, i)] \cap C(\lambda', j) \neq \varnothing, \quad i \neq j \\ 0 & \text{otherwise.} \end{cases}$$

$$\tag{4}$$

The symmetric strong neighbor relation, or simply, the *neighbor relation* is a symmetric variant of relation (4). It is defined as

$$N[C(\lambda, i), C(\lambda', j)]$$

$$= \begin{cases} 1 & \text{if } N^s[C(\lambda, i), C(\lambda', j)] + N^s[C(\lambda', j), C(\lambda, i)] \geq 1 \\ 0 & \text{otherwise.} \end{cases}$$

$$\tag{5}$$

The reaction network on a potential energy surface is defined in terms of a graph. In general, a graph consists of a set of points and a set of lines; the points are called *vertices*, some of which are pairwise interconnected by lines called *edges*.

For a reaction graph g on a potential surface, the vertex set $\mathbf{V}(g)$ may be interpreted as the family of catchment regions,

$$\mathbf{V}(g) = \{C(\lambda, i)\}, \tag{6}$$

and the edge set $\mathbf{E}(g)$ as the set representing the neighbor relations of the catchment regions themselves,

$$\mathbf{E}(g) = \{[C(\lambda, i), C(\lambda', i')] : N[C(\lambda, i), C(\lambda', i')] = 1\}. \tag{7}$$

For simplicity in the notations, a catchment region $C(\lambda, i)$, as a vertex of the graph g, may be replaced by a general vertex symbol v_i. Similarly, the kth pair of neighbor catchment regions, representing an edge $[C(\lambda, i), C(\lambda', i')]$ of the graph g, may be denoted by the edge symbol $e_k(v_i, v_{i'})$.

Additional information, for example, direction may also be incorporated in a graph g, turning it into a *digraph* d. In a digraph the lines have direction and are called *arcs*. Direction may be chosen by various criteria. In the reaction network model we use energy relations between the critical points of catchment regions, together with graph g, to define a digraph d. In what follows we shall use the conventional notation $\in$ for expressing the fact that an element a belongs to a set A, $a \in A$. In the digraph d each *edge* $e_k(v_i, v_j) \in \mathbf{E}(g)$ $[v_i, v_j \in \mathbf{V}(g)]$ of graph g is provided with a *direction*, based on the criterion of nonincreasing energy, that turns the edge into an *arc* of the digraph d. That is, by interpreting an energy function $E(v_i)$ for the vertices v_i of graph g as the energy $E[K(\lambda, i)]$ of the critical point $K(\lambda, i)$ of the corresponding catchment region $C(\lambda, i)$, the following relation holds for the *arc* $a_k(v_i, v_j)$ of digraph d: If

$$E(v_i) \geq E(v_j), \tag{8}$$

then the direction is from vertex v_i to v_j.

If the equality holds in the above inequality (8) then we have two arcs, one from v_i to v_j and another from v_j to v_i. The nonnegative energy difference ΔE_{ij}, defined as

$$\Delta E_{ij} = E(v_i) - E(v_j), \tag{9}$$

may be used to label the arcs of digraph d. The *reaction network* is defined as the labeled digraph d.

Consider the typical case of the catchment regions of two molecules, $C(0, k)$ and $C(0, l)$, separated by a transition structure $C(1, j)$. Neither of the two molecules is a strong neighbor of the other and only two of the three catchment region pairs, as possible edges, are in the edge set $\mathbf{E}(g)$:

$$e(v_k, v_j), e(v_l, v_j) \in \mathbf{E}(g), \tag{10}$$

and

$$e(v_k, v_l) \in \mathbf{E}(g). \tag{11}$$

Consequently, in graph g and in digraph d, two vertices representing stable molecules cannot be connected without involving the vertex representing a formal unstable species, for example, that of a transition structure. This is an important consideration if two stable chemical species can be inter-converted into one another by two or more direct processes, without involving any additional stable species (intermediate). This possibility arises if there are two or more transition structures along the common boundary of the two stable species in the configuration space M. The reaction graph g and reaction network d describe these possibilities in sufficient detail, suitable to distinguish the various reaction mechanisms.

In some instances one is interested only in the sequence of stable chemical species occurring along a chemical reaction, and the transition structures that are actually involved in the interconversion processes are of secondary importance. In these cases simplified reaction graphs and reaction networks are sufficient, involving only the catchment regions $C(0, k)$ of stable species as vertices, and using a different neighbor relation, the *n-neighbor relation*:

$$n[C(0, i), C(0, j)]$$

$$= \begin{cases} 1 & \text{if } \textit{closure } [C(0, i)] \cap \textit{closure } [C(0, j)] \neq \varnothing, \quad i \neq j \\ 0 & \text{otherwise} \end{cases} \tag{12}$$

The vertex set $\mathbf{V}[g'(0)]$ of the resulting graph $g'(0)$ is the family of catchment regions of stable species,

$$\mathbf{V}[g'(0)] = \{C(0, i)\}, \tag{13}$$

and the edge set $\mathbf{E}[g'(0)]$ is the set representing their n-neighbor relations,

$$\mathbf{E}[g'(0)] = \{[C(0, i), C(0', i')] : n[C(0, i), C(0, i')] = 1 \}. \tag{14}$$

The zero in the parentheses in the notation $[g'(0)]$ indicates that only the catchment regions of stable chemical species (molecules), that is, catchment regions with a critical point index $\lambda = 0$, are considered.

The corresponding reaction network is defined as the labeled digraph $d'(0)$ obtained from graph $g'(0)$, by assigning directions to the arcs according to condition (8), and by taking labels according to condition (9). Reaction graph $g'(0)$ and reaction network $d'(0)$ describe the n-neighbor relations of stable chemical species along a potential energy surface. By design, they contain no information on transition structures and on the associated finer details of reaction mechanisms.

The most important elementary relations among various chemical species along the potential energy hypersurface are given by the neighbor relation matrix $\mathbb{N}(g)$ and the n-neighbor relation matrix $\mathbb{n}[g'(0)]$. The elements of these matrices are defined by the corresponding neighbor relations,

$$N_{ij} = N[C(\lambda, i), C(\lambda', j)], \tag{15}$$

and

$$n_{ij} = n[C(0, i), C(0, j)], \tag{16}$$

respectively. These matrices are also called the *adjacency matrices* of the corresponding vertices, and the following alternative notations may be used:

$$\mathbb{N}(g) = \mathbb{A}(g), \tag{17}$$

and

$$\mathbb{n}[g'(0)] = \mathbb{a}[g'(0)], \tag{18}$$

respectively.

A special aspect of fivefold symmetry of reaction mechanisms on potential surfaces may be formulated in terms of reaction graphs, reaction networks, and corresponding adjacency matrices. In Figure 4 a graph of fivefold symmetry is shown. This graph is denoted by K_5, and is known as Kuratowski's K_5 graph [27]. Graph K_5 is a *complete* graph, meaning that each of its vertices is connected by an edge to every other vertex of the graph. Since K_5 has five vertices, the fivefold symmetry is a direct consequence of the above property.

It is easily shown that in any family of chemical processes involving a relaxed potential energy surface with only two active coordinates, the Kuratowski graph K_5 can never occur as a reaction graph g or $g'(0)$, and K_5 cannot occur even as a part (subgraph) of any reaction graph g or $g'(0)$.

The above conclusion is a consequence of a well-known theorem on planar graphs, stating that no planar graph may contain K_5 as one of its subgraphs. A planar graph is a graph that is embeddable in a two-dimensional plane (or, by an equivalent condition, on a two-dimensional sphere). We know that both reaction graphs g and $g'(0)$ of any potential energy surface

Figure 4. The Kuratowski graph K_5 of fivefold symmetry cannot occur as a part of any quantum chemical reaction network d involving a relaxed cross section C of two active coordinates.

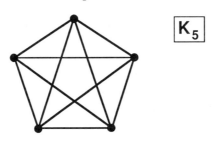

of a relaxed cross section C involving two active coordinates must be planar graphs. This follows, since the corresponding catchment regions provide a partitioning of a two-dimensional surface C, where C is topologically equivalent to an open set of a two-dimensional plane. Consequently, the graphs describing the neighbor relations of these catchment regions must also be embeddable in the same two-dimensional surface. Hence all such graphs are planar; that excludes the possibility of reaction graphs g and $g'(0)$ containing K_5 as a subgraph. We conclude that the special realization of fivefold symmetry for reaction mechanisms that is represented by the graph K_5, is not possible for relaxed cross sections C involving only two active coordinates.

This statemen can be formulated in terms of adjacency matrices. The adjacency matrix $\mathbb{A}(K_5)$ of Kuratowski graph K_5 is given by

$$\mathbb{A}(K_5) = \begin{vmatrix} 0 & 1 & 1 & 1 & 1 \\ 1 & 0 & 1 & 1 & 1 \\ 1 & 1 & 0 & 1 & 1 \\ 1 & 1 & 1 & 0 & 1 \\ 1 & 1 & 1 & 1 & 0 \end{vmatrix}. \tag{19}$$

In terms of matrix $\mathbb{A}(K_5)$ our statement may be rephrased as follows: No reaction graph g or $g'(0)$ of a relaxed potential energy surface of two active coordinates may have an adjacency matrix $\mathbb{A}$ or $\mathbb{a}$, respectively, that can be converted by simultaneous row and column permutations into a matrix that contains $\mathbb{A}(K_5)$ as one of its diagonal blocks.

Note, however, that simpler reaction networks and graphs of fivefold symmetry may occur, if some of the edges of K_5 are not present, for example, if only those edges are retained that appear as the peripherial edges of the representation of Figure 4. The resulting cycle has fivefold symmetry, but the graph is no longer K_5. Furthermore, in higher dimensions, for example, in reaction graphs and networks of reaction mechanisms on relaxed potential surfaces involving three or more active coordinates, all realizations of fivefold symmetry are possible, since K_5 may be a subgraph of nonplanar reaction graphs.

Acknowledgments

The research leading to the developments reviewed above has been supported by both operating and strategic research grants from the Natural Sciences and Engineering Research Council of Canada.

REFERENCES

1. H. Jahn and E. Teller, "Stability of polyatomic molecules in degenerate electronic states. I. Orbital degeneracy," *Proc. R. Soc. London, Ser. A,* **161**, 220–235 (1937).

2. E. B. Wilson, Jr., J. C. Decius, and P. C. Cross, *Molecular Vibrations*. McGraw-Hill, New York, 1955.

3. R. McWeeny, *Symmetry, an Introduction to Group Theory and Its Applications*. Pergamon Press, New York, 1963.

4. R. M. Hochstrasser, *Molecular Aspects of Symmetry*. Benjamin, New York, 1966.

5. (a) J. N. Murrell and K. Laidler, "Symmetries of activated complexes," *Trans. Faraday Soc.*, **64**, 371–377 (1968), (b) J. N. Murrell and G. L. Pratt, "Statistical factors and the symmetry of transition states," *Trans. Faraday Soc.*, **66**, 1680–1684 (1970).

6. D. Bishop, *Group Theory in Chemistry*. Oxford Univ. Press, London, 1973.

7. R. E. Stanton and J. W. McIver, Jr., "Group theoretical selection rules for the transition states of chemical reactions," *J. Am. Chem. Soc.*, **97**, 3632–3636 (1975).

8. T. D. Bouman, C. D. Duncan, and C. Trindle, "Group theory of reaction mechanisms: Permutation theoretic prediction and computational support for pseudorotation modes in $C_2H_5^+$ and $C_5H_5^+$ rearrangements," *Int. J. Quant. Chem.*, **11**, 399–413 (1977).

9. K. Ishida, K. Morokuma, and A. Komornicki, "The intrinsic reaction coordinate. Ab initio calculation for $HNC \rightarrow HCN$ and $H^- + CH_4 \rightarrow CH_4 + H^-$," *J. Chem. Phys.*, **66**, 2153–2156 (1977).

10. S. Kato and K. Fukui, "Reaction ergodography, methane-tritium reaction," *J. Am. Chem. Soc.*, **98**, 6395–6397 (1976).

11. P. Pechukas, "On simple saddle points of a potential surface, the conservation of nuclear symmetry along paths of steepest descent, and the symmetry of transition states," *J. Chem. Phys.*, **64**, 1516–1521 (1976).

12. C. A. Mead and D. G. Truhlar, "On the determination of Born–Oppenheimer nuclear motion wavefunctions including complications due to conical intersections and identical nuclei," *J. Chem. Phys.*, **70**, 2284–2296 (1979).

13. G. S. Ezra, *Symmetry Properties of Molecules*. Springer-Verlag, Berlin, 1982.

14. I. Hargittai and M. Hargittai, *Symmetry Through the Eyes of a Chemist*. VCH, Weinheim, 1986; paper: New York, 1987.

15. P. G. Mezey, *Potential Energy Hypersurfaces*. Elsevier, Amsterdam, 1987.

16. P. G. Mezey, "Reaction topology and quantum chemical molecular design on potential energy surfaces," in *New Theoretical Concepts for Understanding Organic Reactions*, J. Bertran and I. G. Csizmadia (eds.). Kluwer Academic Publ., Dordrecht, 1989, pp. 55–76.

17. P. G. Mezey, "Three-dimensional topological aspects of molecular similarity," in *Concepts and Applications of Molecular Similarity*, G. M. Maggiora and M. A. Johnson (eds.). Wiley, New York, 1990.

18. P. G. Mezey, "A global approach to molecular symmetry: Theorems on symmetry relations between ground and excited state configurations," *J. Am. Chem. Soc.*, **112**, 3791–3802 (1990).

19. P. G. Mezey, "Molecular point symmetry and the phase of the electronic wavefunction: Tools for the prediction of critical points of potential energy surfaces," *Int. J. Quant. Chem.*, in press.

20. P. G. Mezey, "The topology of energy hypersurfaces. II. Reaction topology in Euclidean spaces," *Theor. Chim. Acta*, **63**, 9–33 (1983).

21. P. G. Mezey, "The metric properties of the reduced nuclear configuration space," *Int. J. Quant. Chem.*, **26**, 983–985 (1984).

22. Ref. 15, p. 367.

23. P. G. Mezey, "Catchment region partitioning of energy hypersurfaces. I," *Theor. Chim. Acta*, **58**, 309–330 (1981).

24. Ref. 15, p. 76.

25. C. C. Lee, E. C. Hass, C. A. Obafemi, and P. G. Mezey, "A theoretical study on the protonation of cycloalkanes C_nH_{2n} ($n=3$ to 6)," *J. Comput. Chem.*, **5**, 190–196 (1984).

26. P. G. Mezey, "Quantum chemical reaction networks, reaction graphs and the structure of potential energy hypersurfaces," *Theor. Chim. Acta*, **60**, 409–428 (1982).

27. F. Harary, *Graph Theory*. Addison-Wesley, Reading, MA, 1969.

14 *Buckminsterfullerene, Part A: Introduction*

D. J. Klein and T. G. Schmalz

1. Article Series

The present article is a part of a series on the C_{60} carbon cluster called buckminsterfullerene. This series has five sections:

A: D. J. Klein and T. G. Schmalz, Introduction.

B: A. Léger, L. d'Hendecourt, L. Verstraete, and W. Schmidt, Polyhedral Carbon Ions as Good Candidates to a Solution of the Longest Unsolved Spectroscopic Problem in Astronomy.

C: E. Brendsdal, S. J. Cyvin, B. N. Cyvin, J. Brunvoll, D. J. Klein, and W. A. Seitz, Hückel Energy Levels.

D: V. Elser, E. Brendsdal, S. J. Cyvin, J. Brunvoll, B. N. Cyvin, and D. J. Klein, Kekulé Structures.

E: E. Brendsdal, J. Brunvoll, B. N. Cyvin, and S. J. Cyvin, Molecular Vibrations.

2. Overview

In 1985 at Rice University (of Houston, Texas) Kroto et al. found that laser vaporization of graphite in a (relatively) high-pressure supersonic nozzle produces a persistent C_{60} species, which they then proposed [1] takes the form of a truncated icosahedron. This multiply fivefold symmetric structure with carbons at the vertices and σ-bonds along the edges, and delocalized π-bonding is shown in Figure 1. This "uniquely elegant" soccer-ball-like structure they termed "buckminsterfullerene." This discovery and proposal engendered first, widespread news comments (see, e.g. [2]); second, an avalanche of about 100 purely theoretical papers to date; and third, further experimental work and interpretation. Even earlier Yoshida and Osawa [3], Bochvar and Galpern [4], Davidson [5], and Castells and Serratosa [6] had published comments on this hypothetical species, while Jones [7] had considered a general class of structures of which C_{60} is a member. Further there are rumors of early unpublished work on the species. In the experimental realm there was earlier work [8] that under conditions of laser fluence, pressure, and time of flight differing from those of Kroto et al. [1] showed only a slightly lesser preponderance of C_{60}, but carbon cages were not conceived as a possibility, so that no widespread interest was generated.

Though the work to date has not yielded macroscopic quantities of C_{60}, the species may be very important. O'Brien et al. [9] at Rice University argued that such cage-like clusters might be the seeds for nucleation sites in soot formation, and Iijima [10] recalled electron micrographic evidence in support of this. Gerhardt et al. [11] found C_{60} species in hydrocarbon flames. The possibility of C_{60} in interstellar space was suggested [12]. It is further discussed in Part B. Other discussion of work concerning soot, its relevance in space, and further molecular-beam experiments may be found in companion articles [13, 14] in *Science*.

The theoretical work is even more extensive and has yielded additional insight. The work based upon the Hückel molecular-orbital model is reviewed

Figure 1. A view of the buckminsterfullerene structure showing one of its six equivalent fivefold axes of rotation. This truncated icosahedron is in fact one of Archimedes' semiregular polyhedra.

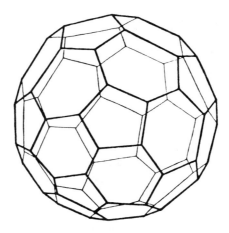

in Part C. That based on resonance-theoretic ideas is reviewed in Part D and that concerning vibrational features is covered in Part E. Especially as regards electronic structure, there is much other work, involving both semi-empirical and *ab initio* computations. In the present part we focus on some simpler rationale.

3. Stability

There is [15] a chain of considerations, thereby indicating the singular nature of the C_{60} buckminsterfullerene structure, and rationalizing some experimental observations. This is developed as a sequence of criteria to be satisfied if one seeks maximal stability of a finite cluster of carbon atoms:

1. Satisfy the tetravalency of carbon. This avoids the high cost of dangling bonds.

2. Form the system into a closed surface. This avoids edges where dangling bonds would arise, and ends up implicating (planar) sp^2-type hybridization in the surface, with delocalized π-bonding accounting for the remaining carbon valence.

3. Make the closed surface homeomorphic to a sphere. This avoids excess curvature strain, and implicates via Euler's relation rings of size 5 or less, which introduces curvature into an otherwise graphitic sheet.

4. Allow only five- and six-membered rings. This minimizes the presence of unstable three- and four-membered rings, while maximizing the number of more stable aromatic six-membered rings.

5. Avoid abutting five-membered rings. This eliminates unstable (pentalenic) fused pairs of pentagons with an antiaromatic eight-membered boundary.

6. Choose higher symmetry cages. This often occurs in repeating a more stable local structure.

Significantly these criteria all follow using straightforward mathematical and chemical arguments, which are described in more detail in [15], [16], and [17]. Notably the smallest structure satisfying criteria 1, 2, 3, and 4 as well as 5 or 6 is that of C_{60} buckminsterfullerene. Experimentally C_{60} stands out [1] with a C_{70} species of secondary prominence. Again notably the next smallest structure satisfying the above criteria is C_{70}, as the structure of Figure 2, also exhibiting a fivefold symmetry. Of tertiary prominence in the long-time high-pressure mass spectra [1, 8, 13] for large carbon clusters are those with other even numbers of carbons, while those with an odd number appear to be absent. This absence also is implied by criteria 1 and 2.

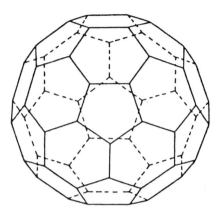

Figure 2. A view of the proposed C_{70} polyhedron from a point of view slightly offset from its single fivefold symmetry axis.

4. Other Structures

A study of other structures with yet more carbons or with a relaxation of different criteria can be made. Several other C_{60} structures have been treated, most diversely in [15]. There all structures satisfying criteria 1–4 with fewer than ~ 86 carbons and symmetry (point) groups of order greater than 12 have been identified, listed, and treated. All structures in this same class up to ~ 252 carbons with symmetry groups of order greater than 8 have been identified and listed [18]. The maximally symmetric icosahedral case has been completely characterized [16, 19] as the class of so-called Goldberg [20] polyhedra, and structures of up to over 200 carbons have been treated. Kroto and McKay [21] have termed these species "fullerenes" and suggested that in soot formation secondary shells around the nucleation site may approximate, e.g., the C_{240} structure of Figure 3. Such Goldberg polyhedra also correlate with the structures of icosahedrally symmetric viruses [22], as well as to the duals of Buckminster Fuller's geodesic domes. The realization of the first member of this sequence is the regular dodecahedron and goes back at least to Plato. The second member is the truncated icosahedron of Figure 1 and is one of the Archimedean semiregular solids.

5. Formation

A key puzzle concerns the mechanistic path by which the C_{60} buckminsterfullerene structure might be formed either in the plasma in the laser-vaporization molecular-beam experiments [7, 8, 13] or in the common flame [11] arising on combusting hydrocarbons in oxygenic at-

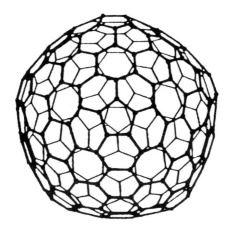

Figure 3. A view of the icosahedral C_{240} structure.

mospheres. To date one study [23] focuses solely on smaller clusters (with 25 or fewer carbons) while another [24] utilizes constructional rules with little attention to energetics. On the other hand Slanina et al. [25, 26] consider concentration ratios of a few selected cages assuming simply thermodynamic equilibrium. It seems evident that thermodynamic equilibrium does not strictly apply for otherwise all of what must be an infinite number of larger more stable structures would totally dominate. But presumably the ideas involved in our stability criteria play a crucial role for otherwise a random collection of structures would surely arise, with many different carbon numbers. A rough outline of what seems a plausible scenario follows: First chains and then single rings form up to perhaps $n \approx 20$ carbons. Then as n increases through collisions involving smaller fragments cross-linking between nonadjacent carbons in a ring sets in to generate somewhat graphitic fragments. But to minimize the number of dangling bonds around the edge, five-membered rings form to introduce curvature and give a bowl-like shape (with less edge). The five-membered rings tend not to form adjacent to one another, in consonance with criterion 5. Thence cages finally result at perhaps sizes $n \approx 40$, with the first that strictly obeys criterion 5 being the C_{60} buckminsterfullerene structure. In the bowl-like stage the edges would be very reactive and subject to rearrangement. With curvature (from five-membered rings) not introduced so rapidly larger cages such as C_{70} of Figure 2 would result.

There finally is the question of rational synthesis, to obtain macroscopic samples. Here ideas seem to be directed to the possibility of adjoining larger fragments together, e.g., two "semibuckminsterfullerene" bowls of McKee and Herndon [27]. Fabre and Rassat [28] have succeeded in synthesizing several conceivable precursor hydrocarbons, with up to 45 carbons and 9 pentagons. The ultimate preparation of buckminsterfullerene would surely be a major synthetic triumph.

REFERENCES

1. H. W. Kroto, J. R. Heath, S. C. O'Brien, R. F. Curl, and R. E. Smalley, "C_{60}: Buckminsterfullerene," *Nature (London)*, **318**, 162–163 (1985).

2. (a) R. M. Baum, "Laser vaporization of graphite gives stable 60-carbon molecules," *Chem. Eng. News*, Dec. 23, 20–22 (1985); see also the cover of this issue. (b) B. Coulter, "What is a "Buckyball"? Rice scientists find new molecule that may be shaped like soccer ball," *Houston Chronicle*, Jan. 26, Sect. 7, pp. 1, 4 (1987).

3. Z. Yoshida and E. Osawa, *Aromatic Compounds* (in Japanese), pp. 174–175. Kagakudojin, Kyoto, 1971.

4. D. A. Bochvar and E. G. Galpern, "Hypothetical systems: Carbododecahedron, s-icosahedron, and carbo-s-icosahedron," *Dokl. Akad. Nauk SSSR*, **209**, 610–612 (1973); English translation pp. 239–241.

5. R. A. Davidson, "Spectral analysis of graphs by cyclic automorphism subgroups," *Theor. Chim. Acta*, **58**, 193–231 (1981).

6. J. Castells and F. Serratosa, "Goal! An exercise in IUPAC nomenclature," *J. Chem. Educ.*, **60**, 941 (1983).

7. D. E. H. Jones, *The Inventions of Daedalus*, pp. 118–119. Freeman, San Francisco, 1982.

8. E. A. Rohlfing, D. M. Cox, and A. Kaldor, "Production and characterization of supersonic carbon cluster beams," *J. Chem. Phys.*, **81**, 3322–3330 (1984).

9. Q. L. Zhang, S. C. O'Brien, J. R. Heath, Y. Liu, R. F. Curl, H. W. Kroto, and R. E. Smalley, "Reactivity of large carbon clusters: Spheroidal carbon shells and their possible relevance to the formation and morphology of soot," *J. Phys. Chem.*, **90**, 525–528 (1986).

10. S. Iijima, "The 60-carbon cluster has been revealed!," *J. Phys. Chem.*, **91**, 3466–3467 (1987).

11. P. Gerhardt, S. Löffler, and K. H. Homann, "Polyhedral carbon ions in hydrocarbon flames," *Chem. Phys. Lett.*, **137**, 306–310 (1987).

12. H. W. Kroto, "Chemistry between the stars," *Proc. R. Inst.*, **58**, 45–72 (1986).

13. R. F. Curl and R. E. Smalley, "Probing C_{60}," *Science*, **242**, 1017–1022 (1988).

14. H. W. Kroto, "Space, stars, C_{60} and soot," *Science*, **242**, 1139–1145 (1988).

15. T. G. Schmalz, W. A. Seitz, D. J. Klein, and G. E. Hite, "Elemental carbon cages," *J. Am. Chem. Soc.*, **110**, 1113–1127 (1988).

16. D. J. Klein, W. A. Seitz, and T. G. Schmalz, "Icosahedral-symmetry carbon-cage molecules," *Nature (London)*, **323**, 703–706 (1986).

17. T. G. Schmalz, W. A. Seitz, D. J. Klein, and G. E. Hite, "C_{60} carbon cages," *Chem. Phys. Lett.*, **130**, 203–207 (1986).

18. P. W. Fowler, J. E. Cremona, and J. I. Steer, "Systematics of bonding in non-icosahedral carbon clusters," *Theor. Chim. Acta*, **73**, 1–26 (1988).

19. P. W. Fowler, "How unusual is C_{60}? Magic numbers for carbon clusters," *Chem. Phys. Lett.*, **131**, 444–450 (1986).

20. M. Goldberg, "A class of multi-symmetric polyhedra," *Tohoku Math. J.*, **43**, 104–108 (1937).

21. H. W. Kroto and K. McKay, "The formation of quasi-icosahedral spiral shell carbon particles," *Nature (London)*, **331**, 328–331 (1988).

22. D. L. D. Caspar and A. Klug, "Physical principles in the construction of regular viruses," *Cold Spring Harbor Symp. Quant. Biol.*, **27**, 1–24 (1962).

23. J. Bernholc and J. C. Phillips, "Kinetics of aggregation of carbon clusters," *Phys. Rev.*, **33B**, 7395–7398 (1986).

24. D. J. Wales, "Closed-shell structures and the building game," *Chem. Phys. Lett.*, **141**, 478–484 (1987).

25. Z. Slanina, J. M. Rudziński, and E. Ōsawa, "$C_{60}(g)$ and $C_{70}(g)$: A computational study of the pressure and temperature dependence of their populations," *Carbon*, **25**, 747–750 (1987).

26. Z. Slanina, J. M. Rudziński, and E. Ōsawa, "$C_{60}(g)$, $C_{70}(g)$, saturated carbon vapour and increase of cluster populations with temperature: A combined quantum-chemical and statistical-mechanical study," *Coll. Czech. Chem. Commun.*, **52**, 2831–2838 (1987).

27. M. L. McKee and W. C. Herndon, "Calculated properties of C_{60} isomers and fragments," *J. Mol. Struct. (Theochem).*, **153**, 75–84 (1987).

28. C. Fabre and A. Rassat, "Découpages Géometriques du footballène C_{60} en fragments en C_{15}, C_{30}, ou C_{45}—"Coupe du roi" de l'icosaèdre tronqué— Préparation de molécules voisines en C_{30} et C_{45}," *C. R. Acad. Sci. Paris*, **308**, Ser. II, 1223–1228 (1989).

15 Buckminsterfullerene, Part B: Polyhedral Carbon Ions as Good Candidates to a Solution of the Longest Unsolved Spectroscopic Problem in Astronomy

A. Léger, L. d'Hendecourt,
L. Verstraete, and W. Schmidt

1. Introduction: Spectroscopic Problems in Astronomy

Mankind has made many advances in physics through confrontation with astronomical observations: Newton mechanics—motion of planets; special relativity—Michelson–Morley experiment; general relativity—motion of the Double Pulsar, etc.

In spectroscopy, there are examples of lines that were first observed in astronomy and, after some time, identified in the laboratory as a part of the spectrum of a newly discovered atom or ion. This is one of the strongest arguments in favor of the universality of the science established on Earth, as opposed to theories stating that the physical "constants" ($\hbar, k, G \ldots$) can change with time and distance.

As an example, the spectrum of an unknown element was observed in the sunlight in 1869, and 26 years later this element was isolated on Earth and named "helios" or "helium."

Now we are facing a problem of this kind, but with the difference that it has remained persistently unsolved for over three quarters of a century: what is the origin of the so-called "diffuse interstellar bands" in the visible spectrum?

2. The Diffuse Interstellar Bands

When men had been able to combine spectrographs and telescopes, they began to analyze the light received from stars. Lines were found, and astronomers classified the stars into different types according to the kind of atomic and ionic structures they showed. Combined with the assumption that all the stars of a given type have the same properties (e.g., intrinsic luminosity) this empirical classification became a very powerful tool of investigation, in particular to establish the cosmic distance scale. After the luminosity calibration of the spectral types by triangulation on nearest objects, the distance of a given star results from its apparent luminosity and its spectral type.

Another application of this classification has been the discovery of the diffuse interstellar bands (DIBs). The basic observation is the following: when distant stars of a given type are observed, the farther they are the clearer a new set of absorption bands appears in their spectra, together with a general reddening of their emission. Both phenomena are attributed to absorption of the starlight by the interstellar medium. The strongest of the DIBs, located near 4430 Å, was discovered in the period 1910–1920 and recognized as interstellar in 1930. Since that date, the mystery of their origin remained complete although many physicists, including several Nobel price winners, have worked hard on the subject.

Over 40 such bands are presently known in the 4400–6670 Å range [1] plus about 29 newly found ones near 6800 Å [2]. Their widths spread in a large range (0.5–30 Å), and their central optical depths are between 2 and 40% for a typically reddened star (see Fig. 1 and Table 1).

The shapes of the bands tell us something about their origin. They are too broad to be some exotic atomic ions, but probably too narrow for absorption in interstellar solids, which are expected to be rather amorphous solids—at least with the information we have on them presently. On the other hand, the absence of fine structure, revealed by high-resolution studies, seems to exclude transitions in small molecules that would have resolved rotation–vibration structures. Such a reasoning points to large molecules whose electronic transitions could give the observed widths.

This idea, that large polyatomic molecules could give the correct line shape, was first proposed by Douglas [3]. However, the molecules he suggested—carbon chains like HC_nN, which are independently observed in radioastronomy in dense and dark clouds, would probably be photodis-

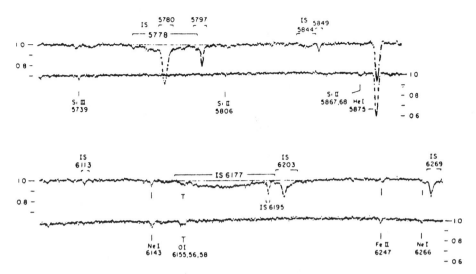

Figure 1. How the diffuse interstellar bands (DIBs) appear: Curves 1 and 3 from top: spectrum of a distant star whose light has traveled through the interstellar medium: curves 2 and 4: spectrum of a nearby star of the same spectral type (from Herbig [1]). Numbers refer to wavelengths in angstroms. Atomic and ionic lines originating in the star photosphere such as He I and Fe II are present in both spectra; the diffuse interstellar bands (DIBs) noted IS are present only in the distant star spectrum.

sociated by the hard UV photons (5–13.6 eV) present in the diffuse interstellar medium where the DIBs are seen.

A new category of interstellar molecules has recently been evidenced through their IR emission: the polycyclic aromatic hydrocarbons (PAH) [4]; they are large (50–200 atoms) and very abundant, involving about 10% of the cosmic carbon. Naturally, soon after their detection, two groups simultaneously proposed them as the carriers of the DIBs [5, 6]. In fact, the PAHs are really good candidates because they fulfill the following required criteria: they are (1) abundant, (2) stable against photo- and photothermodissociation, (3) when large enough or ionized, they have transitions in the visible region, and (4) being large, they would not exhibit resolved rotational structures.

In spite of the above arguments, the main difficulty with PAHs is their great variety. Due to the versatility of organic chemistry there is an impressive number of different PAH species within any reasonable molecular mass range that can be assigned to the interstellar features. In terrestrial natural mixtures of PAHs, this variety has been demonstrated [7]. However, it is possible that a strong selection mechanism originates from the physical processes in the interstellar medium that favors few species whose absorptions

Table 1. Diffuse Interstellar Bands in the Star HD 183143[a]

Wavelength (Å)	Central Depth (%)	Width at Half Depth (Å)
4428.0	16.0	20.0
4501.8	6.5	3.0
4726.0	5.0	5.0
4754.9	3.0	5.6
4763.0	5.0	5.3
4779.7	3.5	1.8
4882.0	6.0	17.0
5362.0	2.5	4.4
5404.3	5.5	1.0
5420.0	2.0	11.0
5449.0	4.5	14.0
5487.31	6.0	4.4
5493.8	5.0	0.8
5535.0	2.5	23.0
5544.6	3.5	0.8
5705.12	7.5	3.5
5778.3	6.0	17.0
5780.41	37.0	2.6
5794.96 ⎱ 5797.03 ⎰	22.0	1.3
5844.1	3.2	4.5
5849.79	9.5	1.0
6010.9	4.0	4.2
6042.0	2.0	14.0
6113.0	5.0	0.85
6177.1	7.0	30.0
6195.95	14.0	0.70
6203.06 ⎱ 6206.49 ⎰	16.0	2.3
6269.77	17.0	1.4
6283.91	38.0	3.8
6314.0	4.0	19.0
6353.5	3.0	3.1
6376.8	5.7	1.5
6379.3	16.0	0.86
6425.7	4.0	1.1
6597.4	3.0	0.6
6613.63	34.0	1.1
6660.71	—	—

[a] From Herbig [1].

give the DIBs. But, which are the happy few is not clear at the present state of our knowledge. This creates an operational difficulty because we do not know which ions should be selected to be studied in the laboratory.

3. The Case of Polyhedral Carbon Ions

In 1985, Kroto et al. [8] gave strong, although circumstantial, evidences for the presence of carbon polyhedral molecules (C_{50}, C_{60}, C_{70}, ...) in the laboratory. They proposed that such species, or their ions, could be abundant in the interstellar medium and responsible for different observations including the DIBs (see [9–11]).

Let us collect and discuss the arguments in favor of these candidates (see Léger et al. [12] for more details).

Let us first consider to what extent the arguments in favor of PAHs also apply to the polyhedral carbon (PC) species and especially their ions:

1. The stability of these remarkable species is expected to be very high as they are aromatic and closed structures. This is supported by a host of theoretical calculations at both the semiempirical and *ab initio* level (see Kroto [9] and references cited therein). Accordingly, these polyhedral clusters do not split into fragments after ionization in the mass spectrometer, and they are relatively resistant to chemical attack by reactive species such as NO, SO_2, and O_2 [13]. Finally, these clusters contain more than 20 atoms, which has been shown [6] to be an important requirement for thermal stability in the interstellar medium.

2. If present as free molecules in the interstellar medium, they would emit in the IR by the same process as PAHs: the energy of absorbed UV photons would be converted into vibrational energy and radiated at wavelengths characteristic of the molecule. However, PC cannot be the major carrier of the observed IR bands because (1) some bands (3.3, 11.3 mm) are typical of CH bonds while PC have no H atoms, (2) the available IR data on corannulene whose C−C bonds probably are relevant for the PC species (presence of a five-membered ring) give a poor fit to the astronomical data: the observed 6.2 mm band is missing and the PC species have 5.87 and 7.0 mm features that are not observed [14]. This is confirmed by a recent theoretical calculation of the IR active modes of C_{60} [15]: modes are found at 7.0−8.9−16.2 and 21.2 μm that do not correspond to the astronomical observations.

So, presently we have no independent estimate of the abundance of the PC species in the interstellar medium. This is a weak point but, on the other hand, they are made from carbon, which is the most abundant heavy element. It is quite possible that a fraction of 10^{-4} of the interstellar carbon is involved in these species; that would already fulfill the abundance criterion for the DIB carriers, but make their IR emission three orders of magnitude weaker than those of PAHs and therefore undetected in the presently available data. In the future, if a precise IR spectrum of C_{60} or C_{60}^+ can be derived experimentally or theoretically, an estimate of the PC abundance, based on a careful study of the astronomical IR emission, should be possible.

3. As for PAHs, electronic transitions are expected for PC species in the visible range.

For the neutral C_{60}, Heath et al. [16] have recently measured in the laboratory the first transition at ~ 3860 Å. It is weak ($f = 0.004$) and has a width of ~8 Å. The experimental setup does not definitely exclude that absorption extends into the visible range. However, the authors have experimentally searched for the 4430 Å DIB in the C_{60} spectrum and have not found it.

For the ionized C_{60}^+ species, the situation is more favorable. The optical spectrum of an aromatic cation in the visible range can be deduced from the photoelectron spectrum of the neutral species. The latter is not known for C_{60}^+, but the π-IPs (ionization potentials) of interest can be predicted using the four-parameter Hückel scheme described by Clar et al. [17]. The first five IPs, corresponding to ionization from the h_u, g_g, h_g, g_u, and t_{2u} orbitals (counted from top to bottom), are calculated to 7.12, 8.17, 8.17, 9.86, and 10.29 eV. Now, within some approximations [18], the optical transitions in the C_{60}^+ ion are simply given by the differences $\Delta IP = IP_i - IP_1$, where $i = 2, 3, 4, 5$. Bands in the near-IR and visible are therefore expected at 11,800 (two accidentally degenerate transitions), 4500, and 3900 Å. The latter two are forbidden, but become partially allowed due to Jahn–Teller forces that lower the symmetry of the ion. Accordingly, this ion has transitions that are in the wavelength range of the DIBs. Specifically, the transition calculated at 4500 Å could be actually at 4430 Å and responsible for the strongest observed DIB. In the UV range, the density of transitions for the C_{60}^+ ion is expected to be very high, and the resulting bands are likely to merge into a continuum.

The precision of the above calculation is about ±20%. If we are correct, this attribution of some of the DIBs to C_{60}^+ would have an important and testable implication: a strong absorption band should be present in the near-IR part of the extinction curve. To our knowledge, this cannot be excluded with the presently available data. We are planning new observations to look for such features.

4. PC ions are large enough for the rotational structure of their transition being compatible with the width of the narrowest DIBs (0.6 Å).

Accordingly, these species appear to be as good candidates as the PAHs except for the second criterion. This point is balanced by the absence of difficulty resulting from the large number of species. In contrast to PAHs, there are only a *reduced number of species* that are excepted to have a high stability, mainly C_{50}^+, C_{60}^+, C_{70}^+, . . . , both as theoretically expected [9] and as observed in the laboratory. This selection effect has been clearly pointed out by Kroto [9, 10]. It would be coherent with the presence of several noncorrelated families of DIBs [19]. However, it is not clear whether the PC ions could be the only carriers of the bands. The rather large range of bandwidths (0.6–30 Å) seems to argue against it, although it is possible that the lowest energy transition of a given species is narrow, the next ones broader, and those occurring at still higher energy so broad that it could explain the absence of resolved DIBs in the UV part of the interstellar extinction curve. Such progressive broadening of electronic transitions is a common feature in molecular spectroscopy and is due to the coupling of higher electronic levels with the vibrational structure of the lower ones [20].

The attribution of some DIBs to the C_{60}^{+} ion could have another testable astronomical implication as pointed out by Jura [21]. Although a precise determination of the degree of ionization of C_{60} would require a better knowledge of its ionization cross section, we estimate it around 50% in H_I regions. Consequently, some neutral C_{60} would also be present in the ISM, and its laboratory measured absorption at 3860 Å should be a part of the extinction curve. The strength of this expected band can be constrained. It depends on the f value of the transition, the abundance of C_{60}^{+} and the degree of ionization. The first quantity has been measured ($f \sim 0.004$) and the second one would result if a DIB (the 4430 Å one?) was attributed to a C_{60}^{+} transition ($\sim$4500 Å?) with a known f value. Unfortunately, the known Hückel scheme is unable to give reliable oscillator strengths. As it is forbidden in the first approximation, we can only assert the $f^{+} < 10^{-1}$, but it can be much less. From this upper limit, a minimum equivalent width ($W_\lambda = \tau_\lambda \, \Delta \, \lambda$) of the expected absorption can be derived. With the above attributions, one finds $W_{3860/A_v} > 25$ mÅ mag^{-1}.

This value is very small (for comparison, $W_{4430/A_v} = 800$ mÅ mag^{-1}) and does not provide a very useful constraint, the more as DIBs are much more difficult to observe in the UV because of the numerous stellar features in that part of the spectrum. However, we think that it is worthwhile to search for this absorption in astronomical spectra because it can be markedly stronger than the above limit.

4. Conclusion

The polyhedral carbon ions appear to be excellent candidates for the carriers of the DIBs. The laboratory determination of their visible absorptions would be most useful as it would provide a decisive criterion. It is a difficult experimental task, although a remarkable result has already been obtained on the neutral C_{60} molecule [16].

We also point out the interest to search for DIBs in the IR (1.2 μm) and UV (3860 Å) parts of astronomical spectra because such bands are predicted within this hypothesis.

The future will probably tell us if the polyhedral carbon species, which are based on the fivefold symmetry carbon cycle, are the clue to the longest spectroscopic enigma in astronomy. Anyway, trying to solve this problem should be fun!

REFERENCES

1. G. H. Herbig, "The diffuse interstellar bands," *Astrophys. J.*, **196**, 129–145 (1975).

2. G. H. Herbig, "The diffuse interstellar bands—VI—New features near 6800 Å," *Astrophys. J.*, **331**, 999–1003 (1988).

3. A. E. Douglas, "Origin of diffuse interstellar lines," *Nature (London)*, **269**, 130–132 (1977).

4. A. Léger and J. L. Puget, "Identification of the "unidentified" IR emission features of interstellar dust," *Astron. Astrophys. Lett.*, **137**, L5–L8 (1984).

5. G. P. van der Zwet and L. J. Allamandola, "Polyclyclic aromatic hydrocarbons and the diffuse interstellar bands," *Astron. Astrophys.*, **146**, 76–80 (1985).

6. A. Léger and L. d'Hendecourt, "Are PAHs the carriers of the diffuse interstellar bands in the visible?," *Astron. Astrophys.*, **146**, 81–85 (1985)

7. P. A. Peaden, M. L. Lee, Y. Hirota, and M. Novotny, "High-performance liquid chromatographic separation of high-molecular-weight polycyclic aromatic compounds in carbon black," *Anal. Chem.*, **52**, 2268–2271 (1980).

8. H. W. Kroto, J. R. Heath, S. C. O'Brien, R. F. Curl, and R. E. Smalley, "C_{60}: Buckminsterfullerene," *Nature (London)*, **318**, 162–163 (1985).

9. (a) H. W. Kroto, "Chains and grains in interstellar space," in *Polycyclic Aromatic Hydrocarbons and Astrophysics*, A. Léger, L. d'Hendecourt, and N. Boccara (eds.), pp. 197–205. Reidel, Dordrecht, 1987. (b) H. W. Kroto, "The stability of the fullerenes C_n, with n = 24, 28, 32, 36, 50, 60 and 70," *Nature (London)*, **329**, 529–531 (1987).

10. H. W. Kroto and K. McKay, "The formation of quasi-icosahedral spiral shell carbon particles," *Nature (London)*, **331**, 328–331 (1988); H. W. Kroto, "The chemistry of the interstellar medium," *Phil. Trans. R. Soc. London Ser. A*, **325**, 405–421 (1988).

11. P. Gerhardt, S. Löffler, and K. H. Homann, "Polyhedral carbon ions in hydrocarbon flames," *Chem. Phys. Lett.*, **137**, 306–309 (1987).

12. A. Léger, L. d'Hendecourt, L. Verstraete, and W. Schmidt, "Remarkable candidates for the carrier of the diffuse interstellar bands: C_{60}^+ and oher polyhedral carbon ions," *Astron. Astrophys.*, **203**, 145–148 (1988).

13. Q. L. Zhang, S. C. O'Brien, J. R. Heath, Y. Liu, R. F. Curl, H. W. Kroto, and R. E. Smalley, "Reactivity of large carbon clusters: Spheroidal carbon shells and their possible relevance to the formation and morphology of soot," *J. Phys. Chem.*, **90**, 525–528 (1986).

14. W. E. Barth and R. G. Lawton, "Dibenzo [*ghi, mno*]fluoranthene," *J. Am. Chem. Soc.*, **88**, 380–381 (1966).

15. S. J. Cyvin, E. Brendsdal, B. N. Cyvin, and J. Brunvoll, "Molecular vibrations of footballene," *Chem. Phys. Lett.*, **143**, 377–380 (1988).

16. J. R. Heath, R. F. Curl, and R. E. Smalley, "The UV absorption spectrum of C_{60} (buckminsterfullerene): A narrow band at 3860 Å," *J. Chem. Phys.*, **87**, 4236–4238 (1987).

17. E. Clar, J. M. Robertson, R. Schlögl, and W. Schmidt, "Photoelectron spectra of polynuclear aromatics—6—Applications to structural eludication: "Circumanthracene"," *J. Am. Chem. Soc.*, **103**, 1320–1328 (1988).

18. S. Obenland and W. Schmidt, "Photoelectron spectra of polynuclear aromatics—IV—The helicenes," *J. Am. Chem. Soc.*, **97**, 6633–6638 (1975).

19. J. Krelowski and G. A. H. Walker, "Three families of diffuse interstellar bands?," *Astrophys. J.*, **312**, 860–867 (1987).

20. J. L. Richards and S. A. Rice, "Radiationless processes in aromatic molecules studied in Shpolskii matrices," *Chem. Phys. Lett.*, **9**, 444–450 (1971).

21. M. Jura, Private communication, 1988.

16 Buckminsterfullerene, Part C: Hückel Energy Levels

**E. Brendsdal, S. J. Cyvin, B. N. Cyvin,
J. Brunvoll, D. J. Klein, and W. A. Seitz**

1. Introduction

The Hückel molecular orbital (HMO) electronic levels for the π-system of buckminsterfullerene have been analyzed by many investigators [1–15]. In the first of these works Bochvar and Galpern [1] reported numerical eigenvalues for this C_{60} cluster, which they termed a carbo-s-icosahedron. Next Davidson [2] independently made a graph-theoretic analysis, identified the orbital symmetries, and succintly concluded: "Should such structures or their higher homologs ever be rationally synthesized or obtained by pyrolytic routes from carbon polymers, they would be the first manifestation of authentic, discrete, three-dimensional aromaticity." These works [1, 2] were "ahead of their time" before the famous experiment of Kroto et al. [16]; after this the number of theoretical works really exploded. Most of the works [1–15] deduced the eigenvalues numerically, presumably via computer. Figure 1 shows these results with I_h symmetries assigned as in Ref. [4]. Two articles [7, 8] further show how the HMOs on the roughly spherical surface of C_{60} correlate with spherical harmonic wavefunctions on the surface of a sphere, as indicated on the far right in Figure 1. With use of the I_h symmetry Byers Brown [11] utilized his method of "reduction to characters" to obtain the eigenvalues analytically as exact numbers. Independently we have derived these numbers, reported in Section 3, by another method.

257

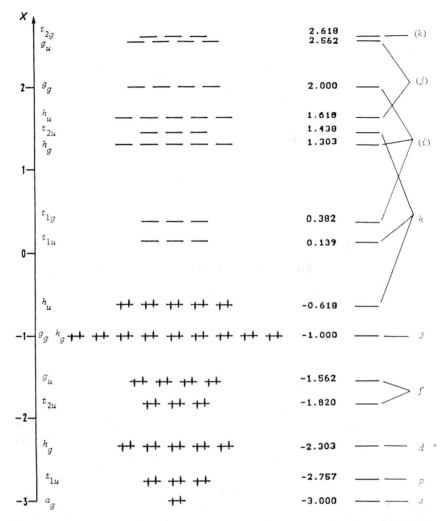

Figure 1. Hückel energy levels of buckminsterfullerene in terms of x values (energies in β units). Levels occupied in the π ground state are indicated by double vertical lines (symbolizing electron pairs). On the right, angular momentum states from which the various levels can arise are indicated. For the three highest angular momenta, where the nodally bounded regions approach the size of an atom, not all the components of an angular momentum manifold are found.

2. Hückel Determinant

In the standard HMO analysis the molecular orbitals are distributed into the irreducible representations of the I_h group according to [2]:

$$\Gamma_\pi = a_g(1) + t_{1g}(3) + t_{2g}(3) + 2g_g(4) + 3h_g(5)$$
$$+ 2t_{1u}(3) + 2t_{2u}(3) + 2g_u(4) + 2h_u(5). \tag{1}$$

The parenthetic numbers indicate degeneracies.

The secular determinant $|A + xI|$, where A is the adjacency matrix and I the unit matrix, is also called the Hückel determinant. This determinant was factorized in consonance with Eq. (1) in a tedious, but elementary way "by hand." The result is presented in Table 1.

3. Hückel Energy Levels

Now it is an easy matter to deduce the exact x values, where x are the conventional dimensionless quantities representing the π-energy levels of the HMO theory. They are reported below. The decimal numbers are given for the sake of easy identification of the levels.

Table 1. The Factored Hückel Determinant for Buckminsterfullerene[a]

	a_g $|x + 3|$		t_{1g} $|x - \phi^{-2}|$	t_{2g} $|x - \phi^2|$		
g_g	$x + \dfrac{1}{2}$	$\dfrac{1}{2} 5^{1/2}$	h_g	$x + \dfrac{9}{5}$	0	$\dfrac{2}{5} 6^{1/2}$
	$\dfrac{1}{2} 5^{1/2}$	$x - \dfrac{3}{2}$		0	$x + \dfrac{1}{2}$	$\dfrac{1}{2} 5^{1/2}$
				$\dfrac{2}{5} 6^{1/2}$	$\dfrac{1}{2} 5^{1/2}$	$x - \dfrac{3}{10}$
t_{1u}	$x + \phi^2$	$-\phi^{-1}$	t_{2u}	x	$-\phi$	
	$-\phi^{-1}$	x		$-\phi$	$x + \phi^{-2}$	
g_u	x	2	h_u	x	1	
	2	$x - 1$		1	$x - 1$	

[a] Here $\phi = \dfrac{1}{2}(5^{1/2} + 1)$ is the "golden mean," often implicated in problems involving fivefold symmetry.

$a_g(1)$: -3

$t_{1g}(3)$: $\frac{1}{2}(3 - 5^{\frac{1}{2}}) \approx 0.381966$

$t_{2g}(3)$: $\frac{1}{2}(3 + 5^{\frac{1}{2}}) \approx 2.618034$

$g_g(4)$: $-1, 2$

$h_g(5)$: $-\frac{1}{2}(13^{\frac{1}{2}} + 1) \approx -2.302776, -1, \frac{1}{2}(13^{\frac{1}{2}} - 1) \approx 1.302776$

$t_{1u}(3)$: $-8^{-\frac{1}{2}}(19 - 5^{\frac{1}{2}})^{\frac{1}{2}} - \frac{1}{4}(3 + 5^{\frac{1}{2}}) \approx -2.756598,$

$\qquad 8^{-\frac{1}{2}}(19 - 5^{\frac{1}{2}})^{\frac{1}{2}} - \frac{1}{4}(3 + 5^{\frac{1}{2}}) \approx 0.138564$

$t_{2u}(3)$: $-8^{-\frac{1}{2}}(19 + 5^{\frac{1}{2}})^{\frac{1}{2}} - \frac{1}{4}(3 - 5^{\frac{1}{2}}) \approx -1.820249,$

$\qquad 8^{-\frac{1}{2}}(19 + 5^{\frac{1}{2}})^{\frac{1}{2}} - \frac{1}{4}(3 - 5^{\frac{1}{2}}) \approx 1.438283$

$g_u(4)$: $-\frac{1}{2}(17^{\frac{1}{2}} - 1) \approx -1.561553, \frac{1}{2}(17^{\frac{1}{2}} + 1) \approx 2.561553$

$h_u(5)$: $-\frac{1}{2}(5^{\frac{1}{2}} - 1) \approx -0.618034, \frac{1}{2}(5^{\frac{1}{2}} + 1) \approx 1.618034.$

4. Characteristic Polynomial

In an impressive computational work Balasubramanian and Liu [15] derived the characteristic polynomial,

$$f(\lambda) = |\lambda I - A| = |A - \lambda I|, \tag{2}$$

for buckminsterfullerene. These investigators do not discuss HMO and Hückel energies, but this is actually what they are reporting when they give the roots of $f(\lambda)$, also referred to as the eigenvalues of A. More precisely, one has

$$x = -\lambda \tag{3}$$

and consequently

$$|A + xI| = f(-x). \tag{4}$$

In Figure 2 we show the Hückel determinant obtained by the slight modification of the characteristic polynomial [15] in consistence with Eqs. (2)–(4). When written in this form, the picture has in our opinion an intrinsic beauty, not quite unlike a modern poem.

With the knowledge of the factored Hückel determinant (cf. Table 1) and the exact x values (cf. Section 3) it is not difficult to render the polynomial of Figure 2 in factorized form. This form is included at the bottom of Figure 2. We have not established directly the identity between the two forms of the polynomial, but performed a few simple checks: (1) both forms represent a polynomial of degree 60, (2) the coefficient of x^{59} vanishes by

$$|A + xI| = x^{60} - 90x^{58} + 3825x^{56}$$
$$+ 24x^{55} - 102160x^{54}$$
$$- 1920x^{53} + 1925160x^{52}$$
$$+ 72240x^{51} - 27244512x^{50}$$
$$- 1700640x^{49} + 300906380x^{48}$$
$$+ 28113600x^{47} - 2661033600x^{46}$$
$$- 347208896x^{45} + 19180834020x^{44}$$
$$+ 3327625680x^{43} - 114118295000x^{42}$$
$$- 25376437920x^{41} + 565407465144x^{40}$$
$$+ 156652575440x^{39} - 2346799508400x^{38}$$
$$- 792175427520x^{37} + 8189116955350x^{36}$$
$$+ 3308173115904x^{35} - 24056403184260x^{34}$$
$$- 11466942645600x^{33} + 59443188508110x^{32}$$
$$+ 33076275953760x^{31} - 123163094844616x^{30}$$
$$- 79417625268960x^{29} + 212712221820840x^{28}$$
$$+ 158412719276240x^{27} - 303315997028160x^{26}$$
$$- 261359090670624x^{25} + 351861389316780x^{24}$$
$$+ 354145195147200x^{23} - 324375523213200x^{22}$$
$$- 390055074762240x^{21} + 228227031040884x^{20}$$
$$+ 344185906596720x^{19} - 122654402736360x^{18}$$
$$- 238553091055200x^{17} + 29617003666920x^{16}$$
$$+ 126428882536240x^{15} + 4679380503120x^{14}$$
$$- 49433493646080x^{13} - 8131429397135x^{12}$$
$$+ 13627897407360x^{11} + 3576552321006x^{10}$$
$$- 2527365617120x^{9} - 831616531095x^{8}$$
$$+ 310065067080x^{7} + 108565938200x^{6}$$
$$- 26034025632x^{5} - 7440712560x^{4}$$
$$+ 1566501120x^{3} + 186416640x^{2}$$
$$- 54743040x + 2985984$$
$$= (x+3)(x+1)^{9}(x-2)^{4}(x^{2}+x-3)^{5}(x^{2}-3x+1)^{3}$$
$$\times(x^{2}-x-1)^{5}(x^{2}-x-4)^{4}(x^{4}+3x^{3}-2x^{2}-7x+1)^{3}$$

Figure 2. The Hückel determinant of buckminsterfullerene, a determinant of degree 60.

virtue of the fact that the sum of the eigenvalues (or x values) vanishes, and (3) the constant term is

$$3 \times 1^9 \times (-2)^4 \times (-3)^5 \times 1^3 \times (-1)^5 \times (-4)^4 \times 1^3 = 2985984.$$

$$(5)$$

5. Implications

The HMO picture is usually expected to make reasonable qualitative predictions. The occupied orbitals (for neutral C_{60}) are exactly those that are bonding, and there is a reasonable HOMO–LUMO gap, as may be seen from Figure 1 (HOMO, highest occupied molecular orbital; LUMO, lowest unoccupied molecular orbital). This is an indicator of stability. There are six low-lying (t_{1g} and t_{1u}) nearly nonbonding MOs so that stable anions are conceivable, with charge up to -12 if one were (incorrectly) to ignore electron–electron repulsion. Amusingly all these consequences might be rationalized from purely classical ideas, in place nearly a century ago. That is, locally the structure is comprised of stable conjugated hexagonal and pentagonal rings, and further from the "rule of six" [17] one might naively expect up to additional electon could be added for each of the 12 five-membered rings.

The excited-state predictions of the HMO model are a more delicate matter. The lowest excitations $h_u \rightarrow t_{1u}$ give rise to $5 \times 3 = 15$ (with the neglect of spin multiplicities) system states, which could be well split up under the influence of electron–electron interaction. Being a $u \rightarrow u$ transition it should be optically (dipole) forbidden. The first optically allowed transition $h_u \rightarrow t_{1g}$ involves the same manifold degeneracy and lies very close to the even larger g_g, $h_g \rightarrow t_{1u}$ manifold, from which electron correlation should be even more relevant. These points are addressed in more elaborate models [18–21].

Of course, many further quantities may be computed for the HMO model, e.g., free valences, bond orders, susceptibilities, and polarizabilities. Also many more elaborate computations all the way to the *ab initio* level may be carried out and have. Though for the ground state quantitative aspects have been commendably improved, little of surprise beyond that anticipated from the simpler model seems to have emerged. See, e.g., Lüthi and Almlöf [22] for one of the more elaborate computations.

REFERENCES

1.   D. A. Bochvar and E. G. Galpern, "Hypothetical systems: Carbododecahedron, s-icosahedron, and carbo-s-icosahedron," *Dokl. Akad. Nauk SSSR*, **209**, 610–612 (1973); English translation, pp. 239–241.

2.   R. A. Davidson, "Spectral analysis of graphs by cyclic automorphism subgroups," *Theor. Chim. Acta*, **58**, 193–231 (1981).

3. A. D. J. Haymet, "C_{120} and C_{60}: Archimedean solids constructed from sp^2 hybridized carbon atoms," *Chem. Phys. Lett.*, **122**, 421–424 (1985).

4. D. J. Klein, T. G. Schmalz, G. E. Hite, and W. A. Seitz, "Resonance in C_{60}, buckminsterfullerene," *J. Am. Chem. Soc.*, **108**, 1301–1302 (1986).

5. R. C. Haddon, L. E. Brus, and K. Raghavachari, "Electronic structure and bonding in icosahedral C_{60}," *Chem. Phys. Lett.*, **125**, 459–464 (1986).

6. P. W. Fowler and J. Woolrich, "π-Systems in three dimensions," *Chem. Phys. Lett.*, **127**, 78–83 (1986).

7. M. Ozaki and A. Takahashi, "On electronic states and bond lengths of the truncated icosahedral C_{60} molecule," *Chem. Phys. Lett.*, **127**, 242–244 (1986).

8. A. J. Stone and D. J. Wales, "Theoretical studies of icosahedral C_{60} and some related species," *Chem. Phys. Lett.*, **128**, 501–503 (1986).

9. S. Satpathy, "Electronic structure of the truncated-icosahedral C_{60} cluster," *Chem. Phys. Lett.*, **130**, 545–550 (1986).

10. P. W. Fowler, "How much unusual is C_{60}? Magic numbers for carbon clusters," *Chem. Phys. Lett.*, **131**, 444–450 (1986).

11. W. Byers Brown, "High symmetries in quantum chemistry," *Chem. Phys. Lett.*, **136**, 128–133 (1987).

12. T. Shibuya and M. Yoshitani, "Two icosahedral structures for the C_{60} cluster," *Chem. Phys. Lett.*, **137**, 13–16 (1987).

13. V. Elser and R. C. Haddon, "Icosahedral C_{60}: An aromatic molecule with a vanishingly small ring current magnetic susceptibility," *Nature (London)*, **325**, 792–794 (1987).

14. V. Elser and R. C. Haddon, "Magnetic behavior of icosahedral C_{60}," *Phys. Rev.* **A36**, 4579–4584 (1987).

15. K. Balasubramanian and X. Liu, "Computer generation of spectra of graphs: Application to C_{60} clusters and other systems," *J. Comput. Chem.*, **9**, 406–415 (1988).

16. H. W. Kroto, J. R. Heath, S. C. O'Brien, R. F. Curl, and R. E. Smalley, "C_{60}: Buckminsterfullerene," *Nature (London)*, **318**, 162–163 (1985).

17. J. Thiele, "Ueber Ketonreactionen bei dem Cyclopentadiën," *Ber. Dtsch. Chem. Ges.*, **33**, 666–673 (1900).

18. P. D. Hale, "Discrete-variational-Xα electronic structure studies of the spherical C_{60} cluster: Prediction of ionization potential and electronic transition energy," *J. Am. Chem. Soc.*, **108**, 6087–6088 (1986).

19. I. László and L. Udvardi, "On the geometrical structure and UV spectrum of the truncated icosahedral C_{60} molecule," *Chem. Phys. Lett.*, **136**, 418–422 (1987).

20. S. Larsson, A. Volosov, and A. Rosén, "Optical spectrum of the icosahedral C_{60} — "follene-60"," *Chem. Phys. Lett.*, **137**, 501–504 (1987).

21. G. W. Hayden and E. J. Mele, "π Bonding in the icosahedral C_{60} cluster," *Phys. Rev.* **B36**, 5010–5015 (1987).

22. H. P. Lüthi and J. Almlöf, "*Ab initio* studies on the thermodynamic stability of the icosahedral C_{60} molecule "buckminsterfullerene"," *Chem. Phys. Lett.*, **135**, 357–360 (1987).

17
Buckminsterfullerene, Part D: Kekulé Structures

V. Elser, E. Brendsdal, S. J. Cyvin, J. Brunvoll, B. N. Cyvin, and D. J. Klein

1. Introduction

The number of Kekulé structures, or Kekulé structure count, K, has long been recognized as an important quantity for conjugated hydrocarbons. See, e.g., current textbooks in organic chemistry, or a recent monograph [1] devoted to this topic.

For C_{60} buckminsterfullerene there are exactly $K = 12{,}500$ Kekulé structures. Perhaps the evident "fiveness" of this count $K = 5^5 2^2$ is some sort of manifestation of the fivefold symmetries of the cage. The count was made [2] early via the transfer-matrix method (applied in two different ways). The number was checked at the Computer Centre of the University of Düsseldorf [3]. Schmalz et al. [4] further utilized the transfer-matrix technique to treat *all* "permissable" C_{60} cages with fivefold or higher symmetries. The K numbers have also been computated for larger icosahedrally symmetric C clusters, e.g., C_{120} [3, 5], C_{180} [5], and C_{240} [5]. Independently Hosoya [6] derived K for buckminsterfullerene as a part of his study of "Z-counting" polynomials for regular and semiregular polyhedra. The first analytical derivation of K without computer aid, using the method of fragmentation, is more recent [7]. In the following we briefly review two of these computational methods, as well as a general "Pfaffian" technique.

2. Transfer Matrix Method

This method is especially applicable in counting different types of subgraphs in a parent graph G, which has cyclic or translational symmetry. For G being the buckminsterfullerene structure there clearly is a fivefold cyclic symmetry, so that this structure may be divided into five equivalent unit cells. Such a division is indicated in Figure 1. Now to specify a Kekulé structure on G we need only specify the occupancy of the bonds on the "fore" and "aft" sides of each unit cell. Let σ_i be 0 or 1 as the ith bond on one of these sides of a unit cell is unoccupied or occupied (by a π-bond of the Kekulé structure under consideration). The occupancy pattern for the five bonds on one side of the present unit cell is abbreviated to $\sigma(5) \equiv (\sigma_1, \sigma_2, \sigma_3, \sigma_4, \sigma_5)$, for which we may conceive of $2^5 = 32$ possibilities. Further let $(\sigma'(5)|\mathbf{T}|\sigma(5))$ denote the number of ways that the occupancy patterns $\sigma(5)$ and $\sigma'(5)$ may be accommodated on the fore and aft sides of a unit cell. For our present G these take only the values 0 or 1, and the transfer matrix $\mathbf{T}$ of which they are elements is of size 32×32, though $\mathbf{T}$ falls into two 16×16 diagonal blocks containing all nonzero elements. The total Kekulé structure count K involves the sum over all fivefold products of the matrix elements such that the occupancy patterns of adjoined fore and aft sides of successive cells match. But such matching is just what occurs in ordinary matrix products, so that we have

$$K = \mathrm{Tr}\ \mathbf{T}^5 \tag{1}$$

After construction of $\mathbf{T}$ this then is readily implemented [2] and indeed also has been used [4] in treating over 50 other cages as well. Applications

Figure 1. A partitioning of the buckminsterfullerene graph into five equivalent cells. The middle sections of the bonds to be cut are missing.

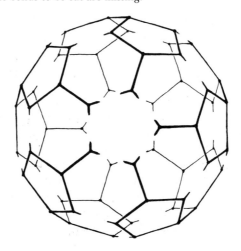

for Kekulé structures go back to Fowler and Rushbrooke [8] in 1937, while applications for a variety of (implicitly graph-theoretic) lattice-gas models were discussed by Montroll [9] in 1941, and a development for chemical electronic structure problems was emphasized only more recently [10]. In the present context the method is readily amended to obtain Pauling bond orders and conjugated circuits, for the present cages [2, 4] for polymers [10, 11], and even for two-dimensional lattices [12].

We can present **T** more or less explicitly, as

$$(\sigma'(5)|\mathbf{T}|\sigma(5)) = \sum_{\sigma_0}^{0, 1} (\sigma'_1 \, \sigma'_2 \, \sigma'_3 |\tau| \alpha_1 \, \sigma_2 \, \sigma_0)(\sigma'_5 \, \sigma'_4 \, \sigma'_0 |\tau| \sigma_5 \, \sigma_4 \, \sigma_3),$$

(2)

where the τ matrix (with entries in the natural order if the row and column labels are viewed as binary numbers) is

$$\tau = \begin{bmatrix} 1 & 0 & 0 & 1 & 0 & 1 & 1 & 0 \\ 0 & 0 & 0 & 0 & 1 & 0 & 0 & 1 \\ 0 & 0 & 1 & 0 & 1 & 0 & 0 & 0 \\ 1 & 0 & 0 & 1 & 0 & 1 & 0 & 0 \\ 0 & 1 & 1 & 0 & 0 & 0 & 0 & 0 \\ 0 & 0 & 0 & 1 & 0 & 0 & 0 & 0 \\ 1 & 0 & 0 & 0 & 0 & 0 & 0 & 0 \\ 0 & 1 & 0 & 0 & 0 & 0 & 0 & 0 \end{bmatrix}.$$

(3)

To obtain this result one considers a subdivision of the unit cell intwo two (equivalent) pieces, divided apart at the bond marked with σ_0 in Figure 2. Here τ is simply a (reduced) transfer matrix for one of these pieces. That

Figure 2. A single unit cell (cf. Section 2) with "occupancy labels" $\sigma(5)$ and $\sigma'(5)$ for the bonds to adjacent cells on the left and right.

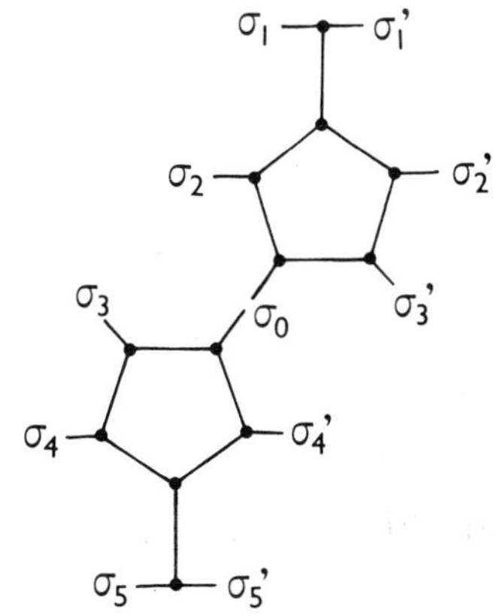

is, $(\sigma'(3) \mid \tau \mid \sigma(3))$ denotes the number of ways the occupancy patterns $\sigma(3)$ and $\sigma'(3)$ for the fore and aft sides of such a piece may be accommodated on the "reduced" cell. The formula of Eq. (2) arises as the algebraic analog of adjoining two such reduced cells together as indicated in Figure 2. The ideas here as indeed in the overall transfer-matrix method are much like those of the method of "fragmentation," but framed so that the powerful tools of matrix algebra may be applied.

3. Pfaffian Method

In a molecule in which each carbon atom belongs to exactly one double bond, a proper Kekulé structure is equivalent to a close-packed dimer configuration on the corresponding graph [13]. The elegant analyses of Temperley and Fisher [14], Fisher [15] and Kasteleyn [16, 17], applied originally to two-dimensional lattice graphs, show that the number of dimer configurations may be expressed as the Pfaffian of an appropriate skewsymmetric adjacency matrix,

$$K = \text{Pf}\mathbf{S}. \tag{4}$$

Kasteleyn [17] observes that this prescription will work whenever the graph is planar. Specifically, it is only necessary to orient the edges of the graph with arrows such that surrounding every even-sided polygon, an odd number of arrows have the clockwise (and counterclockwise) sense. A suitable orientation for buckminsterfullerene is given in Figure 3. The elements of the skew symmetric matrix $\mathbf{S}$, viz. $(\mathbf{S})_{ij}$, are now (1) $+1$ whenever an arrow is directed from vertex number i to vertex number j, (2) -1 whenever the arrow is directed from j to i, and (3) 0 otherwise, i.e., when the vertices are nonadjacent. From the relation $(\text{PF}\mathbf{S})^2 = \det \mathbf{S}$, we then have

$$K = \sqrt{\det \mathbf{S}}. \tag{5}$$

The Pfaffian method for enumeration of Kekulé structures was originally developed in the field of statistical mechanics. Only very recently it was adopted in mathematical chemistry for systematic studies of polyhexes [18]. It was found very useful in the treatment of coronoids (polyhexes with holes), where the well-known method based on the determinant of the adjacency matrix [19–21] may fail. This determinant method is generally applicable to benzenoids.

The number $K = 12{,}500$ for buckminsterfullerene was reproduced by the Pfaffian method. It gives

$$(1/60) \ln K \approx 0.1572, \tag{6}$$

which may be compared with graphite where the (maximal) $N \to \infty$ limit of $[(1/N) \ln K_N]$ is ≈ 0.1615 [12, 17].

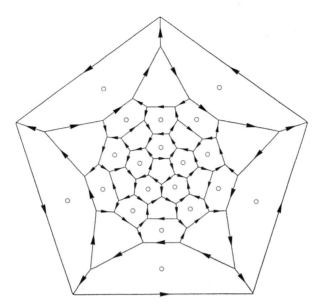

Figure 3. The orientation of edges in C_{60} buckminsterfullerene, represented as a planar graph. For the sake of convenience a small circle is inscribed in each hexagon.

The Pauling bond order [22] in a conjugated hydrocarbon is associated with a carbon–carbon bond or an edge in the corresponding chemical graph. It is defined as the number of Kekulé structures where this particular bond is double, divided by the total number of Kekulé structures (K). For benzenoids it has been proved that the Ruedenberg bond order [23] is identical with the Pauling bond order, since both of them are equal to the appropriate elements of the inverse adjacency matrix [24, 25]. A bond order was defined in analogy with the Ruedenberg bond order by [26]

$$P_{ij} = |\, (S^{-1})_{ij} \,|\,, \tag{7}$$

where S is the skewsymmetric adjacency matrix as in (4) and (5). Now we conjecture that P_{ij} coincides with the Pauling bond order whenever Eqs. (4) and (5) are valid. This hypothesis has been tested by numerous examples for coronoids [26] without detecting a single counterexample.

We have used Eq. (7) to compute the bond orders (P_{ij}) for the two carbon–carbon bonds in buckminsterfullerene to find

$$P_a = 7/25, \qquad P_b = 11/25 \tag{8}$$

in agreement with a previous result [2]. Here a refers to the edges separating pentagons and hexagons, while b refers to edges separating two hexagons. This analysis gives further support to our conjecture.

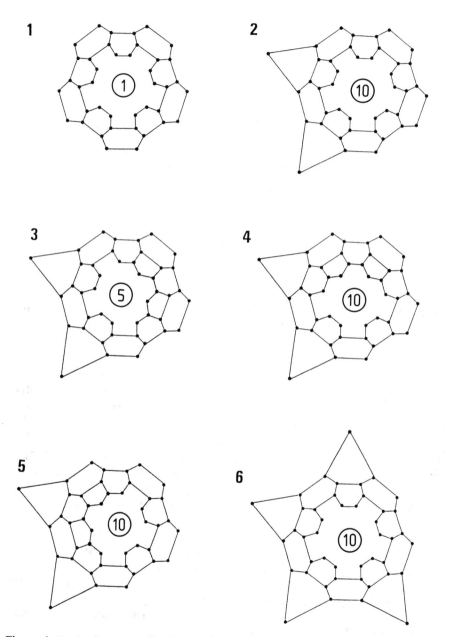

Figure 4. Twelve fragments of buckminsterfullerene. The encircled numbers are multiplicities.

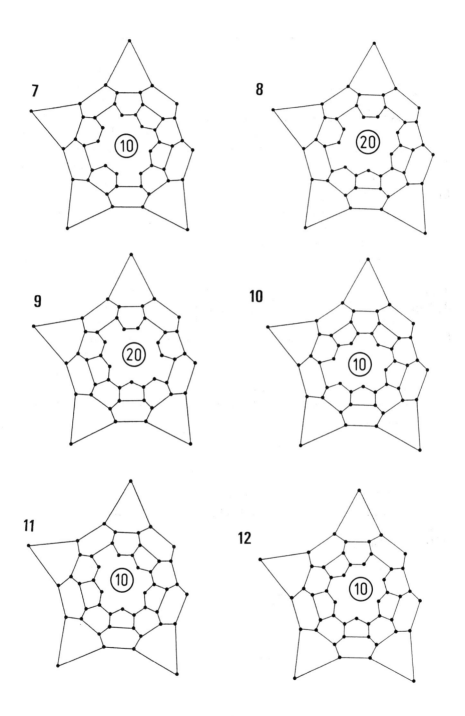

4. Method of Fragmentation

The method of fragmentation, as formulated by Randić [27], is very useful for determination of Kekulé structure counts of chemical graphs by hand [1]. In this method one edge of the graph is considered at a time, assigning to it a double and a single bond, successively. Different modifications of the method are possible by considering the bonding schemes in a set of more than one edge of the graph. The symmetry-adapted method of fragmentation [28–30] is such a modification, which has proved to be very useful. It also gave the clue to the systematization and counting of the Kekulé structures of buckminsterfullerene [7].

Consider two opposite pentagonal rings in the buckminsterfullerene model. The five edges in one pentagon have only three bonding schemes: 0, 1, or 2 double bonds. But when the two pentagons are considered together one must account for 11 bonding schemes in each of them. That gives rise to $11^2 = 121$ fragments of buckminsterfullerene. However, many of these fragments are equivalent (isomorphic). They were inspected with the result that there exist only 12 different fragments among the whole set. The number of times each fragment occurs in the set is called the multiplicity. Figure 4 shows planar graphs of the 12 fragments in question. Their multiplicities are indicated. All of these fragments contain 10 angularly annelated hexagons in a macrocyclic arrangement. In addition (apart from the first fragment, viz. No. 1 in Fig. 4), they have two, four, six, or eight pentagons embedded into bay regions. The number of Kekulé structures of buckminsterfullerene is now

$$K = K_1 + 10(K_2 + K_4 + K_5 + K_6 + K_7 + K_{11} + K_{12})$$
$$+ 5(K_3 + K_{10}) + 20(K_8 + K_9), \tag{9}$$

where $K_I(I = 1, 2, \ldots, 12)$ is the Kekulé structure count of the fragment No. I according to the numbering of Figure 4. The coefficients (viz. 1, 10, 5, and 20) are the multiplicities. The K numbers were determined by means of further fragmentations with the results:

$$K_5 = 75 \tag{10}$$

$$K_2 = K_3 = K_4 = K_6 = K_7 = K_8 = K_9 = 100 \tag{11}$$

$$K_1 = K_{10} = K_{11} = K_{12} = 125. \tag{12}$$

On inserting from (10) − (12) in (9) one attains at the result $K = 12500$ as before.

5. Matching Polynomial

Hosoya [31–33] defined for a graph the nonadjacent number $p(k)$ as the number of subgraphs consisting of k isolated edges. He introduced the Z-counting polynomial as

$$Q(x) = \sum_{k=0}^{[n/2]} p(k) \, x^k, \tag{13}$$

where n is the number of vertices. Here the coefficient of the highest power of x appears to be equal to K. Hosoya deduced the Z-counting polynomials for various graphs including buckminsterfullerene [6]. The Hosoya topological index (Z) is the sum of the coefficients of (13); for buckminsterfullerene:

$$Z = Q(1) = 1417036634543488. \tag{14}$$

The matching polynomial of a graph is strongly related to the Z-counting polynomial and perhaps more familiar. This polynomial was introduced under different names by Aihara (reference polynomial [34] and by Gutman et al. (acyclic polynomial [35]). Some references are given to further advances of the theory of matching polynomials [36–39]. Let the matching polynomial be denoted by $\alpha(x)$. Then [6]

$$\alpha(x) = \sum_{k=0}^{n} (-1)^k \, p(k) \, x^{n-2k}. \tag{15}$$

Figure 5. The matching polynomial of C_{60} buckminsterfullerene.

$$
\begin{aligned}
\alpha(x) = \; & x^{60} - 90x^{58} \\
& + 3825x^{56} - 102120x^{54} \\
& + 1922040x^{52} - 27130596x^{50} \\
& + 298317860x^{48} - 2619980460x^{46} \\
& + 18697786680x^{44} - 109742831260x^{42} \\
& + 534162544380x^{40} - 2168137517940x^{38} \\
& + 7362904561730x^{36} - 20949286202160x^{34} \\
& + 49924889888850x^{32} - 99463457244844x^{30} \\
& + 165074851632300x^{28} - 227043126274260x^{26} \\
& + 256967614454320x^{24} - 237135867688980x^{22} \\
& + 176345540119296x^{20} - 104113567937140x^{18} \\
& + 47883826976580x^{16} - 16742486291340x^{14} \\
& + 4310718227685x^{12} - 783047312406x^{10} \\
& + 94541532165x^{8} - 6946574300x^{6} \\
& + 269272620x^{4} - 4202760x^{2} \\
& + 12500
\end{aligned}
$$

Now the *K* number appears to be the constant term of $\alpha(x)$. The matching polynomial of buckminsterfullerene is given in Figure 5. It coincides exactly with the same polynomial given recently by Aihara and Hosoya [40].

REFERENCES

1. S. J. Cyvin and I. Gutman, *Kekulé Structures in Benzenoid Hydrocarbons* (Lecture Notes in Chemistry 46). Springer-Verlag, Berlin, 1988.

2. D. J. Klein, T. G. Schmalz, G. E. Hite and W. A. Seitz, "Resonance in C_{60}, buckminsterfullerene," *J. Am. Chem. Soc.*, **108**, 1301–1302 (1986).

3. N. Trinajstić, D. J. Klein, and M. Randić, "On some solved and unsolved problems of chemical graph theory," *Int. J. Quant. Chem. Symp.*, **20**, 699–742 (1986).

4. T. G. Schmalz, W. A. Seitz, D. J. Klein, and G. E. Hite, "C_{60} carbon cages," *Chem. Phys. Lett.*, **130**, 203–207 (1986).

5. D. J. Klein, W. A. Seitz, and T. G. Schmalz, "Icosahedral symmetry carbon cage molecules," *Nature (London)*, **323**, 703–706 (1986).

6. H. Hosoya, "Matching and symmetry of graphs," *Comp. Math. Appl.*, **12B**, 271–290 (1986); reprinted in *Symmetry Unifying Human Understanding*, I. Hargittai (ed.). Pergamon Press, New York, 1986.

7. E. Brendsdal and S. J. Cyvin, "Kekulé structures of footballene," *J. Mol. Struct. (Theochem.)*, **188**, 55–66 (1989).

8. R. H. Fowler and G. S. Rushbrooke, "An attempt to extend the statistical theory of perfect solutions," *Trans. Faraday Soc.*, **33**, 1272–1294 (1937).

9. E. W. Montroll, "Statistical mechanics of nearest neighbor systems," *J. Chem. Phys.*, **9**, 706–721 (1941).

10. D. J. Klein, G. E. Hite, and T. G. Schmalz, "Transfer-matrix methods for subgraph enumeration," *J. Comput. Chem.*, **7**, 443–456 (1986).

11. W. A. Seitz, G. E. Hite, T. G. Schmalz, and D. J. Klein, "Resonance in poly-polyphenanthrenes: A transfer matrix approach," in *Graph Theory and Topology in Chemistry*, R. B. King and D. H. Rouvray (eds.), pp. 458–465. Elsevier, Amsterdam, 1987.

12. D. J. Klein, G. E. Hite, W. A. Seitz, and T. G. Schmalz, "Dimer coverings and Kekulé structures on honeycomb lattice strips," *Theor. Chim. Acta*, **69**, 409–423 (1986).

13. V. Elser, "Solution of the dimer problem on a hexagonal lattice with boundary," *J. Phys.*, **A17**, 1509–1513 (1984).

14. H. N. V. Temperley and M. E. Fisher, "Dimer problem in statistical mechanics—An exact result," *Phil. Mag.*, **6**, Ser. 8, 1061–1063 (1961).

15. M. E. Fisher, "Statistical mechanics of dimers on a plane lattice," *Phys. Rev.*, **124**, Ser. 2, 1664–1672 (1961).

16. P. W. Kasteleyn, "The statistics of dimers on a lattice—I—The number of dimer arrangements on a quadratic lattice," *Physica*, **27**, 1209–1225 (1961).

17. P. W. Kasteleyn, "Dimer statistics and phase transitions," *J. Math. Phys.*, **4**, 287–293 (1963).

18. S. J. Cyvin, J. Brunvoll, and B. N. Cyvin, "Kekulé structure counts in coronoid hydrocarbons—A general solution," *Struct. Chem*, in press.

19. M. J. S. Dewar and H. C. Longuet-Higgins, "The correspondence between the resonance and molecular orbital theories," *Proc. R. Soc. London Ser. A*, **214**, 482–493 (1952).

20. D. Cvetković, M. Doob, and H. Sachs, *Spectra of Graphs—Theory and Application*. Academic Press, New York, 1980.

21. R. L. Brown, "Counting of resonance structures for large benzenoid polynuclear hydrocarbons," *J. Comput. Chem.*, **4**, 556–562 (1983).

22. L. Pauling, *The Nature of the Chemical Bond*. Cornell University Press, Ithaca, 1939.

23. K. Ruedenberg, "Free-electron network model for conjugated systems—V— Energies and electron distributions in the FE MO model and in the LCAO MO model," *J. Chem. Phys.*, **22**, 1878–1894 (1954).

24. N. S. Ham, "Mobile bond orders in the resonance and molecular orbital theories," *J. Chem. Phys.*, **29**, 1229–1231 (1958).

25. I. Samuel, "Matrices inverses des matrices alternantes de Hueckel," *Comp. Rend. Acad. Sci. Paris*, **252**, 1795–1797 (1961).

26. J. Brunvoll, B. N. Cyvin, and S. J. Cyvin, "Pauling bond orders in coronoid hydrocarbons—A general solution," *Struct. Chem.*, submitted.

27. M. Randić, "Enumeration of the Kekulé structures in conjugated hydrocarbons," *J. Chem. Soc., Faraday Trans. II*, **72**, 232–243 (1976).

28. S. J. Cyvin, J. L. Bergan, and B. N. Cyvin, "Benzenoids and coronoids with hexagonal symmetry ("snowflakes")," *Acta Chim. Hung.*, **124**, 691–705 (1987).

29. S. J. Cyvin, B. N. Cyvin, and J. Brunvoll, "Trigonal benzenoid hydrocarbons," *J. Mol. Struct. (Theochem.)*, **151**, 271–285 (1987).

30. B. N. Cyvin, S. J. Cyvin, and J. Brunvoll, "Number of Kekulé structures for circumkekulene and its homologs," *Monatsh. Chem.*, **119**, 563–569 (1988).

31. H. Hosoya, "Topological index—A newly proposed quantity characterizing the topological nature of structural isomers of saturated hydrocarbons," *Bull. Chem. Soc. Jpn.*, **44**, 2332–2339 (1971).

32. H. Hosoya, "Graphical enumeration of the coefficients of the secular polynomials of the Hückel molecular orbitals," *Theor. Chim. Acta*, **25**, 215–222 (1972).

33. H. Hosoya, "Topological index and Fibonacci numbers with relation to chemistry," *Fibonacci Q.*, **11**, 255–266 (1973).

34. J. Aihara, "A new definition of Dewar-type resonance energy," *J. Am. Chem. Soc.*, **98**, 2750–2758 (1976).

35. I. Gutman, M. Milun, and N. Trinajstić, "Non-parametric resonance energy of arbitrary conjugated systems," *J. Am. Chem. Soc.*, **99**, 1692–1704 (1977).

36. I. Gutman and H. Hosoya, "On the calculation of the acyclic polynomials," *Theor. Chim. Acta*, **48**, 279–286 (1978).

37. I. Gutman, "The matching polynomial," *Commun. Math. Chem. (Match)*, **6**, 75–91 (1979).

38. E. J. Farrell, "An introduction to matching polynomials," *J. Comb. Theory*, **B27**, 75–86 (1979).

39. C. D. Godsil and I. Gutman, "On the theory of the matching polynomial," *J. Graph Theory*, **5**, 137–144 (1981).

40. J. Aihara and H. Hosoya, "Spherical aromaticity of buckminsterfullerene," *Bull. Chem. Soc. Jpn.*, **61**, 2657–2659 (1988).

18 Buckminsterfullerene, Part E: Molecular Vibrations

E. Brendsdal, J. Brunvoll, B. N. Cyvin, and S. J. Cyvin

1. Introduction

The analysis of the molecular vibrations of the C_{60} buckminsterfullerene model represents a great challenge, which has been grasped by a few research groups. An experimental assignment of the vibrational frequencies is not yet available, but would be very welcome. Haymet [1], for instance, regarded vibrational analysis as a decisive tool in his discussion of the stability of buckminsterfullerene versus "graphitene," a regular hexagonal (D_{6h}) model of C_{60}. If the vibrational spectrum of a C_{60} compound could be recorded, it would indeed give a decisive identification of the structure. For buckminsterfullerene the selection rules predict an exceedingly simple infrared spectrum with only 4 fundamentals and a Raman spectrum with 10 fundamentals. Such low numbers are unheard of among 60-atomic molecules. For graphitene, which also has considerably high symmetry, the predictions are 25 infrared-active and 39 Raman-active fundamental frequencies. More precisely, these spectroscopic activities have been compared from the distributions of the normal vibrations among the different irreducible representations (species) [2, 3]:

$$\Gamma_{vib} \text{ (buckminsterfullerene)} = 2a_g(\text{Ra,p}) + 3t_{1g} + 4t_{2g} + 6g_g$$
$$+ 8h_g(\text{Ra,dp})$$
$$+ a_u + 4t_{1u}(\text{IR}) + 5t_{2u} \tag{1}$$
$$+ 6g_u + 7h_u$$

$$\Gamma_{vib} \text{ (graphitene)} = 10a_{1g}(\text{Ra,p}) + 9a_{2g} + 3b_{1g} + 7b_{2g} + 9e_{1g}(\text{Ra,dp})$$
$$+ 20e_{2g}(\text{Ra,dp}) + 3a_{1u} + 6a_{2u}(\text{IR}) + 10b_{1u}$$
$$+ 10b_{2u} + 19e_{1u}(\text{IR}) + 10e_{2u}. \tag{2}$$

Here the applied abbreviations are dp, depolarized; IR, infrared-active; p, polarized; Ra, Raman-active. The fundamentals of the unmarked species are spectroscopically inactive. It should be mentioned that the symmetric structure (1) has been deduced several times [4−6] in addition to the above citation [2, 3].

2. Survey of Previous Works

Haddon et al. [4] discussed, on the basis of an HMO analysis for nonplanar conjugated molecules, the vibrational spectroscopy of buckminsterfullerene in qualitative terms without reporting frequency values. The first (rudimentary) results of vibrational frequencies for buckminsterfullerene are from two *ab initio* calculations: (1) Disch and Schulman [5], using STO-3G, who reported the two a_g frequencies, viz. 1772 and 570 cm^{-1} and (2) Newton and Stanton [7], using MNDO with full geometry optimization, who reported the range of normal frequencies of molecular vibrations from 186 to 1217 cm^{-1}, including the totally symmetric (a_g) stretching mode at 1179 cm^{-1}. One of the first complete vibrational analyses of buckminsterfullerene is due to Wu et al. [6], who used a spectroscopic approach, but varied their force constant parameters to obtain agreement with quoted *ab initio* (MNDO) calculations. Table 1 includes the results. This analysis did not make use of the symmetry factoring of the vibrational matrices, but implied the diagonalization of a 180 × 180 matrix. The same year (1987) Coulombeau and Rassat [8] also published a complete set of vibrational frequencies for buckminsterfullerene (cf. Table 1) using a simple geometric potential (PSG) in conjunction with calculations of electronic properties. These two works were followed by different approaches of three research groups not knowing about each other, who published their results (cf. Table 1) the same year (1988). Brendsdal and Cyvin with collaborators

Table 1. Calculated Vibrational Frequencies (in cm^{-1}) for Buckminsterfullerene

Species*	a	b	c	d	e	f
a_g (Ra, p)	1627	1423	1830	1442	1798	1409
	548	676	510	513	660	388
t_{1g} (ia)	1464	1047	1662	1398	1459	1282
	811	486	1045	975	904	834
	567	418	513	597	623	686
t_{2g} (ia)	1665	1169	1900	1470	1502	1443
	927	496	951	890	908	1082
	726	423	724	834	867	734
	525	420	615	637	611	525
g_g (ia)	1765	1316	2006	1585	1652	1549
	1590	1195	1813	1450	1467	1384
	1174	846	1327	1158	1169	1054
	673	427	657	770	852	918
	530	419	593	614	628	502
	456	406	433	476	376	452
h_g (Ra, dp)	1831	1377	2068	1644	1816	1601
	1688	1300	1910	1465	1686	1468
	1399	1149	1575	1265	1442	1221
	1160	937	1292	1154	1325	1004
	780	764	828	801	830	743
	552	422	526	691	730	645
	428	410	413	440	463	435
	272	338	274	258	284	218
a_u (ia)	1084/970	502	1243	1206	1018	999
t_{1u} (IR)	1655	1406	1868	1437	1753	1434
	1374	1049	1462	1212	1405	1119
	551	676	618	637	776	622
	491	416	478	544	574	472
t_{2u} (ia)	1720	1322	1954	1558	1799	1510
	1309	1193	1543	1241	1384	1187
	1019	873	1122	999	1224	864
	627	421	526	690	782	755
	362	394	358	350	379	323
g_u (ia)	1765	1351	2004	1546	1728	1537
	1620	1183	1845	1401	1481	1402
	958	918	1086	1007	1298	1071
	756	485	876	832	904	827
	702	422	663	816	829	684
	374	393	360	358	515	331
h_u (ia)	1830	1296	2086	1646	1787	1602
	1579	1144	1797	1469	1500	1380
	1290	955	1464	1269	1390	1154
	770	460	849	812	990	863
	578	422	569	724	782	652
	492	418	470	531	573	519
	355	318	405	403	432	324

* Abbreviations: dp, depolarized; ia, inactive; IR, infrared-active; p, polarized; Ra, Raman-active.
[a] Wu et al. (1987) [6]; the values are rounded off to whole cm^{-1}. Assumed misprint in one species designation.
[b] Coulombeau and Rassat (1987) [8].
[c] Weeks and Harter (1988) [10]; the preferred frequency set out of two.
[d] Negri et al. (1988) [11].
[e] Slanina et al. (1989) [14]. The finer distribution into the species t_{1g}, t_{2g}, and t_{2u} was tentatively executed in the present work.
[f] Present calculations; small corrections of Cyvin et al. (1988) [2] and Brendsdal et al. (1988) [3].

[2, 3, 9] (recalculated here), as well as Weeks and Harter [10] utilized the symmetry to the effect that an 8×8 matrix, pertaining to species h_g, was the largest one to be diagonalized; cf. Eq. (1). Negri et al. [11] applied quantum-chemical (QCFF/PI) computations. The first communication on the AM1 semiempirical computations, a new parametrization of the MNDO method, of Slanina et al. appeared in 1987 [12]. Herein the vibrational frequencies of buckminsterfullerene are mapped in terms of a histogram.

More explicit information on the frequency values appeared 1 year later [13]. The vibrational frequencies were reported to range from 284 to 1816 cm^{-1}, and the number of frequencies below 500 cm^{-1} was revealed to be 22. Finally, in 1989, a full set of frequencies was published [14] (cf. Table 1), apparently from the same AM1 analysis. The range of frequencies agrees with the previous report [13], and there are 22 frequencies below 500 cm^{-1} if they are counted several times according to their degeneracies. (It seems better to say that 22 normal modes have frequencies below 500 cm^{-1}.)

3. Discussion of Vibrational Frequencies

Disch and Schulman [5] stated about their two a_g frequencies (see above) that the values are expected to be overestimated to a considerable extent; the lower frequency could be as small as 400 cm^{-1}. Our analysis confirms this suspicion, giving a value somewhat below 400 cm^{-1}; see set f (Table 1). With regard to the highest a_g frequency the two rudimentary reports from 1986 [5, 7], curiously enough, represent very nearly the two extremes: the frequency value 1179 cm^{-1} from Newton and Stanton [7] is the absolutely lowest ever reported for this mode, while 1772 cm^{-1} of Disch and Schulman [5] was overbid in only two later works (cf. Table 1). Our highest a_g frequency (set f) has the lowest value of the reported complete sets (see Table 1), but is supported by Negri et al. [11] (set d) as to its order of magnitude. With regard to the lowest a_g frequency we are alone with a value below 400 cm^{-1}, while all the other authors report values above 500 cm^{-1} (but see the statement of Disch and Schulman at the beginning of this section). On the other hand, the lowest a_{1g} frequency in coronene is of the same order of magnitude (observed 480 cm^{-1}, calculated 380 cm^{-1} [15]) as our value in question, while those of the two regular hexagonal forms of hexabenzocoronene are substantially lower (calculated 271 and 222 cm^{-1} [16]).

On comparing the complete sets of calculated frequencies (Table 1) in a general way one finds much of the same trends and similar orders of magnitude, but far from a quantitive agreement. The general trends are violated most significantly by set b. The lowest frequency in each of the other sets, for instance, is found in species h_g. Furthermore, the highest frequency is found in h_g or h_u except for set b.

The highest frequencies of sets a, c, and e (Table 1) seem to have unreasonably large values. The highest values of set d and set f are in good agreement with each other. This is also the case for most of the intermediate frequencies. For the lowest frequencies our values (set f) display in general (apart from species a_g) the greatest similarity with set a.

Our calculations (set f) contain the absolutely lowest frequency of all the sets (Table 1). It occurs in species h_g. It is comparable with the lowest e_{1g} frequency in coronene (observed 225 cm^{-1}, calculated 218 cm^{-1} [15], while the calculated lowest e_{1g} frequencies in the hexabenzocoronenes again are substantially lower (144 and 104 cm^{-1} [16]). A citation from Newton and Stanton [7] is relevant: "Its [buckminsterfullerene's] normal mode frequencies are all real and range from 186 to 1217 cm^{-1}. Included is a totally symmetric 'double-bond' stretching mode at 1179 cm^{-1}, intermediate between that of benzene (990 cm^{-1}) and ethylene (1623 cm^{-1})." In this statement the values seem to be somewhat underestimated.

The above discussion should not give the impression that we consider our set of vibrational frequencies for buckminsterfullerene to be superior to all the others (see also below). In conclusion, we wish to repeat the desirability of a synthesis of macroscopic amounts of buckminsterfullerene, which would make an experimental assignment of the frequencies feasible.

4. Vibrational Analysis

4.1. Introductory Remarks

Among the analyses of molecular vibrations referred to above our [2, 3, 9] seems to be the most "symmetry-oriented" one, perhaps in addition to the one of Weeks and Harter [10]. We adhered to a most widely used method often referred to as the Wilson **GF** matrix method (but see below). The approximation of small harmonic vibrations is adopted.

Two versions of a secular-equation method for solving the problem of molecular vibrations were launched independently by Eliashevich [17] and by Wilson [18, 19]. This started the two schools that had and still have an immense impact on the research in vibrational spectroscopy, and each resulted in a classical monograph of great importance: "Molecular Vibrations" by Wilson et al. [20] and "Kolebaniya molekul" (new, revised edition) by Volkenshtein et al. [21]. In the construction of symmetry coordinates of molecular vibrations, as defined by Wilson et al. [20], we employed the modern techniques of projection operators [22].

4.2. Structural Parameters

The buckminsterfullerene model (truncated icosahedron) belongs to the symmetry group I_h and is determined by two bond distances: D belonging to a pentagon and a hexagon and R belonging to two hexagons.

Several values of D and R have been proposed [2–5, 7, 11, 13, 23–27] as derived by different methods. Table 2 shows a collection of them, ordered according to decreasing differences $D - R$, which range from about 9 to about 2 pm.

The presently adopted values [2, 3] are near an extreme end ($D - R = 2.3$ pm); cf. Table 2. They were obtained from calculated Coulson bond orders (P^c) by means of Coulson's [28] empirical formula in the explicit form [29]:

$$X[\text{pm}] = 153.6 - \frac{19.2 P^c}{P^c + 0.765(1 - P^c)} \tag{3}$$

where $X = D$ for $P^c = 0.476$ and $X = R$ for $P^c = 0.601$.

It is interesting that an estimate of bond distances in buckminsterfullerene, seemingly as good as any, is obtainable from the Pauling bond orders (P^P). They lend themselves to a linear relationship between the bond order and bond distance, in contrast to the Coulson bond orders. Specifically, when choosing (among similar variants) the formula from Herndon [30], we have

$$X[\text{pm}] = 146.4 - 12.5 P^P. \tag{4}$$

With $P^P = 7/25 = 0.28$ and $P^P = 11/25 = 0.44$ one obtains $X = D = 142.9$ pm and $X = R = 140.9$ pm, respectively. These values agree very well with those near the bottom of Table 2 and yield $D - R = 2$ pm.

In conclusion it should be mentioned that the accuracy of structural parameters is not crucial in a vibrational analysis. Therefore, for that purpose alone much of the discussion in this paragraph is superfluous.

Table 2. Theoretical C–C Bond Distances (in pm) of Buckminsterfullerene: D(Pentagon/Hexagon), R(Hexagon/Hexagon)

D	R	Reference
146.5	137.6	Disch and Schulman (1986) [5]
145.3	136.9	Lüthi and Almlöf (1987) [25]
146.4	138.5	Schulman et al. (1987) [27]
146.4	138.5	Rudziński et al. (1988) [13]
143.6	136.0	Marynick and Estreicher (1986) [24]
147.4	140.0	Newton and Stanton (1986) [7]
147.1	141.1	Negri et al. (1988) [11]
144.4	139.3	Rudziński et al. (1988) [13]
147.6	143.3	Coulombeau and Rassat (1987) [8]
143.9	139.8	László and Udvardi (1987) [26]
143.4	140.3	Ozaki and Takahashi (1986) [23]
143.2	140.9	Present [2, 3]
142.6	140.5	Haddon et al. (1986) [4]

4.3. Symmetry Coordinates

A complete set of independent symmetry coordinates without redundancies was developed in consistence with the symmetric structure (1). These 174 mutually orthogonal linear combinations of valence coordinates for buckminsterfullerene are far too voluminous to be reproduced here. They are specified in details elsewhere [9].

4.4. Inverse Kinetic Energy Matrix

The inverse kinetic energy matrix or Wilson's **G** matrix [20] in terms of the constructed symmetry coordinates [9] was computed.

4.5. Force Field

A simple force field established for polycyclic aromatic hydrocarbons and referred to as the five-parameter approximation [15, 31–33] was employed. It actually contains five parameters for in-plane and five parameters for out-of-plane vibrations. The numerical values were transferred without modifications to the (nonplanar) model of buckminsterfullerene. Because of the lack of hydrogen atoms the number of parameters altogether reduces to five. The two force constants for the CC stretchings, counted as one parameter, were computed from the Coulson bond orders according to an adaptation of Badger's rule [34], which resulted in the formula [29]

$$f[N/m] = 179.3 \left[\frac{23.5 P^c + 76.5}{0.916 P^c + 65.48} \right]^3 . \tag{5}$$

The result is [3] $f = 421.9$ N/m and $f = 463.4$ N/M for the two stretchings pertaining to D and R, respectively. The other force constants are 40 N/m for the bending, 15 N/m for the out-of-plane bending, 5 N/m for "boat" torsions, and finally 2 N/m for torsion–torsion interactions involving the two torsions about the same bond with respect to two adjacent rings.

The five-parameter force-field approximation has been successfully applied to a number of polycyclic aromatic hydrocarbons, including some with five-membered [35, 36] (and seven-membered [36]) rings. Yet it is an open question how well this approximation works for a nonplanar carbon cluster (without hydrogens) of a cage structure like buckminsterfullerene. Therefore we cannot claim with confidence the superiority of set f to the other proposed sets of Table 1.

The force field, which is defined by the parameters specified above, determines the potential energy (force constant) matrix, **F** [20]. These parameters were transferred to the symmetry force constant matrix, i.e., the **F** matrix in terms of the symmetry coordinates [9].

5. Mean Amplitudes of Vibration and Related Quantities

The mean amplitudes of vibration (l) [37, 38] are temperature-dependent molecular constants of great importance in modern gas electron diffraction structure investigations [39]. Every interatomic distance, bonded or nonbonded, is associated with an l value. In terms of the generalized mean-square amplitudes [37, 40] one has

$$l^2 = \langle \Delta z^2 \rangle \tag{6}$$

where the right-hand side is the mean-square parallel amplitude, Δz signifying the parallel interatomic displacement. The angled bracket indicates both a quantum-mechanical and statistical-mechanical mean value. Among the generalized mean-square amplitudes are also the mean-square perpendicular amplitudes, viz. $\langle \Delta x^2 \rangle$ and $\langle \Delta y^2 \rangle$, where Δx and Δy are the perpendicular interatomic displacements related to the interatomic distance in question. Let the interatomic (equilibrium) separation be denoted by X. Then

Table 3. Equilibrium Interatomic Distances (in pm), Mean Amplitudes of Vibration (l in pm), and Perpendicular Amplitude Correction Coefficients (K in pm) at Absolute Zero, 298.15 K, and 773.15 K

	Distance		$T = 0$		$T = 298.15$		$T = 773.15$	
No.	(Multiplicity)	Equil.	l	K	l	K	l	K
1	(30)	140.9	4.656	0.229	4.722	0.254	5.328	0.405
2	(60)	143.2	4.754	0.232	4.816	0.261	5.456	0.429
3	(60)	231.7	5.551	0.170	5.809	0.201	7.252	0.358
4	(120)	246.0	5.563	0.162	5.825	0.194	7.269	0.349
5	(60)	284.0	5.914	0.142	6.318	0.173	8.195	0.317
6	(120)	355.4	5.987	0.115	6.448	0.143	8.446	0.266
7	(60)	366.5	6.095	0.110	6.583	0.138	8.670	0.256
8	(120)	408.6	6.306	0.097	6.935	0.123	9.334	0.228
9	(120)	447.4	6.425	0.087	7.142	0.110	9.710	0.205
10	(60)	459.6	6.359	0.086	7.059	0.109	9.569	0.203
11	(60)	481.4	6.523	0.080	7.331	0.102	10.06	0.189
12	(60)	515.7	6.699	0.072	7.600	0.091	10.53	0.168
13	(60)	535.2	6.762	0.069	7.721	0.086	10.75	0.158
14	(120)	545.5	6.735	0.068	7.679	0.085	10.67	0.156
15	(120)	575.1	6.886	0.062	7.937	0.078	11.12	0.141
16	(60)	602.8	6.985	0.058	8.110	0.072	11.42	0.129
17	(120)	609.4	6.996	0.057	8.134	0.070	11.47	0.126
18	(60)	645.8	7.163	0.051	8.412	0.062	11.94	0.109
19	(120)	661.2	7.212	0.049	8.499	0.059	12.09	0.103
20	(60)	666.4	7.179	0.049	8.453	0.059	12.01	0.103
21	(60)	690.8	7.295	0.045	8.646	0.054	12.34	0.092
22	(30)	691.3	7.325	0.045	8.693	0.053	12.42	0.091
23	(30)	705.5	7.359	0.043	8.754	0.051	12.53	0.086

$$K = \frac{\langle \Delta x^2 \rangle + \langle \Delta y^2 \rangle}{2X} \tag{7}$$

is referred to as the perpendicular amplitude correction coefficient. This K value is also of great importance in gas electron diffraction studies [39].

In buckminsterfullerene there are 23 different interatomic distances, 2 bonded and 21 nonbonded. Their interatomic (equilibrium) separations, computed simply from the two adopted structural parameters, are given in Table 3 along with their multiplicities. These multiplicities add up to $60 \times 59/2 = 1770$.

The developed force field was used to compute the l and K values for buckminsterfullerene. They were presented by J. Brunvoll in a poster session at The Third European Symposium on Gas Electron Diffraction, Isegran near Fredrikstad, Norway, June 18–22 (1989). Table 3 includes the results at the temperatures (T) of absolute zero and 298.15 K. It is expected that an electron diffraction experiment with buckminsterfullerene would have to be conducted at a higher temperature. Therefore we have also listed the same molecular constants at 500°C (773.15 K); see Table 3.

REFERENCES

1. A. D. J. Haymet, "Footballene: A theoretical prediction for the stable, truncated icosahedral molecule C_{60}," *J. Am. Chem. Soc.*, **108**, 319–321 (1986).

2. S. J. Cyvin, E. Brendsdal, B. N. Cyvin, and J. Brunvoll, "Molecular vibrations of footballene," *Chem. Phys. Lett.*, **143**, 377–380 (1988).

3. E. Brendsdal, B. N. Cyvin, J. Brunvoll, and S. J. Cyvin, "Normal coordinate analysis of "footballene" C_{60}," *Spectrosc. Lett.*, **21**, 313–318 (1988).

4. R. C. Haddon, L. E. Brus, and K. Raghavachari, "Electronic structure and bonding in icosahedral C_{60}," *Chem. Phys. Lett.*, **125**, 459–464 (1986).

5. R. L. Disch and J. M. Schulman, "On symmetrical clusters of carbon atoms: C_{60}," *Chem. Phys. Lett.*, **125**, 465–466 (1986).

6. Z. C. Wu, D. A. Jelski, and T. F. George, "Vibrational motions of buckminsterfullerene," *Chem. Phys. Lett.*, **137**, 291–294 (1987).

7. M. D. Newton and R. E. Stanton, "Stability of buckminsterfullerene and related carbon clusters," *J. Am. Chem. Soc.*, **108**, 2469–2470 (1986).

8. C. Coulombeau and A. Rassat, "Calculs de propriétés électroniques et des fréquences normales de vibration d'agrégats carbonés formant des polyèdres réguliers et semi-réguliers," *J. Chim. Phys.*, **84**, 875–882 (1987).

9. E. Brendsdal, "Symmetry coordinates of molecular vibrations of "footballene" C_{60}," *Spectrosc. Lett.*, **21**, 319–339 (1988).

10. D. E. Weeks and W. G. Harter, "Vibrational frequencies and normal modes of buckminsterfullerene," *Chem. Phys. Lett.*, **144**, 366–372 (1988).

11. F. Negri, G. Orlandi, and F. Zerbetto, "Quantum-chemical investigation of Franck–Condon and Jahn–Teller activity in the electronic spectra of buckminsterfullerene," *Chem. Phys. Lett.*, **144**, 31–37 (1988).

12. Z. Slanina, J. M. Rudziński, and E. Ōsawa, "$C_{60}(g)$, $C_{70}(g)$, saturated carbon vapour and increase of cluster populations with temperature: A combined AM1 quantum-chemical and statistical-mechanical study," *Coll. Czech. Chem. Commun.*, **52**, 2831–2838 (1987).

13. J. M. Rudziński, Z. Slanina, M. Togasi, E. Ōsawa, and T. Iizuka, "Computational study of relative stabilities of $C_{60}(I_h)$ and $C_{70}(D_{5h})$ gas-phase clusters," *Thermochim. Acta*, **125**, 155–162 (1988).

14. Z. Slanina, J. M. Rudziński, M. Togasi, and E. Ōsawa, "Quantum-mechanically supported vibrational analysis of giant molecules: The C_{60} and C_{70} clusters," *J. Mol. Struct. (Theochem.)*, **202**, 169–176 (1989).

15. S. J. Cyvin, B. N. Cyvin, J. Brunvoll, J. C. Whitmer, and P. Klaeboe, "Condensed aromatics—Part XX—coronene," *Z. Naturforsch.*, **37a**, 1359–1368 (1982).

16. B. N. Cyvin and S. J. Cyvin, "Molecular vibrations of hexabenzocoronenes," *Spectrosc. Lett.*, **19**, 1161–1173 (1986).

17. M. A. Eliashevich, "Simple method for calculation of vibrational frequencies of polyatomic molecules," *Compt. Rend. Acad. Sci. URSS*, **28**, 604–608 (1940).

18. E. B. Wilson, Jr., "A method for obtaining the expanded secular equation for the vibration frequencies of a molecule," *J. Chem. Phys.*, **7**, 1047–1052 (1939).

19. E. B. Wilson, Jr., "Some mathematical methods for the study of molecular vibrations," *J. Chem. Phys.*, **9**, 76–84 (1941).

20. E. B. Wilson, Jr., J. C. Decius, and P. C. Cross, *Molecular Vibrations—The Theory of Infrared and Raman Vibrational Spectra*. McGraw-Hill, New York, 1955.

21. M. V. Volkenshtein, L. A. Gribov, M. A. Elyashevich, and B. I. Stepanov, *Kolebaniya molekul* (Vibrations of Molecules, in Russian). Izd. "Nauka." Moscow, 1972.

22. F. A. Cotton, *Chemical Applications of Group Theory*, 2nd ed. Wiley-Interscience, New York, 1971.

23. M. Ozaki and A. Takahashi, "On electronic states and bond lengths of the truncated icosahedral C_{60} molecule," *Chem. Phys. Lett.*, **127**, 242–244 (1986).

24. D. S. Marynick and S. Estreicher, "Localized molecular orbitals and electronic structure," *Chem. Phys. Lett.*, **132**, 383–386 (1986).

25. H. P. Lüthi and J. Almlöf, "Ab initio studies on the thermodynamic stability of the icosahedral C_{60} molecule "buckminsterfullerene"," *Chem. Phys. Lett.*, **135**, 357–360 (1987).

26. I. László and L. Udvardi, "On the geometrical structure and UV spectrum of the truncated icosahedral C_{60} molecule," *Chem. Phys. Lett.*, **136**, 418–422 (1987).

27. J. M. Schulman, R. L. Disch, M. A. Miller, and R. C. Peck, "Symmetry clusters of carbon atoms: The C_{24} and C_{60} molecules," *Chem. Phys. Lett.*, **141**, 45–48 (1987).

28. C. Coulson, "The electronic structure of some polyenes and aromatic molecules—VII—Bonds of fractional order by the molecular orbital method," *Proc. R. Soc. London Ser A.*, **169**, 413–428 (1939).

29. J. C. Whitmer, S. J. Cyvin, and B. N. Cyvin, "Harmonic force field and bond orders for naphthalene, anthracene, biphenylene and perylene with mean amplitudes for perylene," *Z. Naturforsch.*, **33a**, 45–54 (1977).

30. W. C. Herndon, "Resonance theory—VI—Bond orders," *J. Am. Chem. Soc.*, **96**, 7605–7614 (1974).

31. A. Bakke, B. N. Cyvin, J. C. Whitmer, S. J. Cyvin, J. E. Gustavsen, and P. Klaeboe, "Condensed aromatics— Part II—The five-parameter approximation of the in-plane force field of molecular vibrations," *Z. Naturforsch.*, **34a**, 579–584 (1979).

32. S. J. Cyvin, B. N. Cyvin, J. Brunvoll, J. C. Whitmer, P. Klaeboe, and J. E. Gustavsen, "Condensed aromatics—Part III—In-plane molecular vibrations of pyrene," *Z. Naturforsch.*, **34a**, 876–886 (1979).

33. B. N. Cyvin, G. Neerland, J. Brunvoll, and S. J. Cyvin, "Condensed aromatics—Part VI—Force-field approximation for the out-of-plane molecular vibrations," *Z. Naturforsch.*, **35a**, 731–738 (1980).

34. R. M. Badger, "A relation between internuclear distances and bond force constants," *J. Chem. Phys.*, **2**, 128–131 (1934).

35. P. Klaeboe, S. J. Cyvin, A. Phongsatha Asbjørnsen, and B. N. Cyvin, "Condensed aromatics—XIV—Fluoranthene," *Spectrochim. Acta*, **37A**, 655–661 (1981).

36. B. N. Cyvin and S. J. Cyvin, "Condensed aromatics—Part XVIII—Azulene," *Spectrosc. Lett.*, **15**, 549–555 (1982).

37. S. J. Cyvin, *Molecular Vibrations and Mean Square Amplitudes*. Universitetsforlaget, Oslo and Elsevier, Amsterdam, 1968.

38. Y. Morino, K. Kuchitsu, and T. Shimanouchi, "The mean amplitudes of thermal vibrations in polyatomic molecules—I—$CF_2=CF_2$ and $CH_2=CF_2$," *J. Chem. Phys.*, **20**, 726–733 (1952).

39. K. Kuchitsu and S. J. Cyvin, "Representation and experimental determination of the geometry of free molecules," in *Molecular Structures and Vibrations*, S. J. Cyvin (eds.), pp. 183–211. Elsevier, Amsterdam, 1972.

40. Y. Morino and E. Hirota, "Mean amplitudes of thermal vibrations in polyatomic molecules—III—The generalized mean amplitudes," *J. Chem. Phys.*, **23**, 737–747 (1955).

19

Centropolyindans: Benzoannelated Polyquinanes with a Central Carbon Atom

Dietmar Kuck

1. Introduction

There are three principal ways to fuse two five-membered rings: by unifying only one member, two adjacent members, or three consecutive members of each. In organic chemistry, these types of bicyclic structures are well known. In the case of the carbocyclic parent system, cyclopentane (**1**), two rings can be combined to give spiro[4.4]nonane (**2**), bicyclo[3.3]octane (**3**), and bicyclo[2.2.1]heptane (norbornane, **4**). All of them are stable, low-strain molecular systems.

Two stereoisomers of bicyclo[3.3.0]octane (**3**) are known, viz. the *cis* and the *trans* form (**3a** and **3b**). Similar to the stereoisomers of decalin (**5**) and hydrindan (**6**)[1], which do not differ in stability, only **3a** is essentially unstrained; but **3b** is by 27 kJ mol^{-1} less stable than **3a** due to the unfavorable *trans* fusion of the two cyclopentane rings [1, 2]. However, its strain energy is about the same as that of **4** [1, 3]. At this point, the unprepared reader may start wondering about the possibilities of fusing more than two cyclopentane rings in a common polycyclic framework.

1   2   3   4

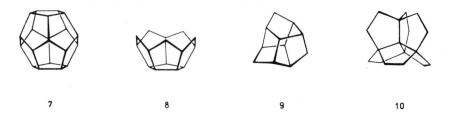

3a   3b   5   6

The chemistry of multiply fused cyclopentane hydrocarbons and their derivatives, termed "polyquinanes" [4–6] has been a very active field of research during the past two decades. In terms of the variety and ease of ring fusion, they indeed represent a borderline class of polycyclic hydrocarbons. Polyquinanes are obviously not as strained as polycyclic small-ring systems (e.g., tetrahedranes, [*l.m.n*]propellanes, cubanes, etc.), yet they are much less known than low-strain polycyclic cyclohexane systems (e.g. [2.2.2]bicyclooctane, twistane, and adamantane) [2, 3]. It appears that the beautiful symmetry of the diamond structure, in particular, has much impacted the understanding of fused cyclohexane hydrocarbons. "Steric fit" and symmetry of multiply fused cyclopentane structures seem less obvious, in part perhaps because of the nonexistence of a diamond modification with fivefold symmetry [7].

Of course, dodecahedrane (**7**), the "Platonic hydrocarbon" among the polyquinanes, is well established as a low-strain polycyclus with perfect molecular symmetry [8]. Several hexaquinanes, e.g. "'[5] peristylane" (**8**), a "C_{15}-hexaquinane," are subunits of the dodecahedrane sphere [4–6, 9]. But consider D_3-trishomocubane (**9**)[10] and **10**, a "C_{17}-hexaquinane," the key polyquinane of this article. Both contain six mutually fused cyclopentane rings as well, but are they "easy-fusing" polyquinane structures?

They are! The following sections will concentrate on various polyquinanes related to **10**. Special emphasis will be given to the benzoannelated analogues, "centropolyindans," a readily accessible and fascinating class of polycyclic organic molecules.

7   8   9   10

2. The Principle of Centropolycyclic Ring Fusion

Indeed, there are many different ways to fuse several cyclopentane rings. Linearly (**A**), angularly (**B**), and axially (**C**) annelated skeletons appear in the exciting field of naturally occurring cyclopentanoids[4–6, 11], and the spherical fusion (**D**) is the structural feature of **7–9** and of other unnatural polyquinanes [3–5, 9]. However, another classification of polycyclic structures, proposed a few years ago [7], includes even more complex, three-dimensional cyclopentanoid frameworks in a particularly clear manner. This array of cyclopentane hydrocarbons has been called *centropolyquinanes* [7]. Centrohexaquinane (**10**) is the most perfect and beautiful among the centropolyquinane hydrocarbons. As a particular feature, the atomic connectivity of such centrohexacyclic species is *topologically nonplanar*; that is, their molecular structure cannot be drawn without at least one mutual crossing of atomic bonds [12, 13].

A B C D

Corresponding to the six edges of a tetrahedron, there is a maximum of six possible bridges between its corners (**E**, Scheme 1)—or between the α substituents of each tetrahedral atomic arrangement derived from methane (**F**). A straightforward way to describe the centrohexacyclic structure of **10** is to bridge *all* pairs of α-C atoms in neopentane (**11**) by ethylene units. Besides the central carbon atom, the four α-C atoms are the only branching points of the polycyclic framework. Branching within the C_2H_4 bridges is not allowed in "regular" centropolyquinanes. All regular centropolyquinanes with less than six rings can be derived from **10**; they are characterized by a central carbon atom that is common to all rings, and by up to four α-C atoms as the only further branching points. Spiro[4.4]nonane (**2**) and *cis*-bicyclo[3.3.0]octane (**3a**) are the simplest centropolyquinanes, viz., centro-*di*quinanes. Norbornane (**4**), however, having a β-branching atom, is not a regular centropolyquinane. The whole array of regular centropolyquinanes with the energetically favorable *cis* fusion of the rings is listed schematically in Scheme 1 [7].

The number of possibilities for connecting a pair of α-C atoms of **11** is equal to that to remove one of the six bridges of **10**. Hence the monocyclus **12** (or **1**) has only one pentacyclic counterpart, centro*penta*quinane (**18**). Twofold bridging gives two centrodiquinanes (**2** and **3a**); correspondingly, removal of two bridges from **10** leads to two centro*tetra*quinanes (**16** and **17**). Both pairs of di- and tetraquinanes consist of one isomer with formal D_{2d} symmetry (**2** and **17**) and another with C_s symmetry (**3a** and **16**). Further closure or, respectively, cleavage of a ring with these four polycycles leads to three centro*tri*quinanes (**13–15**), all of them being accessible from the low-symmetry (C_s), but only one (**14**) from the high-symmetry (C_{2d}) congeners.

The nomenclature characterizing the type of centropolycyclic ring fusion denotes the number of fused C–C bonds of the neopentane core. Except

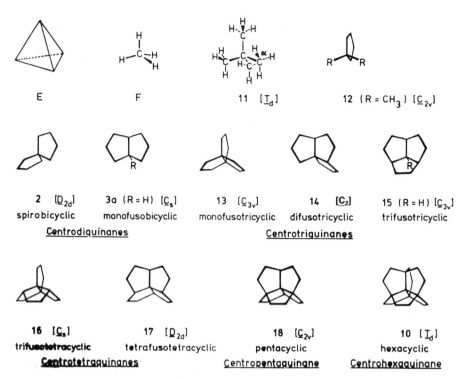

E   F   11 [$\underline{I}_d$]   12 (R = CH$_3$) [$\underline{C}_{2v}$]

2   [$\underline{D}_{2d}$]   3a (R = H) [$\underline{C}_s$]   13 [$\underline{C}_{3v}$]   14   [$\underline{C}_2$]   15 (R = H) [$\underline{C}_{3v}$]
spirobicyclic   monofusobicyclic   monofusotricyclic   difusotricyclic   trifusotricyclic
Centrodiquinanes   **Centrotriquinanes**

16 [$\underline{C}_s$]   17 [$\underline{D}_{2d}$]   18 [$\underline{C}_{2v}$]   10 [$\underline{I}_d$]
trifusotetracyclic   tetrafusotetracyclic   pentacyclic   hexacyclic
Centrotetraquinanes   **Centropentaquinane**   **Centrohexaquinane**

Scheme 1. Regular centropolyquinanes.

for the *spiro*-fused **2**, in which the rings share the central carbon atom only, *mono-*, *di-*, *tri-*, and *tetrafusocyclic* centropolyqinanes can be defined. For example, the three tricycles **13–15** can be distinguished as *monofuso-*, *difuso-*, and *trifuso*centrotriquinanes, according to the number of central C–C bonds used for ring junction.

Of course, the principle of centropolycyclic ring fusion is not restricted to centropolyquinanes. Other ring sizes, in particular the cyclohexane analogues ("centropolysexanes" [7]), mixed "centropolycyclanes," and congeners with remote branching points are possible. The regular centropolyquinanes collected in Scheme 1, however, comprise a particularly concise family of centropolycyclanes.

The geometric conditions for the multiple centroannelation of cyclopentane rings are highly favorable. Different from the cyclohexane analogues, for many of which severe steric interactions have been predicted [7], the much more flexible cyclopentane ring [2] allows the facile fusion of the rings around the central neopentane core. The envelope and half chair conformations of **1** have C–C–C bond angles very close to that of planar **1** (108°) and to those of **11** (109.5°). The steric fit of cyclopentene rings (**19**) [2] and indan units (**20**) should be even better because of less puckering compared to **1**; hence they appear even more promising as bridging entities, in particular for high degrees of annelation (Scheme 2).

From the preparative point of view, indan units are more convenient than nonaromatic C_2 bridges because of their inherent stability and the applicability of the synthetic arsenal of arene chemistry. Indeed, all attempts to synthesize **10** have failed so far. However, structural details of **10** and the corresponding hexaene have been calculated [14]. Most of the *lower* centropolyquinanes are known, but interesting saturated and unsaturated congeners have still eluded their synthesis.

By contrast, the class of benzoannelated centropolyquinanes (centro-polyindans) has been completed by synthetic research performed during the recent years in the Bielefeld laboratory [15–21], including the most beautiful among them, "centrohexaindan" (**21**) [19], the first centrohexacyclic and hence topologically nonplanar hydrocarbon. In most cases, surprisingly efficient multiple C–C bond formation approaches have been developed (Section 4).

Centropolyindans and related arenes comprise a promising new group of aromatic hydrocarbons. The arene rings at the periphery of the centro-polyquinane were expected to increase the molecular stability, allowing for more strain and a higher degree of unsaturation in the polycyclic core as well as for an increased lifetime of transient, e.g., ionic species. Furthermore, the enlarged "wings" of the rigid centropolyquinane core in centropolyindans and analogues with extended aromatic systems offer π-electron donor sites for complexation with metal fragments in sterically well-defined positions [22–24]. Finally, the intermolecular, solid-state coordination of these poly-fused, three-dimensional molecules may be of interest with respect to pin-pointed macroscopic properties of organic materials.

Scheme 2. Centrohexaindan (**21**) as viewed by fusing six indan units (**20**) along the Cartesian coordinates (x, y, z) to give four common C–C bonds along the tetrahedral axes a–d.

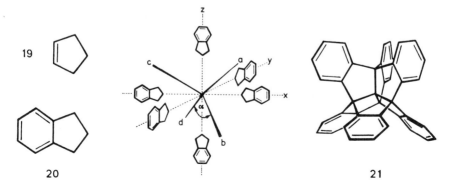

3. Selected Higher Centropolyquinanes

Before discussing the centropolyindans in detail, some few centropolyquinanes will be presented. For comprehensive discussions on polyquinane chemistry, the reader is referred to [4–6, 9, 11]. However, these articles are particularly scarce with respect to benzoannelated centropolyquinanes. Therefore, this section is restricted to *higher* centropolyquinanes, which are of some relevance for the chemistry of the centropolyindans (Scheme 3).

The *monofuso*centrotriquinanes **13** [25] and **23–26**, being prominent members of the well-known class of propellanes [26], have all been synthesized. [3.3.3]propellatriene **22** has apparently not been prepared; but triene **23**, the corresponding trione **24**, and related propellanes have been studied [27]. The trispiro derivatives **25** and **26**, as well as the related mono- and diethers, have been investigated extensively [28–33]. In contrast to the facile conversion of **26** to the isomeric, centrohexacyclic triether **42**, reported in 1981 [28, 29, 33], various attempts to isomerize **25** to the parent hydrocarbon **10** failed [32, 33].

*Trifuso*diquinane **14** has been prepared recently from a mixture of the C_2-symmetrical triene **27** and an unsymmetrical isomer [34].

Scheme 3. Centropolyquinane derivatives.

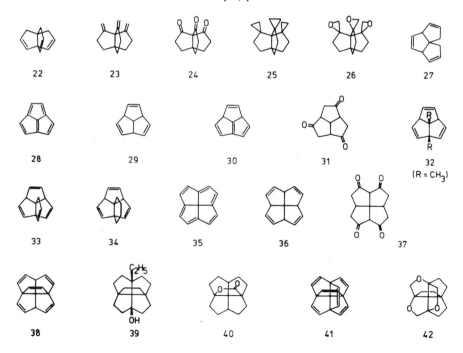

*Trifuso*centrotriquinanes such as **15** and **28–32** have been investigated extensively. Acepentalene (**28**) is the fully unsaturated and yet unknown hydrocarbon of this series. Its reactivity is predicted to be extremely high [35]. The chemistry of triquinacene (**29**) has been investigated extensively by many groups since its first synthesis by Woodward et al. in 1964 [36; 4–6, 9]. In particular, much work has been invested to achieve the face-to-face dimerization of **29** to dodecahedrane (**7**), including the idea to condense two triketones **31** [37]. Of course, the conversion of **29** to **28** has been pursued. This research led to various metal carbonyl complexes of the corresponding tetraenes, e.g., **30** [35]. Some *centro*-substituted triquinacenes (**32**) have been reported very recently [38].

*Trifuso*centrotetracyclic structures are very rare. The parent tetraquinane **16** as well as the tetraene **33** are still unknown, but the cyclohexanotriquinacene **34** has been synthesized very recently as the fist "regular" trifusocentro-tetracyclane [38].

The *tetrafuso*centrotetracyclic congeners have attracted much more interest, being members of the well-recognized class of fenestranes [3, 39–41]. The fully unsaturated hexaene **35** is of high theoretical interest, concerning the planarization of tetracoordinated carbon atoms by embedding it in rigid polycyclic framework. However, **35** seems too strained to be synthetically acessible [39]. The parent all-*cis*-[5.5.5.5]fenestrane (**17**) has been prepared by three independent routes, one of them starting from the tetraketone **37** and involving the tetraene **36** [34].

Centropentaquinane **18** and the pentaene **38** are both unknown. The only known derivative of **18** is the pentacyclic alcohol **39**, which has been obtained in an unsuccessful attempt to convert **25** to **10** [32]. Lactone **40** is a heterocyclic analogue of **18**, from which **17** can be generated by hydrogenolysis [4, 39–41].

The hexaene **41**, like **10**, is still unknown experimentally. However, both of these elusive hydrocarbons have been subject of theoretical studies to predict structural details of the hexacyclic framework [7, 14, 41]. The triether **42** was the first centrohexacyclic, topologically nonplanar molecule reported [28, 29, 33].

4. The Centropolyindans

Benzoannelated centropolyquinanes with various structural features are known. Most of them are collected in Scheme 4. The "regular" centropolyindans **43**, **49**, **55**, **60**, **65**, **71**, **74**, **77** (listed in the left-hand column of Scheme 4) and **21** are all derived from bridging the neopentane core by up to six *ortho*-phenylene units. The whole family of regular centropolyindans is now synthetically accessible, and most members have been reported. Some interesting derivatives are included in Scheme 4. In nonregular centropolyindans, the benzo nuclei are *fused* to the central neo-

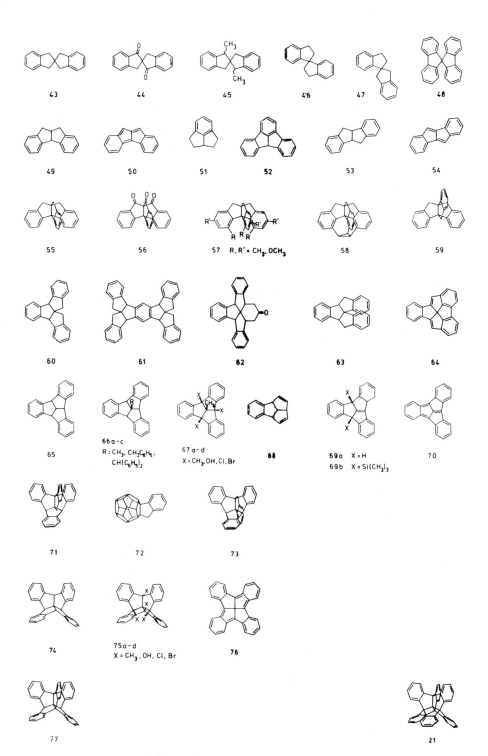

Scheme 4. Centropolyindan derivatives.

pentane core (**46–48**, **51–54**, **59**, **63**, **64**, and **73**). Several of them (**47**, **59**, **64**, and **73**) have not been reported yet.

Substituted 2,2'-spirobiindans derived from the parent hydrocarbon **43**, such as the diketone **44** [42] and the 1,1'-dimethyl derivatives **45** [43], are chiral due to their molecular C_2 symmetry and have therefore been a matter of great interest. The same holds for the 1,1'-spirobiindans (**46**) [44, 45], whereas the mixed isomer **47** is not known. A related group of chiral spiroarenes is derived from 9,9'-bifluorene (**48**), the parent compound of the vespirenes [46; 3].

The *fuso*biindans are derived from the regular centropolyindan **49** [47] and its isomer **53** [48]. Detailed studies have been performed concerning the stability and reactivity of corresponding pairs of unsaturated analogues, e.g., **50** and **54**, the latter being the more stable one [49, 50]. There are a few examples of *peri*-benzoannelated diquinanes such as **51** and **52** [3], in which appreciable strain is imposed by the fusion of two vicinal bonds of the benzo nucleus to the nonplanar diquinane moiety. As a consequence, fluoradene (**52**) is much more acidic than triphenylmethane [51].

Triptindan (**55**) is known for more than two decades [52], but various derivatives including **56** and **57** have been synthesized only recently [17, 18, 21], and a new synthetic access to **55** has been found. The three *endo*-substituents in **57** and related triptindans markedly increase the axial torsion of the propellane skeleton. The *endo*-capped derivative **58** represents a challenging goal because of its fixed bullvalene-type structure. **59**, a non-regularly fused centropolyindan, is not known either.

A very efficient cyclization route has been found to synthesize the angular triindan **60** [15]. Other *difuso*centropolyindans such as **61** [21] and **62** [16] can be prepared by using the same approach. Chiroptical properties and the absolute configuration of **62** and other benzoannelated fenestranes have been determined as well [21]. **63**, an isomer of **60**, was synthesized in an attempt to generate **64** and to study the geometry at the central carbon atom of this elusive, highly strained fenestrane [53].

The *trifuso* framework of tribenzotriquinacene **65** [20, 21] and various *centro* and bridgehead substituted derivatives (**66** and **67**) [15, 21] provides a very rigid framework with a C_{3v} symmetrical "concave" *endo* face bearing three isolated benzenoid π-systems. *Mono*benzotriquinacene (**68**) has also been described [5]. **65** and even **66** can be converted to **69b**, a derivative of the strained, still unknown, tribenzodihydroacepentalene **69a** [20, 23]. The dianion of tribenzoacepentalene (**70**) is formed as an intermediate, suggesting the possibility of generating this elusive, strained polyene [20].

*Trifuso*centrotetraindan (**71**) has been prepared very recently [21]. The only other benzoannelated *trifuso*centropolyquinane reported is indano-dodecahedrane **72** [4]. **73**, the highest unregular centropolyindan, is still unknown.

"Fenestrindan" (**74**) [16] and other *tetrafuso*centrotetraindans [21] have been studied in greater detail because of the possibility of pinpointed moderation of the geometry of tetracoordinated central carbon atom by the [5.5.5.5]fenestrane skeleton [41]. In this context, various bridgehead-substituted fenestrindans (**75**) have been studied [21]. One of the most challenging synthetic targets in this series is "fenestrindene" (**76**), the fully unsaturated congener of **74**.

Centropentaindan (**77**) has been prepared very recently as the last member of the family of regular centropolyindans [21].

Centrohexaindan (**21**) [19] certainly has the most beautiful structure of all centropolyindans since it comprises all of the other regular congeners in its molecular framework. Two efficient, independent syntheses have been developed (Section 4.1.4).

4.1. The Synthesis of the Higher Centropolyindans

4.1.1. The Cyclization Principle The synthesis of the centropolyindans is based on readily available aromatic substrates. Suitably substituted 1-indanones and 1,3-indandiones are used as starting materials, which, in most cases after reducion to the corresponding alcohols, are subjected to acid-catalyzed cyclization with concomitant elimination of water. The two principal cyclodehydration steps are generalized in Scheme 5. Although being the intramolecular variant of one of the classic of electrophilic aromatic substitution reactions (Friedel–Crafts alkylation), this approach proved extremely useful for ring closure in the centropolyindan series. The twofold C–C coupling steps performed in most cases renders most of the preparations highly efficient. One of the two syntheses of centrohexaindan (**21**) involves even a triple cyclodehydration reaction as the final step.

In the beginning of our synthetic efforts, the annelation of up to six benzo bridges around the neopentane core seemed hopeless. Several unfavorable trends had to be considered. (1) The steric crowding of organic ligands at the central, quaternary carbon atom, (2) electrofugic properties of benzylic groups emerging in cationic intermediates, (3) ring cleavage of 1,3-indandiols as substrates for cyclodehydration (Grob-type fragmentation), and (4) steric orientation of the phenyl groups to be cyclized. Indeed, various fragmentation products were obtained in earlier attempts [15, 21].

4.1.2. Triptindan, Difusocentrotriindans, and Tribenzotriquinacenes Thompson's original synthesis of triptindan (**55**) [52] has been simplyfied considerably [21]. By starting from 1,3-indandione (**78**), it can

Scheme 5.

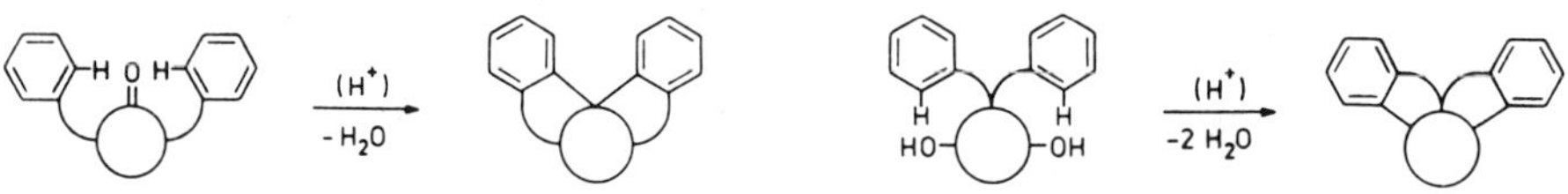

be prepared in three steps with an overall yield of ≃ 60% (Scheme 6). The *difuso* isomer **60** is obtained after reduction of **79** to the 1,3-indandiol **81** and double cyclodehydration in 85% overall yield [15]. Ten Hoeve and Wynberg prepared the nonregular centrotriindan **63** in ∼ 40% overall yield starting from 2-indanone (**82**). Again, a double cyclization was performed in the last step (**84** → **63**).

Centro-substituted tribenzotriquinacenes **66a–66c** can be synthesized from 2-substituted 2-benzhydryl-1,3-indandiones **86a–86c** via the 1,3-indandiols **87a–87c** (Scheme 7) [15, 21]. The twofold cyclodehydration of the benzhydryl group can take place only if the first cyclization leads to an *endo*-oriented intermediate (cf. **89/91**).

Some decades ago, the parent tribenzotriquinacene (**64**) was already considered a possible cyclodehydration product of epimeric diindan alcohols such as **89** [49]. However, the *exo*-orientation of the phenyl groups in **88** and **89**, realized later [54], precluded the cyclodehydration. Therefore, a three-step dehydrogenation–rehydrogenation procedure was developed to give the *endo*-phenyl stereoisomer **90**, which after reduction to **91** affords **65** in good yield (Scheme 7) [20]. This synthesis represents one of the few cases in which a stepwise cyclization procedure was found to be superior to the double cyclodehydration strategy.

4.1.3. The Centrotetraindans and Centropentaindan Trifuso-centrotetraindan (**71**, Scheme 7) has been prepared in a double cyclization step, after introducing a benzyl group in ketone **90** to give **92**. Similar to

Scheme 6.

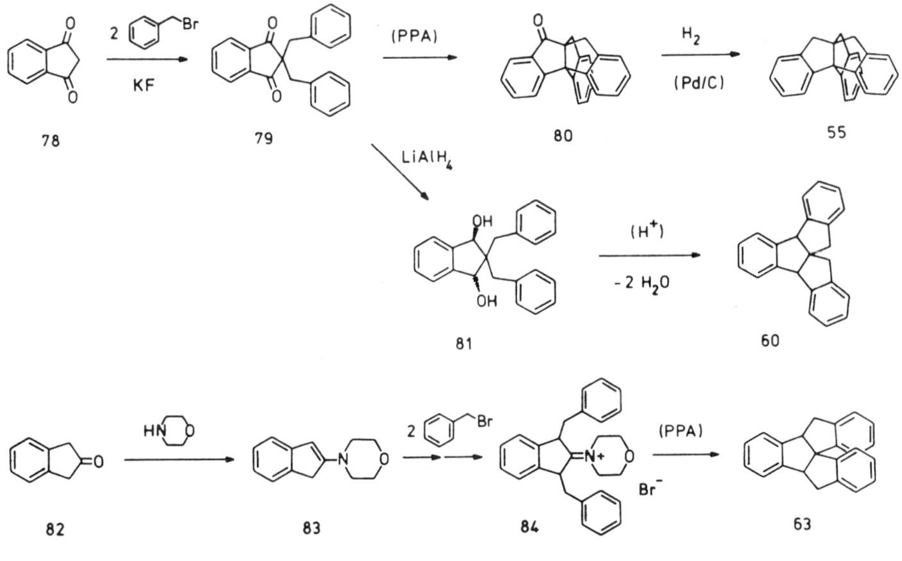

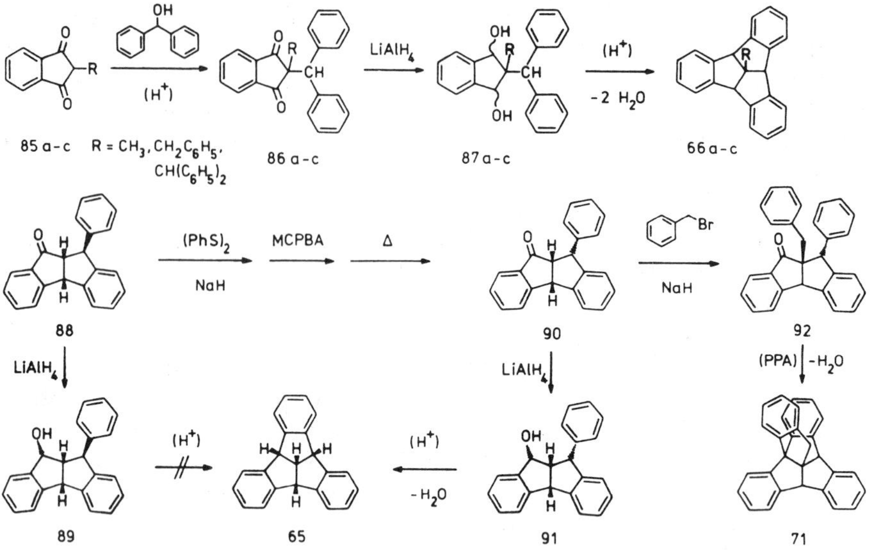

Scheme 7.

91, the *endo*-phenyl group in **92** is fixed in a very favorable orientation, and the benzyl group is probably cyclized *after* the phenyl group [21].

The double cyclodehydration strategy proved extremely convenient as an entry to the benzoannelated fenestranes (Scheme 8), including fenestrindan (**74**) [16]. The *trans*-diphenyl-spirotriketone **93** [55] provided ideal stereochemical conditions for double cyclodehydration after conversion to triol **94**. The [6.5.5.5]fenestranes **95** and **62** are obtained in ~ 50% overall yield. Some further transformations lead to tribenzo[5.5.5.5]fenestrene **96** and fenestrindan **74** [16].

Centropentaindan (**77**) is still the hardest to make of all centropolyindans. Up to now, a maximum of 9% yield has been achieved by *cyclodehydrogenation* of the benzhydryltribenzotriquinacene **66c** (Scheme 8) [21].

4.1.4. Centrohexaindan The synthesis of centrohexaindan (**21**) deserves a separate section. Two synthetic routes have been developed. It is interesting to note that they represent two of the three strategies (**G–I**) envisaged by Simmons [33] for the synthesis of the elusive centrohexaquinane (**10**).

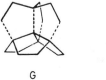

G   H   I

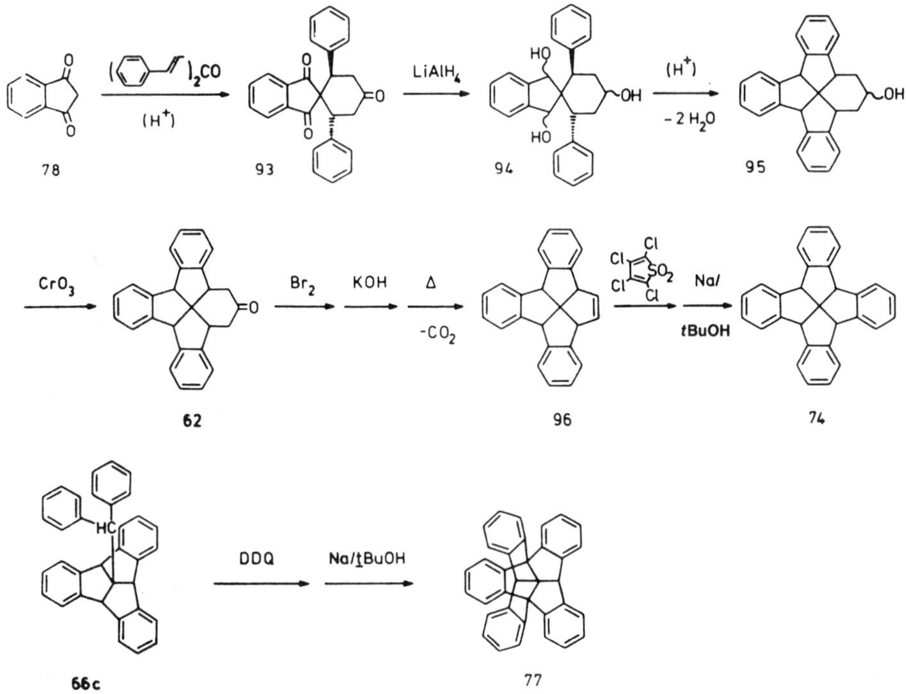

Scheme 8.

The first synthesis of **21** (Scheme 9) [19] was based on fenestrindan **74**. Its tetracyclic framework is particularly resistent to fragmentation. With regard to steric crowding in **75d**, the fourfold bridgehead bromination is surprisingly facile, and the readily occurring incorporation of two further benzene rings by fourfold C–C bond formation, corresponding to approach **H**, is even more so. In total, the synthesis of **21** comprises eleven steps from **78**, with an overall yield of ~ 4%.

The very recently accomplished second synthesis of **21** [21] is much shorter, yielding the hexacyclus in only six steps from **78** with ~ 35% overall yield. Triptindantrione (**56**) is produced in two steps from the monoketone **80**. Surprisingly again, it adds three phenyl groups to give the triol **98**, which yields **21** on threefold cyclodehydration. This synthesis, corresponding to Simmons' approach **I**, is particularly straightforward. It represents the first successful transformation of a [3.3.3]propellane to a centrohexaquinane.

4.2. Some Structural Aspects

Owing to their high molecular symmetry, it is interesting to study structural properties of the now completely accessible group of centropolyindans, both in terms of intra- and intermolecular interactions.

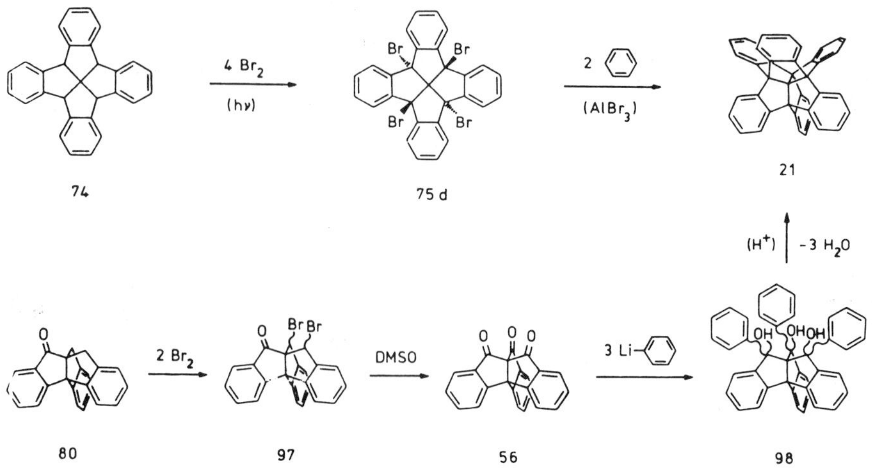

Scheme 9.

At present, spectroscopic and solid-state investigations are underway, a few results of which are discussed in this section.

In saturated centropolyquinanes, even in **10**, a limited conformational flexibility is preserved [2, 6, 14]. Unsaturated derivatives, such as **24** [56], **37** [25], **41** [14], and the centropolyindans are more rigid, and the remaining flexibility depends markedly on the degree and way of ring fusion.

Spirobiindan (**43**) adopts four equivalent ground-state conformations, according to the combination of two envelope forms [43]. Solid-state *fuso*-diindan **49** exists in two equivalent conformations, which equilibrate in solution to give a time-averaged C_s symmetric molecule. The X-ray structure of **49** [47] exhibits dihedral ("torsional") angles $\beta' = 14.8°$, $\beta'' = 16.5°$, and $\beta''' = 17.1°$ at the central C–C bond (**K**, Scheme 10). Similar conformations have been found for various chromium tricarbonyl complexes of **49** and **53** [57].

Triptindan (**55**) exists in two equivalent C_3 symmetric rotamers with, as compared to **49**, increased torsional angles $\beta' (= \beta'' = \beta''') = 23.8 \pm 0.4°$. (**K**, Scheme 10) in the solid state [21]. According to force-field calculations, the torsion is considerably increased by the interaction of three *endo* sub-stituents (e.g., **57**) [18]. X-Ray structural analysis of the *difuso* isomer **60** reveals a similar but slightly unsymmetric torsion of the two central C–C bonds ($22.0 \leq \beta \leq 24.5°$) [21].

By contrast, the trifuso isomer **66a** exists as a single C_{3v} symmetrical conformer. The orientation of the central methyl group is perfectly eclipsed ($\beta''' = 0°$, **L** in Scheme 10), and the six dihedral C–C bond angles within the indan units (e.g., β' and β'') are only 3.6°. Owing to its pronounced cup-like shape, **66a** forms intermolecular stacks along the molecular axis of symmetry with an intermolecular distance of 6.0 Å [21].

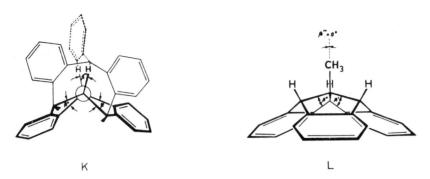

K   L

Scheme 10. Conformational torsion of some centropolyindans. **K**, Newman projections of *fuso*diindan **49** (in bold), triptindan **55** (add dotted line indan unit), and fenestrindan **74** (add, instead, full line indan units). **L**, Projection, inclined to the back, of *trifuso*triindan **66a**.

In turn, solid-state fenestrindan (**74**) [16] exists in two equivalent conformations with S_4 symmetry and two sets of four equivalent central dihedral angles $\beta' = 19.6 \pm 1.0°$ and $\beta'' = 21.1 \pm 0.9°$ (**K**, Scheme 10). A similar but unsymmetrical torsion has been found for **37** [25]. At room temperature, solutions of **74** show rapid equilibration of the rotamers. Similar to **57**, repulsion of the pairs of bridgehead substituents (**75a–75d**) increases the internal torsion. Thus, extreme torsion of the central C–C bonds ($23 \leqslant \beta' \leqslant 28°$ and $25 \leqslant \beta'' \leqslant 33°$) has been estimated from force-field calculations, and solutions of **75c** and **75d** exhibit static behavior of the conformers on the NMR time scale [21].

Obviously, all centropolyindans without *trifuso* junction (**43**, **49**, **55**, **60**, and **74**) exist in at least two equivalent rotamers with mean torsional angles in the order of 15–30°. Conversely, **71** and **77** are predicted to exist in one rotamer only due to the steric locking of the *trifuso* arrangement of (at least) three indan units, as has been found for **66a**. Of course, the X-ray structural analysis of centrohexaindan (**21**) gives the final answer: It shows that, in fact, the four central C–C bonds in this fourfold tribenzotriquinacene are perfectly eclipsed with all central dihedral angles $\beta = 0°$. The two indan units of the three 2,2'-spirobiindan systems are exactly orthogonal to each other, and the dihedral angle between the benzo rings of each of the 12 *fuso*diindan subunits is 120°. Overall, as has been predicted for the hexaene **41** [14], the molecular symmetry of centrohexaindan is perfectly tetrahedral (T_d).

Acknowledgments
The author thanks his co-workers, Dr. Bernd Paisdor and Dipl.-Chem. Thomas Lindenthal, Monika Seifert, and Andreas Schuster, who contributed to the work presented here, for their enthusiastic and skilled collaboration, and Professor Dr. Hans-Fr. Grützmacher for encouraging

discussions and generous support. Thanks is also due to Mr. Dieter Barth for encouraged help in multistep synthetic work. The author is grateful to Professor Dr. Achim Müller, Dr. Hartmut Bögge, Dipl.-Chem. Michael Penk, and Mrs. Beate Neumann for performing the X-ray structural analyses. Financial support from the Deutsche Forschungsgemeinschaft (Ku 663 1-1) and the University Bielefeld (OZ 2194/20) is gratefully acknowledged.

REFERENCES

1. N. L. Allinger, "Conformational analysis. 130. MM2. A hydrocarbon force field utilizing V_1 and V_2 torsional terms," *J. Am. Chem. Soc.*, **99**, 8127–8134 (1977).

2. U. Burkert and N. L. Allinger, *Molecular Mechanics*. American Chemical Society, Washington, D.C., 1982.

3. A. Greenberg and J. F. Liebman, *Strained Organic Molecules*. Academic Press, New York, 1978.

4. L. A. Paquette and A. M. Doherty, *Polyquinane Chemistry, Synthesis and Reactions*. Springer-Verlag, Berlin, 1987.

5. L. A. Paquette, "The development of polyquinane chemistry," *Top. Curr. Chem.*, **79**, 41–165 (1979).

6. L. A. Paquette, "Recent synthetic developments in polyquinane chemistry," *Top. Curr. Chem.*, **119**, 1–163 (1984).

7. P. Gund and T. M. Gund, "How many rings can share a quarternary atom?," *J. Am. Chem. Soc.*, **103**, 4458–4465 (1981).

8. J. C. Gallucci, C. W. Doecke, and L. A. Paquette, "X-ray structure analysis of the pentagonal dodehedrane hydrocarbon $(CH)_{20}$," *J. Am. Chem. Soc.*, **108**, 1343–1344 (1986).

9. P. E. Eaton, "Towards dodecahedrane," *Tetrahedron*, **35**, 2189–2223 (1979).

10. H. Müller, J.-P. Melder, W.-D. Fessner, D. Hunkler, H. Fritz, and H. Prinzbach, "Funktionalisierte, enantiomerenreine [2.1.1]-, [2.2.1]- und [2.2.2]Triblattane," *Angew. Chem.*, **100**, 1140–1143 (1988); *Angew. Chem. Int. Ed. Engl.*, **27**, 1103–1106 (1988).

11. B. M. Trost, "Cyclopentanoids: A challenge for new methodology," *Chem. Soc. Rev.*, **11**, 141–170 (1982).

12. A. T. Balaban (ed.), *Chemical Applications of Graph Theory*. Academic Press, London, 1976.

13. J. Simon, "A topological approach to the stereochemistry of nonrigid molecules," in *Graph Theory and Topology in Chemistry*, R. B. King and D. H. Rouvray (eds.). Elsevier, Amsterdam, 1987, pp. 43–75.

14. O. Ermer, *Aspekte von Kraftfeldrechnungen*. Wolfgang Baur Verlag, München, 1981.

15. D. Kuck, "Ein einfacher Zugang zu benzoanellierten Centropolyquinanen," *Angew. Chem.*, **96**, 515–516 (1984); *Angew. Chem. Int. Ed. Engl.*, **23**, 508–509 (1984).

16. D. Kuck and H. Bögge, "Benzoannelated centropolyquinanes. 2. *All-cis*-tetra-benzotetracyclo[5.5.1.0^{4,13}.0^{10,13}[tridecane, 'Fenestrindan'," *J. Am. Chem. Soc.*, **108**, 8107–8109 (1986).

17. D. Kuck, B. Paisdor, and H.-F. Grützmacher, "Benzoanellierte Centropoly-quinane, 3. Synthese mehrfach substituierter Triptindane (9*H*, 10*H*-4b, 9a-([1,2]Benzenomethano)indeno[1,2-*a*]indene) mit drei Substituenten in der Molekülhöhlung," *Chem. Ber.*, **120**, 589–595 (1987).

18. B. Paisdor, H.-F. Grützmacher, and D. Kuck, "Benzoanellierte Centropoly-quinane, 4. Sterische Effekte in mehrfach substituierten Triptindanen (9*H*, 10*H*-4b, 9a-([1,2]Benzenomethano)indeno-[1,2-*a*]indenen)," *Chem. Ber.*, **121**, 1307–1313 (1988).

19. D. Kuck and A. Schuster, "Die Synthese von Centrohexaindan—dem ersten Kohlenwasserstoff mit topoligisch nicht-planarer Molekülstruktur," *Angew. Chem.*, **100**, 1222–1224 (1988); *Angew. Chem. Int. Ed. Engl.*, **27**, 1192–1194 (1988).

20. D. Kuck, A. Schuster, B. Ohlhorst, V. Sinnwell, and A. de Meijere, "Auf dem Wege zum Tribenzoacepentalen: Tribenzotriquinacen, Tribenzodihydroace-pentalendiid und Tribenzoacepentalen-Radikalanion," *Angew. Chem.*, **101**, 626–628 (1989); *Angew. Chem. Int. Ed. Engl.*, **28**, 595–597 (1989).

21. D. Kuck et al., in preparation.

22. A. Ceccon, A. Gambaro, F. Manoli, A. Venzo, G. Valle, D. Kuck, and T. E. Bitterwolf, submitted.

23. B. Ohlhorst, "Zur Chemie des Dihydroacepentalen-systems: Dianionen, Bi-strimethylsilylderivate und Dimere," Doctoral Thesis, University of Hamburg, 1989.

24. J. Siegel, private communication, 1988.

25. R. Mitschka, J. Oehldrich, K. Takahashi, J. M. Cook, U. Weiss, and J. V. Silverton, "General approach for the synthesis of polyquinanes. Facile generation of molecular complexity via reaction of 1,2-dicarbonyl compounds with dimethyl 3-ketoglutarate," *Tetrahedron*, **37**, 4521–4542 (1981).

26. D. Ginsburg, *Propellanes, Structures and Reactions*. Verlag Chemie, Weinheim, 1975.

27. R. Gleiter, E. Litterst, and J. Drouin, "Interactions in 2,8,9-trifunctional [3.3.3]propellanes," *Chem. Ber.*, **121**, 923–926 (1988).

28. H. E. Simmons, III, and J. E. Maggio, "Synthesis of the first topologically non-planar molecule," *Tetrahedron Lett.*, **22**, 287–290 (1981).

29. L. A. Paquette and M. Vazeux, "Threefold transannular epoxide cyclization. Synthesis of a heterocyclic C$_{17}$-hexaquinane," *Tetrahedron Lett.*, **22**, 291–294 (1981).

30. J. E. Maggio and H. E. Simmons, III, "Trispiro[tricyclo-[3.3.3.0^{1,5}]undecane-2,1':8,1":9,1'''-tris-[cyclopropane]], a chiral fluxional hydrocarbon," *J. Am. Chem. Soc.*, **103**, 1579–1581 (1981).

31. S. A. Benner, J. E. Maggio, and H. E. Simmons, III, "Rearrangement of a geometrically restricted triepoxide to the first topologically nonplanar molecule: A reaction path elucidated by using oxygen isotope effects on carbon-13 chemical shifts," *J. Am. Chem. Soc.*, **103**, 1581–1582 (1981).

32. L. A. Paquette, R. V. Williams, M. Vazeux, and A. R. Browne, "Factors conducive to the cascade rearrangement of sterically congested and geometrically restricted three-membered rings. Facile synthesis of a topologically nonplanar heterocycle," *J. Org. Soc.*, **49**, 2194–2197 (1984).

33. H. E. Simmons, III, "The synthesis, structure, and reactions of some theoretically interesting propellanes: The synthesis of the first topologically non-planar organic molecule," PhD Thesis, Harvard University, 1980.

34. M. Venkatachalam, M. N. Desphande, M. Jawdosiuk, G. Kubiak, S. Wehrli, J. M. Cook, and U. Weiss, "General approach for the synthesis of polyquinenes," *Tetrahedron*, **42**, 1597–1605 (1986).

35. H. Butenschön and A. de Meijere, "The first carbonyl iron complexes with dihydroacepentalene ligands," *Tetrahedron*, **42**, 1721–1729 (1986).

36. R. B. Woodward, T. Fukunaga, and R. C. Kelly, "Triquinacene," *J. Am. Chem. Soc.*, **86**, 3162–3164 (1964).

37. E. Carceller, M. L. García, A. Moyano, M. A. Pericàs, and F. Serratosa, "Synthesis of triquinacene derivatives. New approach towards the synthesis of dodecahedrane," *Tetrahedron*, **42**, 1831–1839 (1986).

38. A. K. Gupta, G. S. Lannoye, G. Kubiak, J. Schkeryantz, S. Wehrli, and J. M. Cook, "General approach to the synthesis of polyquinenes. 8. Synthesis of triquinacene, 1,10-dimethyltriquinacene, and 1,10-cyclohexanotriquinacene," *J. Am. Chem. Soc.*, **111**, 2169–2179 (1989).

39. B. V. Rao and W. C. Agosta, "Fenestranes and the flattening of tetrahedral carbon," *Chem. Rev.*, **87**, 399–410 (1987).

40. K. Krohn, "Fenestrane—Blick auf strukturelle Pathologien," *Nachr. Chem. Techn. Lab.*, **35**, 264–266 (1987).

41. W. Luef and R. Keese, "Angular distortions at tetracoordinate carbon. Planoid distortions in α,α'-bridged spiro[4.4]nonanes and [5.5.5.5]fenestranes," *Helv. Chim. Acta*, **70**, 543–553 (1987).

42. H. Falk, W. Fröstl, und K. Schlögl, "Darstellung, absolute Konfiguration und optische Reinheit von 2,2'-Spirobiindan-1,1'-dion," *Monatsh. Chem.*, **105**, 574–597 (1974).

43. P. Lemmen and I. Ugi, "The chiroptic properties of the diastereomeric 1,1'-dimethyl-2,2'-spirobiindans—a comparative study," *Chem. Scr.*, **27**, 297–301 (1987).

44. J. H. Brewster and R. T. Prudence, "Absolute configuration and chiroptical properties of optically active 1,1'-spirobiindan, 1,1'-spirobiindene, and 1,1'-spirobiindanone," *J. Am. Chem. Soc.*, **95**, 1217–1229 (1973).

45. R. K. Hill and D. A. Cullison, "Dissymmetric spirans. II. Absolute configuration of 1,1'-spirobiindene and related compounds," *J. Am. Chem. Soc.*, **95**, 1229–1239 (1973).

46. G. Haas and V. Prelog, "Optisch aktive 9,9'-Spirobifluoren-Derivate," *Helv. Chim. Acta*, **52**, 1202–1218 (1969).

47. J. M. M. Smits, J. H. Noordik, P. T. Beurskens, W. H. Laarhoven, and F. A. T. Lijten, "Crystal and molecular structure of *cis*-dibenzobicyclo[3.3.0] octa-2,7-diene, $C_{16}H_{14}$," *J. Cryst. Spectrosc. Res.*, **16**, 23–29 (1986).

48. R. S. D. Mittal, S. C. Sethi, and Sukh Dev, "Azulenes and related substances —XV. Azuleno[2,1-*a*]azulene: Reaction of 3,6,7,8-tetrahydrodibenzopentalene with diazomethane; synthesis of 11*H*-indeno[2,1-*a*]azulene," *Tetrahedron*, **29**, 1321–1325 (1973).

49. W. Baker, J. F. W. McOmie, S. D. Parfitt, and D. A. M. Watkins, "Attempts to prepare new aromatic systems. Part VI. 1:2-5:6-Dibenzopentalene and derivatives," *J. Chem. Soc.*, 4026–4037 (1957).

50. M. Randić, "Aromaticity and conjugation," *J. Am. Chem. Soc.*, **99**, 444–450 (1979).

51. H. Dietrich, D. Bladauski, M. Grosse, K. Roth, and D. Rewicki, "Kristallstruktur und Reaktionen des 7b*H*-Indeno[1,2,3-*jk*]*fluorens*," *Chem. Ber.*, **108**, 1807 (1975).

52. H. W. Thompson, "The synthesis of triptindan," *J. Org. Chem.*, **33**, 621–625 (1968).

53. W. Ten Hoeve and H. Wynberg, "Synthetic approaches to planar carbon. 2," *J. Org. Chem.*, **45**, 2930–2937 (1980).

54. D. Kuck, "Single, double, and triple hydrogen rearrangement reactions in ionized 2-benzyl-1-indanols," *Adv. Mass Spectrom.*, **10**, 773–774 (1986).

55. W. Ten Hoeve and H. Wynberg, "Chiral spiranes. Optical activity and nuclear magnetic resonance spectroscopy as a proof for stable twist conformations," *J. Org. Chem.*, **44**, 1580–1514 (1979).

56. T. Prange, J. Drouin, F. Leyendecker, and J.-M. Conia, "X-Ray molecular structure of a highly symmetrical triketone: [3.3.3]Propellane-2,8,9-trione," *J. Chem. Soc. Chem. Commun.*, 430–431 (1977).

57. T. E. Bitterwolf, A. Ceccon, D. Kuck, et al., to be published.

Index